NJU SA 2014-2015

THE YEAR BOOK OF ARCHITECTURE PROGRAM SCHOOL OF ARCHITECTURE AND URBAN PLANNING

南京大学建筑与城市规划学院建筑系 教学年鉴

王丹丹 编 EDITOR: WANG DANDAN

东南大学出版社·南京 SOUTHEAST UNIVERSITY PRESS, NANJING

建筑设计及其理论

Architectural Design and Theory

张　雷 教　授　Professor ZHANG Lei

冯金龙 教　授　Professor FENG Jinlong

吉国华 教　授　Professor JI Guohua

周　凌 教　授　Professor ZHOU Ling

傅　筱 教　授　Professor FU Xiao

钟华颖 讲　师　Lecturer ZHONG Huaying

城市设计及其理论

Urban Design and Theory

丁沃沃 教　授　Professor DING Wowo

鲁安东 教　授　Professor LU Andong

华晓宁 副教授　Associate Professor HUA Xiaoning

胡友培 副教授　Associate Professor HU Youpei

刘　铨 讲　师　Lecturer LIU Quan

尹　航 讲　师　Lecturer YIN Hang

建筑历史与理论及历史建筑保护

Architectural History and Theory,
Protection of Historic Building

赵　辰 教　授　Professor ZHAO Chen

王骏阳 教　授　Professor WANG Junyang

胡　恒 教　授　Professor HU Heng

肖红颜 副教授　Associate Professor XIAO Hongyan

冷　天 讲　师　Lecturer LENG Tian

建筑技术科学

Building Technology Science

鲍家声 教　授　Professor BAO Jiasheng

秦孟昊 教　授　Professor QIN Menghao

吴　蔚 副教授　Associate Professor WU Wei

郜　志 副教授　Associate Professor GAO Zhi

童滋雨 副教授　Associate Professor TONG Ziyu

南京大学建筑与城市规划学院建筑系
Department of Architecture
School of Architecture and Urban Planning
Nanjing University
arch@nju.edu.cn　http://arch.nju.edu.cn

教学纲要
EDUCATIONAL PROGRAM

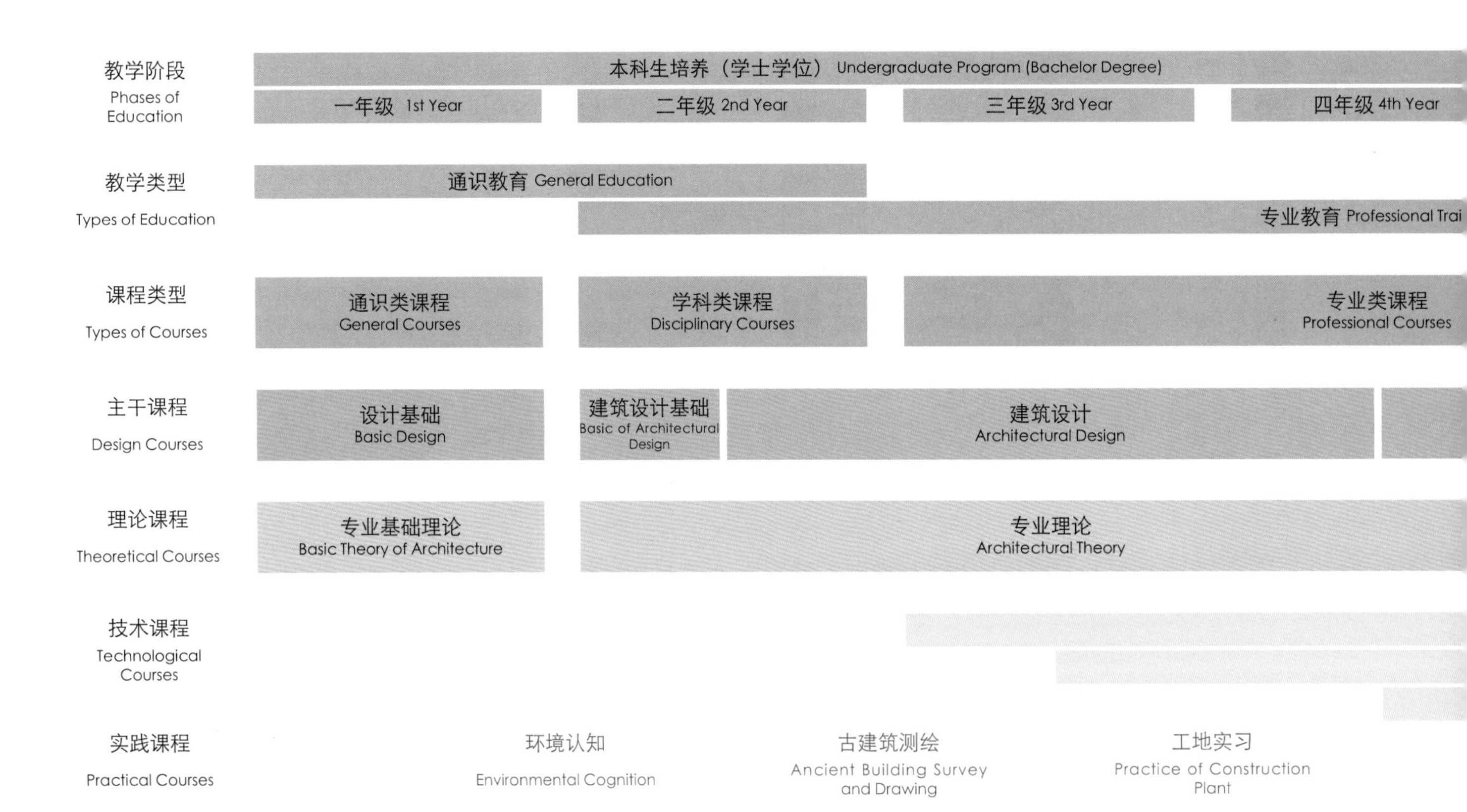

研究生培养（硕士学位）Graduate Program (Master Degree)

一年级 1st Year

二年级 2nd Year

三年级 3rd Year

研究生培养（博士学位）
Ph. D. Program

学术研究训练 Academic Research Training

学术研究 Academic Research

建筑设计研究
Research of Architectural Design

毕业设计
Thesis Project

学位论文
Dissertation

学位论文
Dissertation

专业核心理论
Core Theory of Architecture

专业扩展理论
Architectural Theory Extended

专业提升理论
Architectural Theory Upgraded

跨学科理论
Interdisciplinary Theory

建筑构造实验室 Tectonic Lab

建筑物理实验室 Building Physics Lab

数字建筑实验室 CAAD Lab

产实习
e of Profession

生产实习
Practice of Profession

课程安排

CURRICULUM OUTLINE

	本科一年级 Undergraduate Program 1st Year	本科二年级 Undergraduate Program 2nd Year	本科三年级 Undergraduate Program 3rd Year
设计课程 Design Courses	设计基础 Basic Design	建筑设计基础 Basic Design of Architecture 建筑设计（一） Architectural Design 1 建筑设计（二） Architectural Design 2	建筑设计（三） Architectural Design 3 建筑设计（四） Architectural Design 4 建筑设计（五） Architectural Design 5 建筑设计（六） Architectural Design 6
专业理论 Architectural Theory	逻辑学 Logic	建筑导论 Introductory Guide to Architecture	建筑设计基础原理 Basic Theory of Architectural Design 居住建筑设计与居住区规划原理 Theory of Housing Design and Residential Planning 城市规划原理 Theory of Urban Planning
建筑技术 Architectural Technology	理论、材料与结构力学 Theoretical, Material & Structural Statics Visual BASIC程序设计 Visual BASIC Programming	CAAD理论与实践 Theory and Practice of CAAD	建筑技术（一） 结构与构造 Architectural Technology 1: Structure & Construction 建筑技术（二） 建筑物理 Architectural Technology 2: Building Physics 建筑技术（三） 建筑设备 Architectural Technology 3: Building Equipment
历史理论 History Theory	古代汉语 Ancient Chinese	外国建筑史（古代） History of World Architecture (Ancient) 中国建筑史（古代） History of Chinese Architecture (Ancient)	外国建筑史（当代） History of World Architecture (Modern) 中国建筑史（近现代） History of Chinese Architecture (Modern)
实践课程 Practical Courses		古建筑测绘 Ancient Building Survey and Drawing	工地实习 Practice of Construction Plant
通识类课程 General Courses	数学 Mathematics 语文 Chinese 名师导学 Guide to Study by Famed Professors 计算机基础 Basic Computer Science	社会学概论 Introduction of Sociology 社会调查方法 Methods for Social Investigation	
选修课程 Elective Courses		城市道路与交通规划 Planning of Urban Road and Traffic 环境科学概论 Introduction of Environmental Science 人文科学研究方法 Research Method of the Social Science 美学原理 Theory of Aesthetics 管理学 Management 概率论与数理统计 Probability Theory and Mathematical Statistics 国学名著导读 Guide to Masterpieces of Chinese Ancient Civilization	人文地理学 Human Geography 中国城市发展建设史 History of Chinese Urban Development 欧洲近现代文明史 Modern History of European Civilization 中国哲学史 History of Chinese Philosophy 宏观经济学 Macro Economics 管理信息系统 Management Operating System 城市社会学 Urban Sociology

本科四年级 Undergraduate Program 4th Year	研究生一年级 Graduate Program 1st Year	研究生二、三年级 Graduate Program 2nd & 3rd Year
建筑设计（七） Architectural Design 7 建筑设计（八） Architectural Design 8 本科毕业设计 Graduation Project	建筑设计研究（一） Design Studio 1 建筑设计研究（二） Design Studio 2 数字建筑设计 Digital Architecture Design 联合教学设计工作坊 International Design Workshop	专业硕士毕业设计 Thesis Project
城市设计理论 Theory Urban Design	城市形态研究 Study on Urban Morphology 现代建筑设计基础理论 Preliminaries in Modern Architectural Design 现代建筑设计方法论 Methodology of Modern Architectural Design 景观都市主义理论与方法 Theory and Methodology of Landscape Urbanism	
建筑师业务基础知识 Introduction of Architects' Profession 建设工程项目管理 Management of Construction Project	材料与建造 Materials and Construction 中国建构（木构）文化研究 Studies in Chinese Wooden Tectonic Culture 计算机辅助技术 Technology of CAAD GIS基础与运用 Concepts and Application of GIS	
	建筑理论研究 Study of Architectural Theory	
生产实习（一） Practice of Profession 1	生产实习（二） Practice of Profession 2	建筑设计与实践 Architectural Design and Practice
景观规划设计及其理论 Theory of Landscape Planning and Design 东西方园林 Eastern and Western Gardens 地理信息系统概论 Introduction of GIS 欧洲哲学史 History of European Philosophy 微观经济学 Micro Economics 政治学原理 Theory of Political Science 社会学定量研究方法 Quantitative Research Methods in Sociology	建筑史研究 Studies in Architectural History 建筑节能与可持续发展 Energy Conservation & Sustainable Architecture 建筑体系整合 Advanced Building System Integration 规划理论与实践 Theory and Practice of Urban Planning 景观规划进展 Development of Landscape Planning	

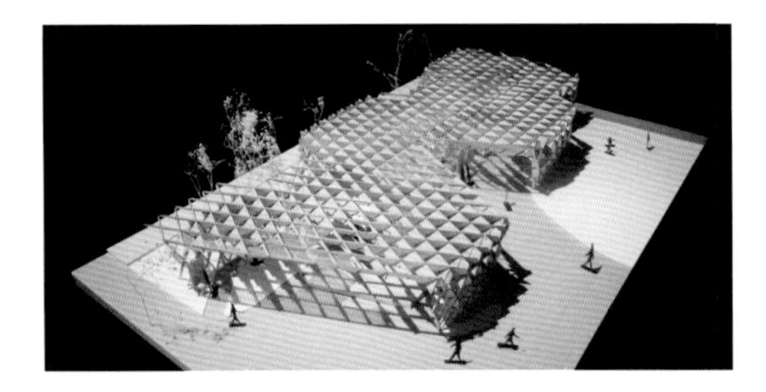

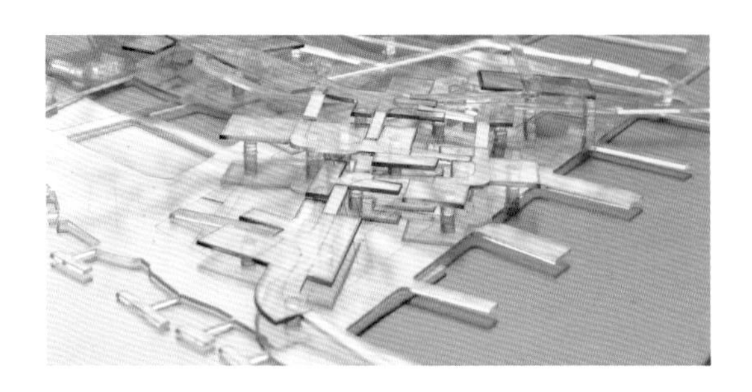

教学论文
ARTICLES ON EDUCATION

过渡与转换——对转型期建筑教育知识体系的思考

TRANSITION AND TRANSFORMATION: THINKING OF THE KNOWLEDGE SYSTEM OF ARCHITECTURAL EDUCATION

丁沃沃

1. 引言

改革开放以来，我国建筑教育得到了突飞猛进的发展，在“量”与“质”两个方面都得到了很大的提升。前些年为了适应国家建设的需要，“建筑学”在办学数量和招生数量两个方面数量大增，近年来提高建筑学教学质量的要求越来越受到重视，不断涌现各类教学改革。在此背景下，作为从事建筑学教育多年的教师，在此试图从学科发展的角度考量建筑学的知识体系及其教育模式。

2. 处于转型之中的建筑学

自法国巴黎美术学院奠定了建筑学教育的学院派教育体系以来，建筑教育在大学教育体系里已经历练了300多个年头[1]，在此过程中，无论在教育理念和教育方法上都经历了若干次大的变动或者说是改革。导致建筑学变革的是建筑学认识论的转变：建筑是“艺术”、建筑是“建造的艺术”、建筑是“居住的机器”以及建筑学必须“自治”，如此等等。本世纪初，当经济发展推动的城市化在全球蔓延之时，建筑的形式借势已经转化成了商品并产生了价值，建筑学的“自治”的概念即刻消解。此时，西方建筑学理论界因支撑建筑形式的理论基础开始变异而担忧，提醒西方建筑学的传统核心价值观正面临挑战。我们开始看到根植于西方古典审美理论的建筑学正面临着重构或者转型，其建筑教育也正处在变化之中。

在中国，我们有着自己几千年的建造文明史，也有着在世界建筑史之林中独树一帜的传统建筑。在中国建筑文化中，建筑被定义为“器”[2]，而非“艺”。然而，就设立在大学体系中的建筑学学科而言，我们却仅有100年左右的历史。不仅如此，目前我国大学中的建筑教育体系和其他许多学科一样是随着上个世纪初“西学东进”的潮流而从国外引进，从“认知”到“方法”都深深地烙上了西方建筑学的烙印。显然，对于作为“艺术”的建筑有着使用的功能，作为“用器”的建筑也有着艺术的价值。建筑是“艺术”或建筑是“用器”是同一事物的两种不同的认知角度，而两种不同的认识论会导致设计方法的不同。经历了近百年的发展，我国的建筑教育体系一方面延续了和西方建筑学可交流的共同认知和方法，另一方面为了适应我国自身来自社会和文化的需求而变化，逐渐形成了自己的模式，培养了具有中国特色的优秀建筑师。尤其是改革开放的30年以来培养的大多数建筑师已经为国家的建设作出了有目共睹的贡献[3]。

当今，科学的发展、技术的进步以及多学科融合所产生的新的知识都给建筑学知识体系更新带来了挑战。对于我国建筑学来说，尽管与西方建筑学的发展路径不尽相同，快速的城市化进程使得我国的建筑学学科过早地面临到与发达国家同样的问题，即：如何理解我们的城市，如何改进我们的城市环境，以及如何使得我们的人造环境进入可持续发展的轨道。因此，对于进入21世纪的中国来说建筑的内涵也已扩展。在城市化进程中，建筑不再仅仅是一个“用器”，城市建筑已经构成了人们的“生活环境”。在城市中，讨论单体建筑已经没有意义，重要的是建筑与建筑的组合方式和建筑之间的城市空间。我们已经注意到了全球范围内的城市化进程似乎给中国建筑学学科的发展又带来了新的机遇，同时也意识到全球化的趋势导致建筑的文化特质比任何时期更加受到重视。由于我们的建筑学学科体系依附于西方建筑学的评价体系，对建筑学与建筑的认知一直纠结于东西方文化差异而难以摆脱，因此重新梳理学科体系成为我们学科发展的必要任务。

当西方建筑学转型之际，我们更应该以我们自己的文化视角重新思考。建筑学是和一个国家和地区的社会发展紧密联系在一起的学科，建筑学学科的认知和发展不可能脱离它所处的社会发展阶段，因此，和学科发展直接相关的建筑教育改革从来都不仅仅源自学科发展的需求，还有来自社会发展的需要，这是重新建构学科核心知识体系的真正动力。因此，面对社会转型的机遇、科学进步的支撑、城市化进程的需求和文化自信的挑战，都促使我们对建筑学学科重新思考。立足本民族的文化，中国建筑学应该自己作出抉择，才会有新的机遇。

3. 建筑学科的知识亟待更新

建筑学科知识体系的构成主要取决于它所培养建筑师应具备的知识体系，目前公认最早的论著是罗马建筑师维特鲁威的《建筑十书》。维特鲁威认为：“建筑师只是要具备许多学科和种种技艺。以各种技艺完成的一切作品都要依靠这种知识的判断来检查。它是由手艺和理论产生的。”[4]以维特鲁威的观点，建筑学中存在两种事物，即被赋予意义的事物和赋予意义的事物（拉丁文版本为Quod Significatur et Quod Significat），而建筑师应该精通这两种事物，“建筑师既要有天赋的才能，还要有钻研学问的本领。因为没有学问的才能或者没有才能的学问都不可能造出完美的技术人员”[4]。维特鲁威的《建筑十书》囊括了建筑师应该掌握的从大尺度的城市到细部的建筑材料、从涉及美学的比例与尺度到物理环境与机械原理等各类学问，奠定了建筑学科的核心知识体系的基础。

意大利文艺复兴是西方建筑学发展的重要时期，意大利建筑师和理论家阿尔伯蒂对建筑师的认识和维特鲁威不尽相同。在阿尔伯蒂看来建筑师应该是一个学者或绅士而不仅仅是一个工匠或手艺人[5]。在阿尔伯蒂建构的建筑学知识体系中主要是建筑形式的艺术及其美学理论，有关建筑建造的技术却被忽略。

欧洲自17世纪以来对建筑师的定位及其知识与技能形成了四种不同的角色，如：学术建筑师（Academic Architect）、手艺人或工匠（Craftsman-Builder）、市政工程师(Civil Engineer)以及稍后形成的社会学家（Social Scientist）。当巴黎美术学院的建筑教育在大学体系里设立之时，主要继承了文艺复兴的建筑学传统，选择了“学术建筑师”作为建筑学的培养目标[6]。作为大学的一个学科，巴黎美术学院的建筑学教育的课程体系秉承了大学的学术传统，建构了完整的课程体系。通过课程体系给未来的建筑师输送5类知识：分析类、科学类、建造类、艺术类、项目类。分析类主要分析或模仿优秀的建筑实例；科学类（要通过考试）包括数学、解析几何、静力学、材料性能、透视学、物理学、化学以及考古学；建造类包括做模型和结构分析；艺术类包括用炭笔画石膏模型的素描、各种装饰细部和临摹雕像；项目类包括建筑相关的各项指标构成[7]。可以看出，进入大学的建筑学延续并完善了维特鲁威时期奠定的建筑学的知识结构，该知识体系旨在将学术建筑师培养成一名有品位的学者，它奠定了西方建筑学知识体系的基础。

梳理历史脉络不难发现，建筑学在西方经历了三百多年，从古典建筑到现代建筑，建筑的认识论发生巨大的转变，认识论的转变带来的是审美观的转变以及设计方法的变化；对自然界认知的更新和技术进步导致了设计知识和媒介的更新，最终都体现在了建筑形式的更替上。整个过程中，建筑学的认识论起到了引领作用[5]。在学科知识构成方面，虽然具体内容在不断更新和扩充，但是建筑学知识体系的构成没有发生根本的变化。概括起来包括了三个主要方面：建筑的认知理论、建筑设计的方法论以及和建筑学相关的科学与技术知识。

当我国在大学里设立建筑学学科之时就引进了西方建筑学的知识构成框架。正因如此，建筑学不再是工匠之手艺活而成为一门学问，仅此一点从根本上改变了我们传统意识中对建筑的认知。虽然大学里的建筑学按知识体系设立了相应的课程，但是知识体系和理论研究与建筑设计之间的关系一直并不十分清晰。实际上，是否沿用西方建筑学体系不是问题，关键在于当转化为“学科”而不再仅仅是“造物”的建筑学学科时，如果没有严谨的研究体系，学科的知识显然难以更新。其结果是在国际交流中，尽管我们能在竞图方面取胜，而在理论的建树方面我们却很少有独立的话语权。反映在建筑教育中，松散的知识构架和陈旧的知识使得学生在学习过程中感受不到知识对于建筑师的重要性。

如果说建筑学的知识结构没有变化的话，那么核心知识的内容应该随着时代的变迁而不断更新。当然，知识的更新需要研究的耕耘。笔者以为，如果建筑学依然作为一个学科在大学里存在的话，重视学科的相关研究、完善学科知识构成体系和内容不仅非常重要而又迫在眉睫。对于完善学科体系，我们主要有两个方面的主要任务：首先，城市化已经使得城市正在成为我们主要的生活场所，我们的城市物质空间的现状要求我们的建筑学亟待扩充城市方面的知识。基于我国的人口基数，高密度的城市物质形态将会成为我们的主要选择，它将给城市建筑学注入新的内涵和要求。其次，由于高密度城市形态将影响城市的整体气候状况，以此，城市物理和城市气候学将会成为建筑学知识体系中的重要组成部分[8]。这方面的研究不仅能够服务于我国的需求，也是对整个建筑学学科的贡献。此外，在全球化的趋势下，建筑文化的地域性特征比任何一个时候都备受关注[9]。理论研究证实，建筑的形式来源于对事物的认知与思考，形式的发生并不源于巧合或偶然[10]。目前就建筑教育而言，我们不缺时尚建筑范例，而缺乏对于形式生成原因的研究。

4. 通识知识和设计能力

既然建筑学是大学中的一个学科，建筑学培养的人才就一定不是作为手艺人的建筑师，而应该是一个有学问的建筑师。何谓学问？建筑学科人才培养主要分为两个方面：其一，广博的知识和思辨能力；其二，形式的规律和设计能力。前者奠定了一个学者的基本素质，后者决定了其专业素养，缺一不可。进入21世纪以来，出于社会的需求和科学发展的需要，国际一流大学纷纷强调学科之间交叉与合作。在人才培养方面，开始强调通识教育，实际上是赋予新时期人才应该具备的共同的知识基础，为未来的发展和变化做好准备。国际一流大学的建筑学也不例外，大学初期用以夯实学生的基础，提高学生一般知识素养，而将专业教育向研究生教育衍生。旨在当社会需求发生变化时，未来的新一代建筑师具备了能够应对社会发展需要的基本知识和专业素质。

建筑设计教学是建筑学教育体系中的重要环节，也是建筑教学体系中最具特色的部分。建筑设计课是建筑教育中的核心课程，在任何院校建筑学教育中都是最受重视的内容。在巴黎美术学院时期，建筑理论教育和设计教学分离，建筑设计教学的任务由学院聘请执业建筑师来承担。设计训练由设计导师(Patron)指导，在导师的工作坊受到训练[11]。美国的宾夕法尼亚大学的帕尔•克瑞（Paul Cret）教授继承了巴黎美术学院的建筑学教育基本理念，同时改进了巴黎美术学院的校外导师工作坊式的建筑设计教学，将建筑设计作为正式课程引入大学，与理论课一样列入课程表[12]。这样，大学中的建筑设计课和工作坊的设计训练任务也更加多元。克瑞认为通过设计课不仅训练设计技能，而且可以通过设计分析来学习建筑理论和历史理论[12]，从此大学里设计课不单纯是设计手法的训练，而且承担了传授知识的平台。建筑设计不仅贯穿于整个建筑学教育过程，而且成为检验学生是否能够获取学位的主要环节——学生独立的毕业设计。随着学科的发展，学校的建筑设计内容变得丰富多彩，建筑设计与理论研究相结合，甚至独立完成的毕业设计开始逐渐被由教师引领的研究性设计所取代[13]。

建筑设计教学在我国的建筑院校里受普遍受到重视，其重要程度居所有课程之首，设计教学的质量往往代表了一个学校的建筑学教学的质量。改革开放以来，由于国家建设快速发展，急需大量的建筑设计人才。为了配合这样的需求，建筑设计教学逐渐以模拟现实实践需求为主，目标是入职后能尽早上手出活。为此，大学短期内的确为市场输入了大量的有用的人才，尽管如此，一方面我们的设计教学始终还是满足不了现实市场对人才素质的要求，另一方面我们又意识到我们的学生在思维训练和创意训练方面远不如欧美大学，在未来竞争力方面显然处于弱势。那么，大学里具有如此重要地位的建筑设计课在建筑学教育中究竟承担怎样的角色？

其实，如果学习建造一个房子，那么最好的学习过程无疑应该在建造工地，只有通过工地的学习，才能真正地体验和理解真实建造的问题。如果为了学习建造建筑或房子同时知晓如何设计，那么可以直接去设计院或事务所从帮助制图开始学。通过设计院或事务所的工作体验可以理解真实的建筑设计，但也学不到建造，所以在工地学习过程依然不可缺失。进而，如果不仅为了建造房子和不仅为了设计一个要造的房子，而是为了理解设计房子或建筑的基本原理、思维方法和设计手法，学习对建筑形态的认定标准和一般规律，那就得进入学校进行专业学习。所以，学校的建筑设计课必须提供建筑设计的核心知识（而非全部）和建筑设计的一般道理，不可能提供的是市场上的设计实践和工地上的建造实践，设计实践和建造实践还要通过设计事务所和工地方能解决。很明显，大学里的建筑设计课程承担的知识传授的任务，不得不有着自身的训练规律和方式。训练的目的是满足未来现实的需要，而不是即刻的需求。欧洲大陆有着最为成熟的职业教育体系，传统上就有因不同需要培养的设计人才。就建筑学而言，就有许多高等专业学校（Hochschule），培养的人才同样进入设计市场，而且进入市场后立马上手。而欧洲大陆的大学建筑学的职业教育的出口通常在研究生层面（Diploma），应对的是不同的社会需求。因此，如果我们的市场急需立马能用的设计人才，不能简单地向研究型大学提要求，而是去多办些高职或大专来训练，这样既好又快。一个研究型大学的设计课程的设置不必纠结是否学生一出校门就会盖房子，也不必因刚出校门的学生不能马上上手施工图而感到惭愧，这些都会通过周而复始的工作得到解决。然而，一个大学倒是应该为没有给一个面向未来的建筑师足够的知识基础、应有的社会责任感和价值判断能力、进入社会后所应该具备的不断学习的能力以及能够应对国际竞争的专业素质而反思。

当下，我国社会发展正处于转型期，很多行业正处于转型之中，建筑教育亦然。就建筑设计课而言，笔者认为应该加强三个方面的训练：

首先，建筑设计课应该加强建造知识的训练。中国传统建筑虽然在意识上没有直接归属为“艺术”，但是它从来都讲究“建造的艺术”。真实的材料以及合情理的诗意表达才使得中国传统建筑具有永恒的魅力。因此，当我们意识到“建构”是中国建筑之魂而不再是简单的“形式符号”的时候，我们自己的独立创作才会开始。此外，融入建造知识的建筑设计训练才能务实地探讨建筑各个层次的形式问题，将建筑的形式问题落到实处。建造训练需要图纸表达，但并不是施工图训练，应该强调的是建造的逻辑如何表达设计的理念。

其次，建筑设计课应该加强思维逻辑训练。建筑学既是实践性很强的学科，又是理论领域较广的学科。然而在现实中，学生总是认为建筑理论听起来很有意思，但实际上并没有真正地重视。实际上，建筑理论的重要性在于帮助学生提高认知世界的能力，因此设计课应该成为一个平台，使学生基于建筑理论训练设计方法。我们通常强调设计过程，过程不仅仅只是简单地为了显示设计方案的从无到有，而是通过设计过程训练设计的思维。设计过程就是理论的思辨和演绎的过程，形式只是最终的结果。当然，建筑理论的教学也应该更加地通俗和明白[14]。

第三，建筑设计课应该加强研究和探索性训练。建筑学学科在其发展过程中一直不断更新自身的知识内容和技术方法，研究的意味着收集新信息、挖掘新和发现新问题[14]。然而在建筑学科里，研究的传统一直没有受到重视，学生也不太重视知识类的课程，误认为知识与设计无关。应该强调的是，虽然知识不能直接推演出建筑的形式，但是新知识的运用往往会带来建筑形式的创新。我们体会到在形式创新方面的落后，但还没有意识到这和我们不重视研究有着密不可分的关系。纵观历史，只有具备知识的设计者才有创造力，设计的创新需要新的知识来支撑。

5. 结语

作为学科的建筑学的任务已经不再是建造几幢能用的物体那么简单，城市化带来的对城市高密度物质空间的挑战已经将前所未遇的问题摆在了我们学科的面前，其中有科学问题值得我们去探索，如高密度的城市物质形态与城市微气候环境的关联性问题，也有人文问题值得我们去思考。因此，建筑学需要研究，建筑设计需要新的知识去支撑。

1.Foreword

Since the reform and opening up, rapid development of architectural education has been achieved in China, which has been substantially improved in terms of both “quantity” and “quality”. In the past years, in order to accommodate the demand of national construction, quantity of both teaching schools and enrolled students of “architecture” increased tremendously, more and more attention was paid to the requirement of improving quality of architectural education, and various reforms of teaching methodology were emerged one after another. In this context, as a mentor carrying out architectural education for so many years, the author intends to examine the knowledge system and educational mode of architecture in the aspect of disciplinary development.

2.Architecture in Transition and Transformation

Since the academic education system of architectural education was established by the Ecole des Beaux-Arts in Paris, architectural education has gone through over 300 years in the system of higher education [1]. Several significant transformations or reforms in other words in terms of education idea and educational methodology have been experienced in this process. It is the evolution of architectural epistemology that has led to the transformation of architecture: architecture is “arts”, architecture is the “arts of construction”, architecture is the “machine for living”, and architecture must be of “autonomy”, and so on. At the beginning of this century, when urbanization driven by economic development was spreading across the world, the form of architecture was transformed into goods and value was created by relying on the trend of times, and the concrete of “autonomy” of architecture was dissolved immediately. For the time being, western theoretical circle of architecture began to worry about the initial variation of the theoretical basis behind the architectural form, alerting that the traditional core value of western architecture was confronted with challenges. We saw that the architecture being rooted in western classical aesthetic theory was faced with restructuring or transformation, and its architectural education was also in the process of transformation.

In China, we have thousands of years of civilization history of architecture, and also the unique traditional architecture in the diversified styles of world architectural history. In the Chinese architectural culture, building is defined as “ware” [2], instead of “art”. However, in term of architectural discipline set up in the system of higher education, we have a history of just 100 years approximately. Moreover, the current architectural education system is still the one introduced in the tide of “eastward propagation of western learnings” like many other disciplines in Chinese colleges and universities, which are deeply stamped with the hallmark of western architecture in the aspects from "cognition" to "methodology". Obviously, the building of "art" has the function of practical use, while the building of "ware" also has the artistic value. The "art" or "ware" of architecture is just the same thing derived from two different cognitive perspectives, and these two different epistemologies would result in different design methodologies. After several hundred years of development, the architectural education system in China, on the one hand, inherited the common cognition and methodology that can be exchanged with western architecture, and on the other hand, evolved and gradually took shape its own mode in order to accommodate the social and cultural demand in the country, and cultivated many outstanding architects with Chinese characteristics. In particular, most architects cultivated in the 30 years since reform and opening up have made universally recognized contributions to the construction of the nation [3].

Today, new knowledge generated through development of science, advance of technologies, as well as integration of multiple disciplines brought challenges to the update of the knowledge system of architecture. For architecture in China, although it has a different path of development in comparison with western architecture, rapid urbanization make Chinese architecture discipline have to face with the same issues in an earlier phase as that faced in developed countries, that is, how to understand our cities, how to improve environment in our cities, and how to enable our man-made environment to step on a track of sustainable development. Therefore, the implication of architecture has been expanded for China into the 21st century. In the course of urbanization, building is not just a "ware", urban buildings constitute the "living environment" of mankind. In a city, discussion of single building does not make sense anymore, what more importance is the combination mode between buildings as well as the urban space among buildings. We have noticed that the world-wide urbanization process appears to have brought new opportunities again for the development of architecture discipline in China, and also realized that the cultural trait of architecture has received more attention than in any time because of the trend of globalization. Our disciplinary system of architecture is attached to the evaluation system of western architecture, and our cognition to architecture and buildings has always been tangled with the difference between eastern and western cultures which cannot be got rid of, so renewing the disciplinary system has become an imperative task for development of this discipline.

In the transition of western architecture, we should be even more thinking from our own cultural perspective. Architecture is a discipline that is closely connected with the social development of a country or a region, and cognition and development of architecture discipline cannot be separated from the social development phase it belongs to, therefore, the architectural education reform that is directly related to disciplinary development never comes just from the demand of disciplinary development, but also from the demand of social development, this is the real momentum for re-establishing the core knowledge system of the discipline. Therefore, being confronted with the opportunity of social transition, support from advance of science, demand of the urbanization process and the challenge of cultural confidence, all make us to rethink the discipline of architecture. Based on our native culture, Chinese architecture should make its own choice, that's the new opportunity.

3.Architectural Knowledge Need Renewal

Composition of knowledge system of architecture discipline mainly depends on the knowledge system required for architects to be cultivated, and currently the recognized earliest literature is the *Ten Books on Architecture* by Vitruvius, a Roman architect. In the opinion of Vitruvius, "What required for an architect is just the acquisition of many disciplines and various kinds of craftsmanship. All works completed with various kinds of craftsmanship must be examined by judging with such knowledge. It is generated from craftsmanship and theories. "[4] According to opinion of Vitruvius, there are two things in architecture, i.e. what is meant and what means (in Latin: quod significatur et quod significat), and an architect should master these two things, "an architect should have not only gifted talent, but also ability of studying knowledge, because neither talent without knowledge nor knowledge without talent may create perfect technicians"[4] . The *Ten Books on Architecture* by Vitruvius covers various kinds of knowledge that should be mastered by an architect, from large-scale city to detailed building materials, from aesthetic scale and dimensions to physical environment and mechanical principles, which laid down foundation for the core knowledge system of the architecture discipline.

Italian Renaissance is an important period for the development of western architecture, and the Italian architect and theorist Alberti has different understanding on architect from that of Vitruvius. In Alberti's opinion, an architect should be a scholar or gentleman, not just a builder or craftsman [5]. Art of architectural forms and their aesthetic theories are major subjects in the architectural knowledge system constructed by Alberti , and techniques about building construction are neglected.

Positioning for architects and their knowledge and skills consists of four different roles in Europe since the 17th century, including: the academic architect, the craftsman-builder, the civil engineer, and later the social scientist. When architectural education of the Ecole des Beaux-Arts in Paris was established in higher education system, it mainly succeeded the architecture tradition of the Renaissance, and selected "the academic architect" as its educational objective [6]. As a discipline in college, curriculum system of architectural education in the Ecole des Beaux-Arts in Paris inherited the academic tradition of colleges, and constructed a complete curriculum system. It conveys 5 categories of knowledge to future architects through the curriculum system: analysis, science, construction, art and project. The category of analysis mainly includes analysis or simulation of outstanding architectural cases; the category of science (examination must be passed) includes mathematics, analytic geometry, statics, material properties, perspective, physics, chemistry and archaeology; the category of construction includes modelling and structure analysis; the category of art includes sketch of plaster models, various decorative details and copy of statues with charcoal pencil; and the category of project includes composition of various index relating to architecture [7]. We can see that the architecture in colleges inherited and improved the knowledge structure of architecture established in the period of Vitruvius, that system aimed to educate academic architect into an elegant scholar, which laid the foundation for the knowledge system of western architecture.

It is not difficult for us to find out by examining historical context that architecture has gone through over 300 years in western countries, from classical architecture to modern architecture, the epistemology of architecture has undergone tremendous transformation, and transformation of epistemology resulted in the evolution of aesthetics as well as changes of design methodology; update of cognition to nature and technical advance led to update of design knowledge and media, which was reflected by supersession of architectural forms ultimately. The epistemology of architecture played a leading role in the whole process [5]. In term of composition of disciplinary knowledge, the specific content is constantly being updated and expanded, but the composition of architectural knowledge system has not gone through any fundamental changes. Generally it consists of three primary aspects: cognitive theories of architecture, methodology of architectural design, as well as scientific and technological knowledge relating to architecture.

The framework of knowledge composition of western architecture was already introduced to China when architecture discipline was established in Chinese colleges. For this reason, architecture is no longer the craftsmanship of builders, but a branch of science, this alone changed the cognition to architecture in our traditional ideology fundamentally. Although relevant courses have been set up for architecture in colleges based on the knowledge system, the relationship between knowledge system as well as theoretical research and architectural design has not been clearly defined all the time. In fact, the question is not about whether to follow the system of western architecture, but that when architectural discipline is transformed into "discipline" other than just "building", if there has no a strict research system, apparently the discipline would be difficult to be updated. As a result, we have the ability to win drawing competition in international exchanges, but we have little independent say in theoretical achievements. When reflected in architectural education, loose knowledge framework and obsolete content of knowledge are unable to make students sense the importance of knowledge to an architect in the process of learning.

If we say that the structure of architectural knowledge has not been changed, the content of core knowledge should be constantly updated over time. Of course, update of knowledge requires painstaking research. In the author's opinion, if architecture is still to exist as a discipline in colleges, focusing on research about this discipline, and improvement of composition system and content of the disciplinary knowledge is not only very important, but also imperative. In order to improve the disciplinary system, we have two primary tasks: first, urbanization has made cities become our main living places, and current situation of material space in our cities requires us to promptly expand urban-related knowledge in the discipline of architecture. Given the population base in China, high-density urban material morphology will be our primary option, which will infuse the urban architecture with new implications and requirements. And second, high-density urban form will affect overall climatic conditions in a city, therefore, urban physics and urban climatology will be important components in the knowledge system of architecture [8]. Research in such fields can not only serve the demand in China, but also make contributions to entire architecture discipline. In addition, with the trend of globalization, regional characteristics of architectural culture attract more attentions than any time [9]. Some theoretical researches show that architectural forms come from our cognition and thinking to objects, and occurrence of forms does not come from coincidence or accident [10]. Currently in term of architectural education, we are not lack of cases of stylish buildings, but of the research on causes for formation of forms.

4.General Knowledge and Design Capability

Now that the architecture is a discipline in colleges, talents cultivated with architecture should certainly be no craftsman-type architects, but architects of scholarship. What does the scholarship mean? Education of talents with architectural discipline mainly includes two aspects: first, extensive knowledge and critical thinking skills; second, rules about forms and design capability. The former establishes basic quality for a scholar, and the latter determines his/her professional quality, neither one is dispensable. Since the beginning of the 21st century, given the demand of society and requirement of scientific development, all first-class international universities have been emphasizing intersection and cooperation among different disciplines. And in term of talent cultivation, they emphasize general education, which in fact is to endow talents in the new era with required general knowledge base, and make them be prepared for future development and changes. It has no exception for architecture in first-class international universities. We should strengthen students' knowledge base, improve their general knowledge cultivation in early stage of higher education, and extend specialized education into postgraduate education. So in case of change of social demand, future new generation of architects will be equipped with basic knowledge and professional quality required to answer the demand of social development.

Teaching of architectural design is an important link in the system of architectural education, and also the most unique part in the system. The course of architectural design is a core course in architectural education, and is the most emphasized content of architectural education in any institution of higher education. In the period of the Ecole des Beaux-Arts in Paris, architectural theory education and design teaching were separated with each other; the task of architectural design teaching was undertaken by practice architects hired by the school, while design training was instructed by patrons, and students were trained at studio of their patrons [11]. Pro. Paul Cret from American University of Pennsylvania succeeded the fundamental idea of architectural education of the Ecole des Beaux-Arts in Paris, and improved the studio-style architectural design teaching by outside Patrons adopted by the Ecole des Beaux-Arts in Paris, and architectural design was introduced into colleges as formal course, which was listed in curriculum same as theoretical courses [12]. In this way, architectural design course and design training tasks in studio are more diversified. Cret believes that design course can not only train students with design skills, but also can allow students to learn architectural theories and historical theories through design analysis [12], and since then design course in colleges is not just the training of design skills, but also the platform for imparting knowledge. Architectural design not only runs through the whole process of architectural education, but also becomes a key link to examine whether students can be qualified for degree award – graduation project completed by students independently. With development of the discipline, content of architectural design in schools is rich and colorful, architectural design is combined with theoretical research, and even independently completed graduation project starts to be replaced gradually by research-oriented design guided by professors [13].

The architectural design teaching is emphasized universally in all architectural colleges and universities in China, its importance overrides all other courses, and the quality of design teaching often represents the quality of architectural education of a school. Since the reform and opening up, with rapid development of national construction, the country was badly in need of a large number of architectural design talents, and in order to accommodate this demand, architectural design teaching was focusing on simulating the need of real-world practice, so that graduates can get started and undertake projects as soon as possible after stepping on their posts. Therefore, colleges did export large number of competent talents to market in a short period, nevertheless, on the one hand our design teaching still cannot meet the requirement of real-world market on quality of talents, and on the other hand, we realized that our students are far behind those from European and American colleges in terms of thinking training and creativity training, and are disadvantaged in term of competiveness in future. So, what role should the architectural design course play in architectural education, given it is so important in higher education?

In fact, if one wants to learn how to build a house, the best learning process is no doubt the learning at construction site, it is only by studying at job site that one can actually experience and understand real-world construction. If the goal is to learn how to construct a building or house while get to know how to design, we can go to a design institute or firm to start from drawing work. We can understand real architectural design with working experience in a design institute or firm, but we cannot learn construction, so the learning process at construction site is indispensable. Furthermore, if the goal is not just to construct a house or design a house to be constructed, but to understand the fundamental principles, thinking methods and design methodology for designing a house or building, and to learn the identification standards and general rules of architectural form, we must study specialized courses in schools. Therefore, architectural design course in schools must offer core knowledge (not all knowledge) and general rationales of architectural design, it is impossible to offer design practice in market or construction practice at job site, the question of design practice and construction practice must be solved at design firms and construction sites. Apparently, architectural design course in colleges should undertake the task of imparting knowledge, so it must have its own training rules and methods. The goal of training is to meet real-world demand in the future, instead of immediate demand. Continental Europe has the most sophisticated vocational education system, and has the tradition of training design talents targeting different demands. In term of architecture, there are many Hochschules, talents trained by them also enter the design market, and they can start work immediately after entering the market. While the outlet of vocational education of architecture in colleges in Continental Europe often lies in the level of Diploma, which answers a different social demand. So, if the market is badly in need of design talents that can be used immediately, we cannot just simply make demand to research universities, but should establish some higher vocational schools or junior colleges to undertake such training task, which is a good and fast way. Setup of design course of a research university should not worry whether students is capable of building a house immediately after graduation, or feel shamed because its fresh graduates are not capable of completing construction drawings immediately after graduation, all these problems will be solved through accumulation in daily work. However, a university should indeed rethink for being failed to offer a future-oriented architect with sufficient knowledge base, required sense of social responsibility and value judgement ability, required ability of continuous learning after graduation, as well as the professional quality of being able to meet international competition.

For the time being, China's social development is in a transitional period, many industries are in the process of transformation, which is also true for architectural education. For the course of architectural design, trainings in three aspects should be strengthened in the author's opinion:

First, architectural design course should enhance training on construction knowledge. Chinese traditional architecture is not classified into "art" in term of ideology, but it always emphasizes the "art of building". Authentic materials and rational poetic presentation render Chinese traditional architecture with everlasting charm. Therefore, when we realized that "construction" is the soul of Chinese architecture instead of simple "symbol of the form" anymore, we just start our own independent creation. Moreover, then the architectural design training infused with construction knowledge can probe into various levels of questions of architectural form in a pragmatic manner, and put the questions of architectural form into practice. Construction training requires presentation with drawings, but it is not the training on construction drawings, but should emphasize how to present design idea with logic of construction.

Second, architectural design course should enhance training on thinking logic. Architecture is not only a discipline of strong practical nature, but also a discipline with broad knowledge domain. In reality however, students always think that architectural theories sound very interesting, but they never pay attention to theories actually. In fact, the importance of architectural theories is helping students increase ability in the world of cognition, so design course should be a platform to train students on design methodology based on architectural theories. We usually stress the process of design; the process is not just to show development of design scheme from nothing, but to train design thought in the process of design. Design process is the process of dialectic thinking and deduction with theories, and the form is just the final result. Of course, teaching of architectural theories should also be more plain and clear[14].

Third, architectural design course should enhance training on research and exploration. The discipline of architecture has been constantly updating its knowledge content and technical methods in the process of development, and research implies collecting new information, exploring new fields and discovering new questions [14]. In the discipline of architecture however, the tradition of research has never been emphasized, and students pay little attention to courses of knowledge, mistakenly assuming that knowledge is not related to design. It is worth emphasizing that architectural form cannot be deduced directly from knowledge, but application of new knowledge often brings with innovation of architectural form. We have become aware of lagging behind in term of form innovation, but still not realized that it is closely connected to the fact that we do not value research. Throughout the history, only designers with knowledge are creative, and the innovation of design requires support from new knowledge.

5.Conclusion

As a discipline, the task of architecture is no longer as simple as constructing several usable buildings, challenge of urban high-density material space brought by urbanization has placed questions never encountered before in front of this discipline, among which there are scientific questions worthy of exploration by us, such as the question of correlation between high-density urban physical form and urban microclimate, as well as humanity question worthy of thinking by us. Therefore, architecture requires research, and architectural design requires support of new knowledge.

参考文献:
[1] EGBERT D D. The Beaux-arts Tradition in French Architecture: Illustrated by the Grands prix de Rome [M]. New Jersey: Princeton University Press，1980: xxi-xxii.
[2] 丁沃沃 . 回归建筑本源：反思中国的建筑教育 [J]. 建筑师，2009(8): 85-92.
[3] 当代中国建筑设计现状与发展课题研究组 . 当代中国建筑设计现状与发展 [M]. 南京：东南大学出版社 ，2014.
[4] 维特鲁威 . 建筑十书 [M]. 高履泰，译. 北京: 中国建筑工业出版社，1986.
[5] HEARN F. Ideas that Shaped Building [M]. Cambridge: The MIT Press，2003: 32.
[6] EGBERT D D. The Beaux-arts Tradition in French Architecture: Illustrated by the Grands prix de Rome [M]. New Jersey: Princeton University Press，1980: 3-4.
[7] WEISMEHL L A. Changes in French architectural education [J]. Journal of Architectural Education，1967，21(3): 1-3.
[8] KHAN A Z. VANDEVYEVERE H. ALLACKER K. Design for the ecological age: Rethinking the role of sustainability in architectural education [J]. Journal of Architectural Education，2013，67(2): 175-185.
[9] 闵学勤，丁沃沃，胡恒. 公众的建筑认知调研分析报告[M]// 当代中国建筑设计现状与发展课题研究组 . 当代中国建筑涉及现状与发展 . 南京：东南大学出版社，2014:132-143.
[10] LLLIES C，RAY N. Philosophy of Architecture，Handbook of the Philosophy of Science. Volume 9: Philosophy of Technology and Engineering Science，Elsevier BV，2009: 1199-1256.
[11] CARLHIAN J P. The Ecole des Beaux-Arts: modes and Manners [J]. Journal of Architectural Education，1979，33(2): 7-17.
[12] WRIGHT G. History for Architects [M]//Wright G. The History of History in American Schools of Architecture，1865-1975. Princeton: Architectural Press，1990: 25.
[13] SALOMON D. Experimental Cultures: On the “end” of the design thesis and the rise of the research studio [J]. Journal of Architectural Education，2013，65(1): 33-34.
[14] TEYMUR N. Architectural Education: Issues in educational practice and policy [M]. London: Question Press，1992: 25-35.

南京大学建筑学教育的基本框架和课程体系概述

THE FRAMEWORK AND CURRICULUM SYSTEM OF ARCHITECTURE EDUCATION IN NANJING UNIVERSITY

周凌 丁沃沃

1. 办学历史

2000年，南京大学成立了建筑研究所，同年9月研究所正式招收首届建筑设计及其理论专业硕士生。2006年，南京大学在建筑研究所基础上成立建筑学院。2007年，建筑学院通过建筑学硕士研究生教育评估，获建筑学硕士学位授权，同年9月开始招收本科生。2010年，建筑学院与本校地理与海洋科学学院的城市与区域规划系合并组建成立南京大学建筑与城市规划学院，下设建筑系和城市规划与设计系。目前建筑学科拥有建筑设计及其理论二级学科博士点，建筑学一级学科硕士点，以及建筑与土木工程领域工程硕士点。

2. 学科发展现状与办学目标

南京大学建筑学专业虽起步较晚，但在办学之初，就将“借鉴国外一流大学建筑学办学模式、紧密结合中国特色”作为办学理念。经过近十五年的努力，已初步摸索出一套既与国际一流建筑学教育体系接轨，又在当代中国切实可行的培养模式。该模式在中国建筑教育中开启了建筑学通识教育和本硕贯通培养之先河，即“2（通识）+2（专业）+2（研究生）”模式。目前，南京大学建筑学专业已经初步建立起从本科到硕士的连续贯通的完整教学体系。在这个体系中，建筑学专业的基础教学在南京大学“三三制”教改体系中统一安排，让学生在大学低年级接受宽基础的通识教育，此举与国际一流大学建筑学专业实施的教学计划与课程设置相同或相似。在“三三制”教改体系中的通识教育基础上，进一步实施分类培养和分阶段培养方案。而建筑学专业研究生教育业已通过国家专业教育评估，获得了建筑学硕士授予权，为建筑学高水平专业学位教育提供了保障。此外，目前国家教育部实施的专业硕士学位政策也鼓励贯彻建筑学专业本硕贯通培养计划，从而为这一体系提供了有力的导向。

南京大学建筑学专业在这一模式和体系上所进行的深入研究和大胆探索，取得了一些经验和成果。自2007年起，这一体系经过了8年的实践，第一届学生已经完成整个本硕贯通教学流程。从教学成果、学生质量、社会反馈等各方面来看，这一体系已初步表现出了旺盛的生命力。这为南京大学建筑学专业进一步深入开展建筑学专业综合改革的探索打下了最重要的体制基础，最大限度地避免了在改革中可能出现的理念重构、学制调整、框架重建乃至队伍重组等问题。

建筑学专业注重学术型、应用型和复合型三类人才的培养。对于不同类型的人才有不同的培养层次或学位相对应：将学术型人才培养定位在博士学位；将应用型人才培养定位在建筑学硕士学位；根据社会和学科发展的需要，在硕士学位阶段培养复合型人才。总体来说，以宽基础的本科教育应对各类型高层次人才的培养，既能满足国家和地方建设对高层次专业人才的需要，又能在学科前沿研究方面和国际一流大学开展合作。

3. 和国际上的比较

除了与学校层面的教学体制改革同步，还充分考虑建筑学专业教育的特点，南京大学建筑学教育的目标是瞄准国际一流的建筑学院和建筑系，建立与国际同步的建筑教育体制和体系，教学质量达到国际上较高水平。

建筑学专业教育分为两个层次，即以培养的学生所获取的专业学位来划分的建筑学学士和建筑学硕士两个不同层次，分别应对不同的社会需求。国外著名建筑院系大多都把培养目标定位在培养建筑领域的高端人才，以美国为例：全美排名前列的著名建筑院校基本上不设本科，如哈佛大学和哥伦比亚大学只有建筑学的研究生教育；或者采取通识的本科教育和研究生的专业学位相结合，如麻省理工大学、普林斯顿大学、耶鲁大学、宾夕法尼亚大学等。麻省理工大学的建筑教育明确了建筑学的本科学位是普通工学学位，如果需要专业学位，就要继续深造，通过研究生阶段的学习获得建筑学的专业学位——建筑学硕士。美国大学的建筑学本科教学模式过去也是以5年制为主，进入21世纪以来，美国的研究型大学为了应对社会发展的需要，对建筑学本科教育进行了改革，由专业教育转为通识教育+专业培养，将专业深造的过程放入研究生教育。这样做的优势在于，利用通识教育夯实优秀人才知识的深度和广度，应对未来变化对专业转型带来的挑战，而专业教育由研究生阶段完成，此时学生已经比较成熟，以广博的学识为基础更加有利于对专业知识的理解。欧洲的一流大学建筑学教育也分为两种类型：一类以英国为代表的建筑学教育模式，另一类是以德国为代表的建筑学教育模式。前者和美国一流大学相类似，而后者的建筑学教育传统就是本硕连读，建筑学教育的直接出口就是研究生。

南京大学建筑学科在办学之初，充分研究分析过国际建筑学办学的类型和发展趋势，决定了“4+2（3）”的本硕贯通模式，本科阶段4年，以工学学位毕业，硕士阶段2～3年，以建筑学硕士毕业。建筑学职业教育的最终目标在硕士、博士阶段完成。

4. 课程体系介绍

南京大学建筑学课程设置方面，既要考虑南京大学通识教育的需要，又要考虑建筑学职业教育和评估的要求。因此，在通识教育的层面上，建筑学本科学制采用“2+2”模式，即2年通识课程加2年专业课程。为了满足职业教育评估要求，在一年级通识课程中加入造型基础课（相当于过去建筑系美术课的内容），在二年级增加“学科通识”类课程，如建筑设计基础课程，因此专业教育时间上其实是3年。研究生分为学术型和专业型，学术型硕士学制3年，以论文的形式毕业；专业型硕士学制2.5年，以毕业设计的方式毕业（图1～图3）。

由于篇幅限制，下面重点介绍本科阶段的一些课程，以反映教学特色。

一年级美术基础由南京大学建筑学院的老师和南京艺术学院的美术老师共同教学，分

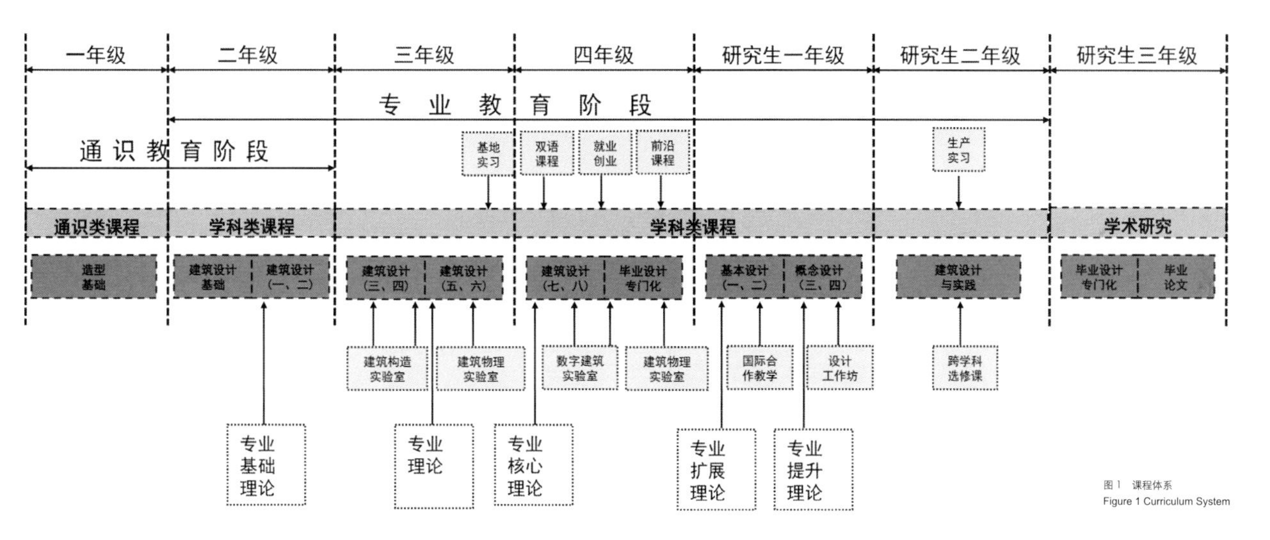

图 1　课程体系
Figure 1 Curriculum System

为美术和建筑两部分内容，美术训练由南京艺术学院的老师指导，在该校完成，建筑训练由南京大学建筑学院的老师指导。造型基础包括三个作业：作业一“动作装置”要求设计一个装置，改变人与场地环境的关系，目的是使学生通过对身体动作的精确分析和有目标的干预，初步认识身体、尺度与环境的相互影响；作业二“折纸空间”要求利用折叠纸板创造一个复合空间，并通过轴测图和拼贴图，使学生初步掌握二维到三维的转化、分析和体验的转化；作业三“覆盖结构”要求建造一个覆盖大尺度空间作展示用途的结构，目的在于初步理解支撑体系和围护体系，通过对力的关系的分析感受结构美和空间美。整个课程希望综合学生的分析和感受能力，为今后对空间和造型的训练建立正确的基本认识。

二年级上学期“建筑设计基础”课程是建筑学专业本科生的专业通识基础课程。本课程的任务一方面让新生从专业的角度认知与实体建筑相关的基本知识，如主要建筑构件与材料、基本构造原理、空间尺度、建筑环境等知识，另一方面通过学习运用建筑学的专业表达方法，如平立剖面图、轴测图、实体与计算机模型等来更好地掌握建筑基本知识。教学通过认知建筑、认知图示、认知环境等环节建立起学生在这两方面的思维联系，为今后深入的专业学习奠定基础。

二年级下学期“界面限定下的空间组织训练”是在学生掌握了基本的建筑专业知识与表达技巧之后，进行的第一个建筑设计训练。学生需要综合运用在建筑设计基础课程中学习的知识点来推进设计，初步体验用建筑的形式语言组织空间、进行设计操作。课程也希望学生在设计学习开始之初，就关注场地条件与建筑生成之间的紧密关系。在初次设计时，学生对各种设计要素的提取和组织能力是十分有限的，因此教案需要进行简化、抽象与限定。课程首先将真实场地环境限定为垂面和坡面两种基本的空间界面条件，各用一个设计练习加以训练。建筑设计（一）强调竖向界面的限定，建筑设计（二）强调坡地斜平面的界面限定。在形式语言上，要求学生用体块和片墙

二年级上学期	二年级下学期		三年级上学期		三年级下学期		四年级上学期		四年级下学期
设计基础	古玩店设计	茶室设计	赛珍珠纪念馆扩建	大学生活动中心设计	社区商业中心设计	观演中心设计	高层办公楼设计	城市设计	毕业设计
	形式与语言 FORM & LANGUAGE		材料与构造 MATERIAL& CONSTRUCTION	空间与场所 SPACE & PLACE	功能与混合 PROGRAM& MIXURE	流线与公共性 CIRCULATION & PUBLICITY	技术与规范 TECHNIQUE & REGULATION	城市与环境 URBANISM& ENVIRONMENT	专门化 SPECIALISM

图 2　本科生常规课程纲要

Figure 2 General Curriculum Outline for Undergraduate Students

两种基本的形式去填充和划分场地空间，在建筑设计（一）中，要求学生在统一场地以两种形式语言分别做方案，建筑设计（二）则可以综合运用体块和片墙。

三年级上学期有两个设计课程——建筑设计（三）和建筑设计（四）。建筑设计（三）以“材”为主题，关注最基本的建造问题，使学生在学习设计的初始阶段就知道房子如何建造起来，深入认识形成建筑的基本条件（结构、材料、构造原理及其应用方法），同时课程也面对场地、环境和功能问题。课程训练的核心是结构、材料、场地，在学习组织功能与场地同时，强化认识建筑结构、建筑构件、建筑围护等实体要素。建筑设计（四）以“空间”为主题，学习建筑空间组织的技巧和方法，训练空间的效果与表达。空间问题是建筑学的基本问题，课题基于复杂空间组织的训练和学习，从空间秩序入手，安排大空间与小空间、独立空间与重复空间，区分公共与私密空间、服务与被服务空间、开放与封闭空间。训练的重点是空间组织，包括空间的秩序、空间的内与外、空间的质感及其构成等。课程以模型为手段辅助推敲，设计分体积、空间、结构、围合等阶段，最终形成一个完整的设计。

三年级下学期两个设计课程——建筑设计（五）和建筑设计（六），分别训练复杂建筑的功能和流线组织，以及大跨度结构、声学、视线问题。

四年级上学期有两个设计课程——建筑设计（七）和建筑设计（八）。前者以“高层建筑”为主题，涉及城市、空间、形体、结构、设备、材料、消防等内容，比较复杂与综合，课题采取贴近真实实践的视角，教学重点与目标是帮助学生理解、消化涉及的各方面知识，提高综合运用并创造性解决问题的技能。后者是城市设计，着重训练空间场所的创造能力，熟练掌握城市设计的方法，熟悉从宏观整体层面处理不同尺度空间问题的能力，并有效地进行图纸表达。教学重点在于使学生通过分析，理解城市交通、城市设施在城市体系中的作用。

四年级下学期课程是毕业设计及专门化，两个方向题目分别针对读研和不读研的同学。继续读研的同学选择专门化主题（数字化建造、建筑节能等），由在相应领域有所专长的教师担任导师。不继续读研的学生完成毕业设计，由设计教师担任指导，完成16张A1幅面的图纸。毕业设计题目要求具有综合性和足够的设计深度。

研究生教学是专业教育的提高阶段。研究生的知识体系中，既要有很强的专业性和规范性，又要有很强的创造性和开拓性，这是建筑设计学科的特点。在学制方面，总体课程设置有一定的特点：研究生进入学校第一学年不分导师，统一授课，统一完成设计课的课程教育，避免学生较早进入导师工作室，并带来知识取向单一化、知识结构不完整的弊病；第二学年进行设计实践，可选择在校内或在校外相关共建基地实习，通过一年的实践，达到建筑学专业教育所需要的实践技能；第三年进行毕业论文或者毕业设计。通过课程教育—实践—毕业设计（或毕业论文）的过程，循序渐进地完成研究生的专业教育，达到建筑学学位评估提出的要求。

在设计课课程体系设置方面，学院做出了一定的探索，既强调研究生的基本技能训练，又重视创造性思维开拓。在研究生一年级集中开设研究生设计课，完成两个阶段的设计课程：第一个阶段可选“基本设计”“概念设计”两者之一，分别锻炼学生的基本技能和概念思维方法；第二个阶段可选“建构设计”“城市设计”两者之一，分别起到深化实践技能和开拓创造性视野的作用。设计课第一阶段“建筑设计研究（一）”，时间9周（研究生一年级上学期），同时开设两个课题：基本设计研究（Basic Design），解决功能、空间、场地、建造等基本问题，深化建筑设计过程与设计方法的基本认识与理解；概念设计研究（Conceptual Design），研究建筑空间创意的方法。设计课第二阶段“建筑设计研究（二）”，时间9周（研究生一年级上学期），同时开设两个课题：建构设计研究（Tectonic Design），深化符合建造意义结构与构造的设计过程与设计方法的认识与理解；城市设计研究（Urban Design），研究城市空间形态创意的方法。两个阶段分别提供4位教师各自命题的工作坊（Studio），分属两大课题类。其中有1～2位外聘教师的工作坊，题目根据每次的实际情况归属不同课题。通过两个阶段的学习，研究生提高了设计的技能，也获得了一定的知识积累，掌握了一定的研究方法，为下一个阶段的学习奠定了基础。

此外，每年春季学期末开设4个国际工作营或联合教学，请国际著名院校教师或建筑师任教，时间1～4周，丰富了课程的结构和层次，开阔了学生的视野。总之，以建筑学理论框架串接多元化设计课模块，不仅有利于研究生认知整体的知识框架，而且多元化的建筑设计课程设计训练模块为提高设计能力提供了保障。

建筑教育是一个长期的过程，需要很长周期才能显示其成果。建筑学专业课程设置也是一个错综复杂的综合课题，尤其是在解决通识教育和专业教育的矛盾方面，仍需要持续关注和探索。

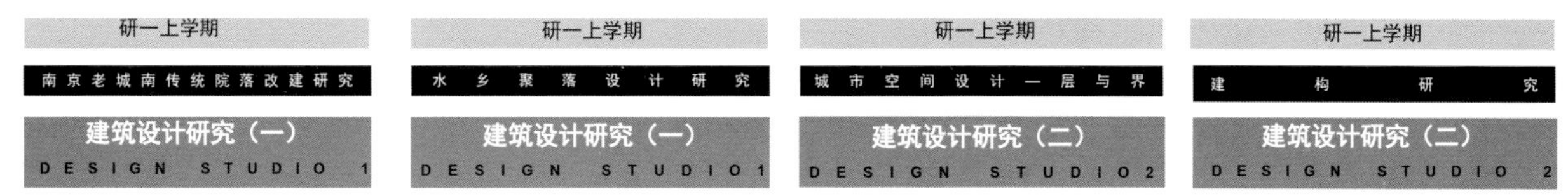

图 3 研究生常规课程纲要
Figure 3 General Curriculum Outline for Graduate Students

1. School History

The Institute of Architecture was founded in Nanjing University in 2000, and the institute started to enroll first batch of postgraduate students of Architectural Design and Theories formally in September. In 2006, the School of Architecture was founded in Nanjing University based on the Institute of Architecture. In 2007, the School of Architecture passed assessment on architectural education for postgraduate students, was approved to grant master degree of architecture, and started to enroll graduate students in September of the same year. In 2010, the School of Architecture and the Department of Urban and Regional Planning of the School of Geographic and Oceanographic Sciences were merged to form the School of Architecture and Urban Planning of Nanjing University, comprising of Department of Architecture and Department of Urban Planning and Design. Currently the school has a sub-disciplinary doctoral program of architectural design and theory, a disciplinary master's degree program of architectural history and theory, and two engineering master's programs of architecture and civil engineering.

2. Current Disciplinary Development and School Objective

Nanjing University started the architectural major relatively late, but since the School of Architecture was founded, it already established the educational philosophy of "borrowing architectural education mode from overseas first-class universities, while closely combining Chinese characteristics". After nearly fifteen years of efforts, it has explored a type of training mode that is not only in line with the architectural education system of international first-class universities, but also feasible in today's China. This mode pioneered a road of general education and undergraduate-postgraduate run-through education for architecture major in China's architectural education, that is, the mode of "2 (general education) + 2 (specialized education) + 2 (postgraduate education)". Currently, a continuous and complete teaching system of architecture from undergraduate to master's program has already taken shape preliminarily in Nanjing University. In this system, fundamental teaching of architecture is arranged in a uniform manner in the "Double-Three" education reform system of Nanjing University, enabling students to receive broadly-based general education in lower grades, which is identical or similar to teaching plan and curriculum implemented for architecture major by international first-class universities. Classified training and staged training are carried out on basis of the general education under the "Double Three" education reform system. And the school's postgraduate education of architecture has passed assessment of national specialized education, is authorized to grant master degree of architecture, which can provide guarantee for advanced level of professional degree education of architecture. In addition, the professional degree policy currently implemented by Ministry of Education also encourages the undergraduate-postgraduate run-through education program of architecture, thus provided strong guidance for this system.

After in-depth research and bold exploration in this mode and system, some experience and achievements have been achieved for architectural education in Nanjing University. Since 2007, after 8 years of practice in this system, students of first enrollment have completed the entire undergraduate-postgraduate run-through teaching process. In respects of teaching achievements, student quality and social feedback, this system demonstrated vigorous vitality preliminarily. It laid down the most important systematic foundation for further exploring in-depth comprehensive reform of architecture discipline in Nanjing University, and avoided problems such as idea reconstruction, educational system adjustment, framework reconstruction as well as restructuring of faculty team that may be encountered in reform to the greatest extent.

Architectural education shall focuses on training of three types of talents, i.e. academic, applied and integrated talents. There are different levels or degrees of education corresponding to different types of talents: training of academic talents is positioned at doctor degree; training of applied talents is positioned at master degree; and integrated talents are trained in the stage of postgraduate education according to demand of social and disciplinary development. In general, by applying broadly-based undergraduate education for training of various types of high-level talents, it can not only meet the demand of national and local construction on high-level professional talents, but also carry out cooperation with international fist-class universities in term of frontier research of the discipline.

3. Comparing with International Universities

In addition to synchronization with educational system reform of the university, characteristics of architectural education are also fully considered. The goal of architectural education of Nanjing University is an internationally first-class architectural school and architectural department, establishing an architectural education mechanism and system in line with international universities, and achieving relatively high education quality among international universities.

Architectural education consists of two levels, that is, training students to obtain two levels of professional degrees of architectural bachelor and architectural master to meet different social demands respectively. Training goals of most overseas distinguished architectural schools and departments are cultivating high-end talents for the field of architecture, in the United States for instance: there has no undergraduate education in top-ranked distinguished American architectural schools, for example, Harvard University and Columbia University only have postgraduate education for architecture major; or apply professional degree education combining general education and postgraduate education, like Massachusetts Institute of Technology, Princeton University, Yale University and University of Pennsylvania. Architectural education of MIT clearly defines undergraduate degree of architecture as a general engineering degree, students must receive further education if they want obtain professional degree, and complete postgraduate education to obtain the professional degree of architecture – master of architecture. Undergraduate teaching mode of architecture in American universities were also mainly a 5-year system in the past, but since the beginning of the 21st century, in order to answer the demand of social development, research universities in the United States carried out reform on undergraduate education, transformed professional education into general education + professional training, and placed further professional training in postgraduate education. One benefit of this reform is that depth and breadth of knowledge of outstanding talents can be consolidated through general education, so as to meet challenge of profession transformation caused by change in the future. And professional education is completed in the stage of postgraduate education when students become relatively mature, and their broad knowledge base can facilitate their understanding on professional knowledge. Architectural education in European first-class universities also consists of two modes: the architectural education mode represented by UK, and another architectural education mode represented by Germany. The former is similar to that of American first-class universities, and the latter is the traditional undergraduate-postgraduate run-through architectural education, which outputs graduate students directly.

When education of architectural discipline was started in Nanjing University, the school analyzed the types and development trend of architectural education in international universities, and determined the "4 + 2 (3)" undergraduate – postgraduate run-through mode: 4 years of undergraduate education, granting engineering degree upon graduation, and 2~3 years of master's degree education, granting master of architecture upon graduation. The final objective of professional education of architecture is completed in master, doctoral stages.

4. Introduction to Curriculum System

For curriculum of architecture in Nanjing University, both the demand for general education in the unversity and requirements for education and assessment of professional education of architecture are taken into account. Therefore, in term of general education, the 2 + 2 mode is adopted for undergraduate education of architecture, that is, 2-year general curriculum plus 2-year specialized curriculum. In order to meet requirements on professional education assessment, basic modeling courses are added in general curricumlum of the first year (equivalent to content of fine arts of architecture in the past), and crouses of "disciplinary general knwledge" are added in the second year, such as fundamental courses of architectural design, so acturally professonal education lasts for 3 years. Postgraduate students include academic and professional students, master's program for academic students lasts for 3 yeas, completed with graduation thesis; master's program for professional students lasts for 2.5 years, completed with graduation project (Figure1~3).

Due to limit of space, only some key courses for undergraduate education are introduced in the following paragraphs to reflect teaching characteristics.

Art basis in the first year is taught jointly by teachers from the School of Architecture of Nanjing University and art teachers from Nanjing University of the Arts, including two parts of arts and architecture; arts training is directed by teachers from Nanjing University of the Arts and completed in the university, while architectural training is directed by teachers from the School of Architecture of Nanjing University. Modeling basis includes three assignments: first assignment is "action device", which requires students to design a device to change relations between human and site environment, aiming to allow students to preliminarily understand relations among body, dimensions and environment through accurate analysis and targeted intervention on body actions; the second assignment is "paper folded space", which requires students to create a composite space with folded carton boards, thus allow students to preliminarily grasp transformation from 2D to 3D, analyze and experience such transformation with axonometric drawing and collage; the third assignment is "covered structure", which requires students to build a large-scale structure of covered space for the purpose of demonstration, aiming to allow students to preliminarily understand supporting system and enclosure system, and to sense the beauty of structure and space through analysis on relations among forces. The whole course intends to integrate analysis ability and sensation of students, and allow them to establish proper fundamental perception to training of space and modeling in the future.

The course of "Basic of Architectural Design" in first semester of the second year is a fundamental course of general knowledge for undergraduate students of architecture. One task of the course is allowing students to perceive fundamental knowledge about physical building from a professional angle, such as main building elements and materials, basic construction principles, spatial scale, architectural environment and other knowledge, and another task of the course is allowing students to better grasp fundamental knowledge of architecture by studying and applying professional presentation methods of architecture, such as horizontal and vertical sectional views, axonometric drawings, physical and computer models. Teaching of this course intends to help students set up associated thinking between these two aspects through perception to architecture, drawings and environment, and lay down foundation for in-depth professional study in the future.

The "Training on Space Organization under Interface Restriction" in the second semester of the second year is the first architectural design training after students have grasp fundamental architectural knowledge and presentation methods. It requires students to push forward design by comprehensively using knowledge learned in fundamental courses of architectural design, preliminarily experience organizing space and to carry out design operation with form language of architecture. The course also intends to allow students to pay attention to the close connection between site conditions and building formation from the beginning of studying design. When carry out their first design task, students have limited ability of extracting and organizing various design elements, so the teaching plan must be simplified, abstract and restricted. In this course, first the real-world site environment is restricted to two basic spatial interface conditions: vertical interface and sloped interface, and one design training is practiced for each condition respectively. Architectural Design 1 focuses on restriction of vertical interface, while Architectural Design 2 focuses on interface restriction of sloped plane. In term of form language, it requires students to fill and divide site space with two basic forms: physical block and unit wall. Architectural Design 1 requires students to complete design at a unified site with these two form languages respectively, and in Architectural Design 2, students can comprehensively apply physical blocks and unit walls.

There are two design courses in first semester of the third year – Architectural Design 3 and Architectural Design 4. Subject of Architectural Design 3 is "material", it focuses on the most fundamental construction issues, allowing students to know how is a building built from the beginning of study phase, obtain in-depth understanding on basic conditions to form a building (structure, materials, construction principles and their application), in the meantime, the course also covers issues such as site, environment and functions. Core issues of the course training are structure, materials and site, while teaches organization of functions and site, it also strengthens understanding of architectural structure, architectural elements, architectural enclosure and other physical factors. Subject of Architectural Design 4 is "space", it teaches students with techniques and methods of space organization, and train them on effect and presentation of space. Space is a fundamental issue of architecture, the course starts from spatial order, based on training and studying of organization of complex spaces, to arrange large and small spaces, independent and overlapped spaces, and to distinguish public and private spaces, spaces serving and being served, as well as open and closed spaces. Key point of the training is space organization, including spatial order, interior and exterior spaces, texture and composition of spaces. With assistance of modeling, the course covers design of partial volume, space, structure, enclosure and other phases, and finally to shape an entire design.

There are two design courses in second semester of the third year – Architectural Design 5 and Architectural Design 6, which aims to train function and flow line organization of complex building, as well as large-span structure, acoustics and sight issues respectively.

There are two design courses in first semester of the fourth year – Architectural Design 7 and Architectural Design 8. The former focuses on "high-rise building", involves city, space, form, structure, utilities, materials, fire protection and other items, it is relatively complex and integrated; the course applies a perspective close to real-world practice, key point and objective of teaching is to help students understand, digest various aspects of knowledge involved, improve ability of integrated application and solving problems in a creative manner. The latter is about urban design, which focuses on training of creative ability in spatial place, mastering methods of urban design, acquiring ability of coping with issues of different scale spaces from a macroscopic and overall perspective, and completing drawing presentation effectively. Key point of teaching is to allow students to understand the role of urban traffic, urban facilities in the urban system through analysis.

Courses in second semester of the fourth year are graduation project and specialized subject, targeting at students who will continue to receive postgraduate education and those will not respectively. Students who will continue to receive postgraduate education choose specialized subjects (digital construction, building energy conservation, etc.), and are tutored by professors from corresponding fields of specialty. Those who will not continue to receive postgraduate education choose to complete graduation project under direction of design teachers, and must complete 16 drawings of A1 sizes. Subject of gradation project must be comprehensive and has sufficient design depth.

Postgraduate education is an advanced stage of professional education. For the knowledge system of postgraduate education, it requires not only strong specialty and standardization, but also strong creativity and pioneering, which are the characteristics of the discipline of architectural design. In term of educational system, the curriculum has some features: postgraduate students will not be assigned to specific tutors in the first year, they take uniform courses, and complete education of design courses in a unified manner, so as to avoid drawbacks such as single knowledge orientation, incomplete knowledge structure due to early access to tutor's studio; design practice is carried out in the second year, students can choose to practice in the school or outside bases jointly established by the school and other organizations, and they will acquire practical skills required for architectural education through one year of practice; the third year is for graduation thesis or graduation project. Through the process of course education – practice – graduation project (or thesis), students will complete postgraduate professional education step by step, and meet requirements for assessment on degree of architecture.

In term of curriculum system, the school made some explorations; it not only stresses basic skill training of postgraduate students, but also emphasizes formation of creative thought. Postgraduate design courses are provided to students in a unified manner in the first year, and students will complete two stages of design courses: students can choose one of the "basic design" and "conceptual design" in the first stage, which aims to train students with basic skills and conceptual thinking methods respectively; students can choose one of the "tectonic design" and "urban design" in the second stage, which aims to deepening practical skills and opening up creative vision of students respectively. The "Architectural Design Research 1" in the first stage of design courses lasts for 9 weeks (first semester of first year of postgraduate education), and consists of two subjects simultaneously: Basic Design, which aims to solve some basic issues such as function, space, site and construction, and to deepen basic cognition to and understanding on design process and design methods; Conceptual Design, which aims to study creative methods about architectural space.

The "Architectural Design Research 2" in the second stage of design courses lasts for 9 weeks (first semester of first year of postgraduate education), and consists of two subjects simultaneously: Tectonic Design, which aims to deepen cognition to and understanding on design process and deign methods in line with structure and construction of architectural meaning; Urban Design, which studies creative methods of urban space morphology. These two stages provide studios with topics assigned by 4 professors respectively, falling into two primary topics. Of which there are studios of 1~2 external teachers, which fall into different topics according to actual conditions each time. After two stages of study, postgraduate students will improve design skills, acquire some knowledge accumulation, master certain research methods, and lay down foundation for study in next stage.

In addition, 4 international working camps or joint teaching are provided in spring semester each year, inviting teachers or architects from international renowned universities, lasting for 1~4 weeks, which can enrich structure and tiers of curriculum, and broaden vision of students. In short, link diversified modules of design courses with framework of architectural theories, which can not only facilitate postgraduate students to cognize the entire knowledge framework, and design training module of diversified architectural design courses can provide guarantee for students to improve design capability.

Architectural education is a long-term process, which requires a long period to see achievements. Curriculum design for architecture major is also a complex, comprehensive subject, especially in the aspect of dealing with the conflict between general education and professional education, which requires persistent attention and exploration.

以地形表达为切入点的低年级设计教学

THE TOPOGRAPHIC MORPHOSIS AS THE START-POINT IN THE ARCHITECTURAL DESIGN COURSE OF FIRST YEAR

刘铨

1. 低年级设计教学中的“场地”与“形式”问题

低年级设计课程首先要让学生理解设计是以解决问题为目的的。如果说“建筑设计初步”的课程让建筑学专业新生接触到了一些基础性的建筑知识和建筑表达方法，那么本科阶段的系列设计课程则是训练他们能够逐步地综合考虑并处理好几个基本的建筑设计问题，即为什么建造（功能与空间）、在哪里建造（场地与环境）以及怎样建造（材料、结构、构造）。

以往的设计课程序列，主要是以功能性空间的组织难易程度进行安排，在场地和建造问题上却缺乏清晰的训练要点和系统安排。特别是作为外在的限制要素的场地与环境问题，在激发建筑空间的生成上和建筑功能有着同等重要的作用，但在低年级的课程设计的任务书中却往往语焉不详，缺少对学生的明确引导，学生容易忽视其重要性。到了高年级，就往往在遇到更为复杂的设计场地与环境条件时，要么视而不见，要么不知道如何入手进行分析。

南京大学建筑与城市规划学院在建筑专业开始的两个建筑设计课程中，明确和强化了场地环境方面的训练要点。虽然课程选择的是城市中的真实场地（一个是老城传统民居地区内的一个地块，另一个是城郊风景区内的一处自然坡地），为学生提供必要的真实情境，但教案将其环境限定因素分别简化为垂面限定和坡面限定。这样，训练的要点可以更加突出，引导性更加明确。既在设计操作过程中引入场地环境要求，使低年级学生了解了如何在在设计中对真实场地环境进行提取分析，却又不会过于增加他们的设计难度（图1）。

其次，低年级设计课程要让学生了解设计的基本过程和方法，形成正确的价值标准。形式不仅体现在设计最终结果中，更是建筑师在解决建筑设计问题过程中使用的基本工具。一个好的建筑师，能够用一个统一的形式策略综合地解决多方面的建筑问题。设计课程必须能够引导学生从解决建筑问题的角度来正确评价形式的“好”与“坏”。

一般来说，在建筑设计入门课程中有各种构成训练来培养学生的空间形式感。但这种单纯的形式训练与建筑设计课程中所要解决的建筑问题往往存在脱节，学生并不知道如何将单纯的空间形式构成原则与建筑问题进行有效地链接。这就往往造成学生的两种设计倾向：一种是陷入以个人好恶为评价基础的形式结果的追求，建筑问题被硬塞到建筑造型的套里；另一种就是对建筑问题的处理缺少整体、清晰的形式逻辑，设计只关注和解决了一部分的问题。同时，形式及其构成规则千变万化，对于低年级学生来说是很难把握的，因此在教学中进行较为严格的限定才能够起到有效的引导作用，使学生更好地聚焦于对建筑问题解决过程的理解。

本文以南京大学建筑与城市规划学院建筑学专业第二个建筑设计课题“风景区茶室设计”为例，以多样化的地形表达为设计切入点，来探讨通过可操作性形式工具的设定，使学生更好地把场地环境的解读与建筑空间的生成联系起来，并推进设计过程的教学方法。

2. “风景区茶室设计”的教案设置

教案的功能设定比较简单，300m^{2}的建筑包括容纳60人的大茶室、30人的雅座和必要的辅助空间，如操作间、厕所、储藏室、值班室等，主要针对基本流线组织与尺度问题进行训练。设计场地为一处城郊风景区内的西向坡地，西侧坡底有南北向景区小路，路西侧为湖面。建筑必须位于山坡之上。课程共8周，第1周除了看场地，学生要在选定的地形表达方向上做一个底板A3尺寸、1:50的局部地形的试验模型。这一周的成果好坏参差，主要的问题是有些学生的形式表达缺乏秩序与尺度的考虑，在材料运用上也不够丰富，缺少想象力。第2~3周通过修改调整，学生基本上完成了对地形的表达。这一阶段教师必须对这些形式表达在后续对建筑生成的可能性上进行把控，帮助学生对形式秩序进行符合建筑需求的调整。例如，调整地形表达单元的尺度，使之适应于建筑内部空间的功能使用要求；调整地形表达单元的方向，以增加景观面、减小体量感使之适应于建筑内外的景观需要等。之后的建筑方案推敲就都基于地形表达模型的推拉、提压等操作之上，方案的修改和推进有了十分明确清晰的可操作性。同时，从山下道路到达建筑入口必须进行室外场地和景观的设计。这就使学生必须在原有地形表达基础上考虑室外场地的景观设计，从而进一步用形式原则将室内外空间的深化联系和统一起来（图2）。

实际上，许多建筑院校课程中都有坡地建筑设计的题目，斜面场地引发了对建筑内部空间标高变化和场地竖向设计的重点训练，但学生对坡地场地条件的理解也多仅限于此。近年来，景观建筑学、地形学等理论的发展，使我们认识到自然地形在创造建筑内外空间逻辑的连续性方面与传统城市空间相比有了更多可以发掘的潜力。

首先，自然地形提供了更多的形态解读可能性。在城市环境中，人工化的环境条件，特别是垂直界面对建筑形态的提示要比自然地形强很多，形态的限定性也就更加大。但自然地形中要植入建筑，建筑师就必须重新将自然地形予以人工化的解读，即“赋形于场地”。以往我们习惯性地使用等高线，不论是图，还是模型。但这一惯用的表达形式其实限制了我们对自然地形的多样化理解。因此，在教案中，为了让学生更容易地找到设计切入点，我们设定了几个不同的地形表达形式——堆叠、阵列、切片、覆盖，通过案例引导学生对场地进行创造性的解读。

其次，这样的解读大大提高了设计过程的可操作性。以往的等高线实体模型（包括计算机模拟地形），在设计过程中的挖、填等操作都十分复杂、费时，使得学生难以真正借助三维模型工具来辅助思考设计方案。新的地形表达除了赋予场地形式秩序，还要求学生充分考虑材料、制作的特性，使设计过程中地形的改变更加易于操作，例如堆叠单元的推、拉，阵列单元的升、降，覆盖单元的伸缩、折叠等。这不仅激发出学生对材料表达可能性的极大兴趣，也使他们获得了对设计的形式操作方法的实在体验。

更为重要的是，形式化的场地为建筑形态的生成提供了有效的提示。原有教学中多运用等高线模型、剖面图等传统手段进行设计训练，因此对于低年级学生来说，很难从场地上直观地提取到对建筑空间形态生成的形态要素。但是形式化了的地形表达，不仅再现了场地，而且产生了形式的秩序。将其作为设计的工具，从场地解读到建筑空间生成的链接就自然形成了。在这一过程中，教师在指导过程中要强调的是场地形式再现中秩序、尺度、方向等需要根据建筑功能要求进行调整、改进的反馈过程（图3）。

3. 成果与讨论

在教案设计阶段，教学组考虑了两个可能的问题。第一个问题是，过强的形式设定会不会限制学生的思维。但由于教学对象是低年级学生，他们对于怎样做设计并不十分清楚，我们将这一阶段教学最主要的目的设定为学习形成设计思路和方法，因此必要的设定有利于教学目标的达成。而且，在地形表达的探索阶段，学生的思路是有很大自由度的。第二个问题是，从地形表达切入，会不会弱化对建筑空间本身的关注。成熟建筑师会从多角度综合考虑建筑问题，但初学者很难做到这一点，所以在教学中需要选择从某一个基本问题入手。因此教案选择了常被低年级学生忽视的场地环境问题作为设计的主要切入点。不过，这也要求教师对形式的建筑空间转化可能性有清晰的认识，才能帮助低年级学生顺利找到可以发展的地形表达形式。这一表达形式只是一个起点，在设计发展过程中也不是不变的，需要根据建筑功能、流线、尺度、建造等问题的反馈进行调整。这一过程也帮助学生更好地理解了形式是如何作为工具来协调和链接不同建筑问题的。

从教学过程和最后的成果来看，有了地形表达阶段的形式设定，学生就能够在这一设计工具引导下顺利、明确地推进和深化设计方案。本教案并不是单纯的教会学生遇到坡地地形时如何进行设计，更多的是让学生学会一种选择和运用适当的形式工具来综合应对多种建筑问题，推进设计过程的工作思路和方法。

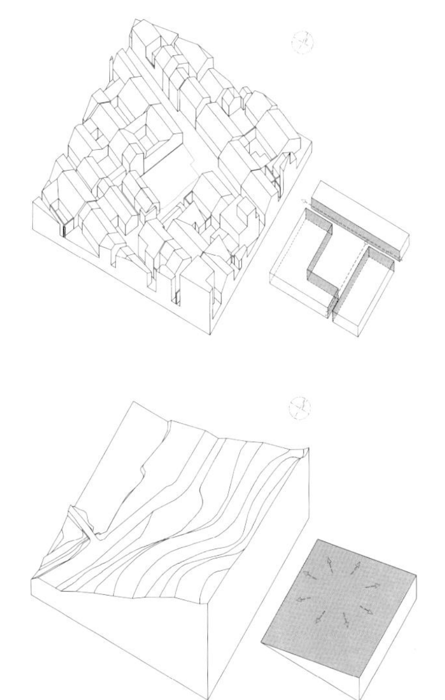

图1　建筑设计（一）与建筑设计（二）课程的场地问题设定

Figure 1 Setting of Site Issue in Architectural Design 1 and Architectural Design 2 Courses

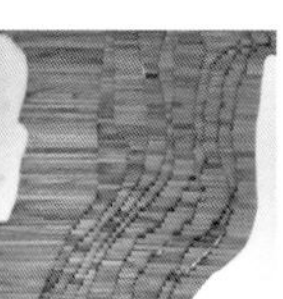
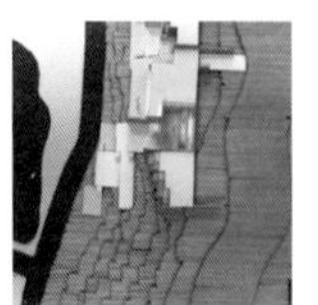

图 2 教学过程中的模型推敲
Figure 2 Refining the Model in the Process of Teaching

1. The "Site" and "Form" in the Architectural Design Course for the Lower Grade Students

A design course for lower grade students should first make students understand that the goal of design is to solve problems. If we say that the course "Preliminary Architectural Design" allows fresh students of architecture to get to know some fundamental architectural knowledge and architectural expressions, then the series of design courses for undergraduates are aiming to educate them gradually to consider and deal with several basic issues of architectural design comprehensively, i.e. why to build (function and space), where to build (site and environment), and how to build (materials, structure, construction).

The sequence of design courses in the past was mainly arranged based on the degree of difficulty for organization of functional spaces, with no clear training focuses and systematic arrangement on the issue of site and building. In particular, site and environment as external restriction factors, which are of equal importance on inspiring the formation of architectural spaces and architectural functions, were not elaborated clearly in the assignment book of course design for lower grade students, and were lack of explicit guidance for students, thus their importance was easily to be neglected by students. When come to higher grades, students often choose to neglect or do not know how to carry out analysis when they are in the face of more complicated design site and environmental conditions.

In two architectural design courses for architecture major started in the School of Architecture and Urban Planning of Nanjing University, training focuses on site and environment are clearly specified and enhanced. Real sites in the city are selected for the course (one plot within the traditional residential area in old town, and a natural slope located in a suburb scenic spot), which can provide students with necessary real-world situations, however, environmental restriction factors in the teaching plan are simplified as vertical-plane restriction and slope restriction. In this way, training focuses can be more prominent, and the guidance can be more explicit. This method not only introduced requirements on site conditions in the process of design, allowing lower-grade students to understand how to extract and analyze real site conditions in design, but also did not increase design difficulty for them (Figure1).

Secondly, design courses for lower grade students should make them understand basic process and methods of design, and form proper value standards. The "form" not only will be reflected in the final results of design, but also is a basic tool frequently used by architects in the process of solving problems of architectural design. A good architect should be able to solve various aspects of architectural problems comprehensively with a uniform form-oriented strategy. A design course must guide students to assess the "Good" and "Bad" properly in the perspective of solving architectural problems.

Generally, there are various formation trainings in introductory courses of architectural design to cultivate students with sense of spatial form. However, these pure form trainings often are separated from architectural problems to be solved in architectural design courses, and students do not know how to effectively link pure formation principles of spatial form to architectural problems. It often leads to two design tendencies among students: one is being trapped in the pursuit of form results with assessment based on personal preference, who often just stuff architectural problems into the jacket of architectural modelling; the other one is only focusing on and solving some problems in the design, with no overall, explicit form logic when handling architectural problems. Meanwhile, forms and their formation rules are constantly changing, which are difficult for lower grade students to grasp, therefore, in order to achieve effective guidance, strict restrictions must be established in teaching program to enable students to better focus on the understanding of the solving process of architectural problems.

In the paper, we took the second architectural design subject "Tea House Design" of the School of Architecture and Urban Planning of Nanjing University as the study case, by using various topographic morphosis as the start point, to search how to make students better connect the interpretation on site conditions to the formation of architectural space by setting up operable form tools, and to advance teaching methods for design process.

2. Setup of Teaching Plan for "Tea House Design"

Function setup in the teaching plan is relatively simple, the 300m^2 building consists of a 60-seat large tea room, a 30-seat private room and necessary supplementary space, such as operation room, restrooms, storeroom and duty room, and trainings were carried out mainly on basic flow organization and dimensions. The design site is a westward slope located within a suburb scenic spot, there has a south-to-north sightseeing footpath at western foot of the slope, and a lake at west side of the footpath. The building must be located on the slope. The course lasted for 8 weeks. During week 1, in addition to field investigation, students were required to make a 1:50 test model of partial topography with A3 base dimensions at selected direction of topographic morphosis. Mixed results were obtained in this week, and the main problem is that the form expression of some students was lack of consideration on order and dimensions, and material application was not rich enough, being lack of imagination. After modification and adjustment in week 2-3, students basically completed topographic morphosis. During this phase, lecturers must control possibility of subsequent building formation based on such topographic morphosis, and help students complete adjustment on form orders according to requirements of building. For example, adjust dimensions of the unit of topographic morphosis to adapt them to functional use requirements of internal space of the building; adjust direction of the topographic morphosis, so as to increase scenic surface, reduce sense of mass, and adapt it to the scenic requirements inside and outside the building. Subsequent refinement of building scheme was all based on operations of the topographic morphosis model such as push-and-pull, and lift and press, so specific and clear operability was rendered to modification and advancement of the scheme. Meanwhile, outdoor pavement and landscape must be designed from

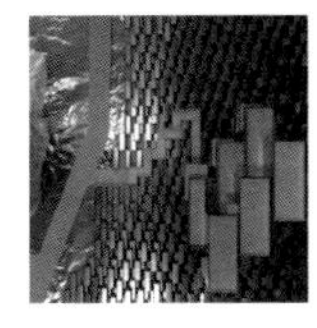  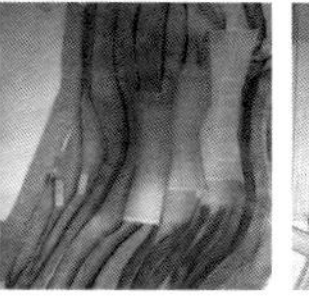 

图 3 地形表达与建筑方案
Figure 3 Topographic Morphosis and Building Scheme

hill-foot path to main entrance of the building. So students must take into account landscape design for outdoor space based on the existing topographic morphosis, thus further linked and unified in-depth connection between indoor and outdoor spaces with form principles (Figure 2).

As a matter of fact, the subject of slope-site building design is included in courses of many schools of architecture, and slope site triggered focused trainings on change of elevations within buildings and on vertical design of the site, however, understanding of students on slope site conditions was often limited in such scope. With development of landscape architecture, topography and other theories in recent years, we realized that greater potential may be explored for natural topography in term of creating continuity of logic of internal and external building spaces in comparison with traditional urban space.

First, natural topography provides more possibilities for form interpretation. In urban environment, artificialized environmental conditions, especially the implication of vertical interface on architectural form is much stronger than natural topography, with greater restriction on the form accordingly. However, in order to implant a building into natural topography, the architect must carry out artificialized interpretation on natural topography, that is, "offering a form to the site". In the past we were used to use contours, no matter in drawing or models. However, this conventional expression in fact limited our diversified understanding on natural topography. Therefore, in this teaching program, in order to help students find start point for design more easily, we set up several different forms of topographic morphosis – stack, array, slices and overlapping, and guided students to perform creative interpretation on site with cases.

Secondly, interpretation in this way increases operability of design process substantially. For traditional contour physical models (including computer simulation of topography), digging, filling and other operations in design process are very complicated and time-consuming, and it is difficult for students to think over design scheme as a supplementary way by really utilizing 3D modeling tools. In addition to offering the site with form order, this new topographic morphosis also requires students to fully take into account features of materials and fabrication, and the operation for change of topography in design process is easier, for instance, the pushing, pulling of stack units, the lifting, lowering of array units, expansion and contraction, folding of overlap units. It not only inspired tremendous interest of students in possibility of material expression, but also offered them with real experience on method of form operation in design.

More importantly, formalized site provides effective implications for the formation of architectural form. In existing teaching plans, traditional methods such as contour models and profiles are frequently used for design training, so it is difficult for lower grade students to extract form factors generated from architectural space form visually at the site. However, formalized topographic morphosis not only represented the site, but also generated order of forms. When it is used as a design tool, the link from site interpretation to formation of architectural space would be formed naturally. In this process, what the lecturer should emphasize in instruction process is the feedback process of modification, improvement on order, dimensions and directions in representation of site form that are required so according to functional requirements of the building (Figure 3).

3. Results and Discussions

In the design phase of teaching program, the teaching group considered two possible questions. The first question is that if the over-strong form setup confines the thought of students. However, the teaching program is targeting at lower grade students, they are not very clear about how to carry out design work, and the primary teaching objective of our program in this phase is how to shape ideas and methods of design, so this necessary setup can facilitate the achievement of teaching objectives. In addition, in the exploratory stage of topographic morphosis, thinking of students is still quite free. The second question is that if attention on architectural space will be weakened by making topographic morphosis as the start point. A mature architect will consider architectural issues comprehensively from many angles, but it is hard to do so for a learner, so it is necessary to choose one basic issue as start point in the teaching program. To this end, the issue of site conditions that is often neglected by lower grade students is selected as the primary start point in this teaching program. Nevertheless, it requires lecturers to have clear understanding on possibility of architectural space transformation for the form, so they can help lower grade students successfully find a form of topographic morphosis that can be developed. This form of morphosis is just the beginning, and it is not invariable in the development process of design, which requires adjustment based on feedback on such aspects as building functions, flow lines, dimensions and construction. This process can also assist students to better understand how does the form coordinate and link different architectural issues as a tool.

According to teaching process and final results, with this form setup in the phase of topographic morphosis, students can successfully and explicitly advance and detail design scheme under guidance of this design tool. This teaching program is not purely to teach students how to design when they are in the face of slope topography, more importantly, it is to enable students to learn how to choose and apply proper form tools to handle various architectural issues from an overall perspective, and to advance working ideas and methods in the process of design.

参考文献：

[1] 丁沃沃，刘铨，冷天．建筑设计基础［M］．北京：中国建筑工业出版社，2014.
[2] 丁沃沃．过渡与转换——对转型期建筑教育知识体系的思考［J］．建筑学报，2015（5）：1-4.
[3] 华晓宁．地形建筑［J］．现代城市研究，2005（8）：64-72.
[4] 刘铨．建筑设计基础课程中的城市空间认知教学［M］//全国高等学校建筑学科专业指导委员会．2012全国建筑教育学术研讨会论文集．北京：中国建筑工业出版社，2014：320-324.

引入建构的构造课教学——南京大学建筑与城市规划学院构造课教学浅释

INTRODUCING TECTONICS IN CONSTRUCTION COURSES: TEACHING CONSTRUCTION IN THE SCHOOL OF ARCHITECTURE AND URBAN PLANNING OF NANJING UNIVERSITY

傅筱

在我国建筑学教育中，构造课很像是给学生开设的“中药铺”，内容繁杂，难上难学，究竟难在哪里？如何改观？南京大学建筑与城市规划学院教学组对此进行了深入的思考和实践，尝试从构造课教学评价标准、构造课知识架构以及构造教学体系化等方面进行探索，通过近几年来的教学实践，取得了较好的教学效果，借此机会，浅释成文，以飨同行。

在阐述南京大学的构造课教学之前，有必要讨论一个关键问题，那就是在建筑学教育中该如何定位构造课，定位的不同，必然带来教学观和教学方式方法的差异。回顾构造课在我国建筑学教育中的定位，基本上可以从两个角度来检视。第一个角度是教研室体系划分方式。构造课基本上都是划归技术教研室；第二个角度是构造教材编写。我国通行的构造教材均以详述各种技术做法为主。由此可见构造课在建筑学中被看做纯技术课程。与建筑设计相关课程相比，纯技术课程历来是学生逃课的首选，枯燥的技术原理让学生生厌，庞杂的技术细节让学生无所适从，如何让构造课生动而不枯燥、系统而不庞杂，我们认为解决问题的关键是构造课的合理定位。换言之，就是必须将构造课纳入建筑学范畴，而不只是工程技术范畴，因此我们进行了将建构（Tectonic）引入构造教学的尝试。建构作为一种联系建筑学与工程学的实践操作理论和方法，将让我们以更为全面的视角来看待构造教学，构造将源于技术而超越技术，从纯技术视野进入建筑学的视野，这必然会带来构造教学相应的改变。

1. 构造课教学评价标准的改变

1.1 原理课与设计课之分

在教学时，构造课教师经常会被问及，“学生上完你的构造课，为何很多大样还是不会画？”这类问题反映出构造课的评价是以实用性和技术性为标准，如不仔细甄别，似乎并无不妥。然而，构造课为原理课，如同住宅设计原理课一样，原理课强调理论认知，设计课强调动手实践，二者教学内容及目标虽有关联却也差异甚大。如果以设计课的标准来衡量构造课，教师容易掉进职业技巧训练的陷阱，既担心学生对某个具体节点技术的掌握程度，又担心是否遗漏了某个知识点的灌输。相反，如果以原理课的标准来衡量构造课，教师授知的重点将是从庞杂的技术细节中，有意识地进行筛选，并将之合理串联，从而加强学生对构造整体性原理的认知。

1.2 技术原理与建构原理

事实上，让原理课去承担实训目的是高校构造课的授知误区，更何况构造职业技巧的掌握远非几十节课程就能解决之事，由此南京大学的构造课明确定位为“原理课”，但是对于“原理”二字的理解却与以往不同。在南京大学的构造课中有两种描述原理的关键词，关键词一是自然力、材料、构件和连接；关键词二是沿革、意图、语言和表达。前者是纯粹的技术范畴，而后者属于建筑学范畴。对原理进行区分，其理论基础就是“建构”。关键词一包含的技术原理是构造课必须面对的基本问题，关键词二所包含的内容则是对基本问题的升华，其概念接近于建构中“建造的诗学”之意，与纯粹的“技术原理”相对应，我们也可以姑且称之为“建构原理”以便行文（图1）。

通过原理课与设计课之分，技术原理与建构原理之分，我们将构造课的评估标准从技术实训与理论认知的纠缠中廓清，教师可以将精力从职业技巧实训中解放出来，在有限的课时内教会学生技术原理分析，而建构理论的引入，则让学生理解构造技术原理与设计的关联关系，以及如何在设计中运用构造进行表达，所以，当面对上述提问，我们完全可以这样回答，“学生虽然画得不太好，但是他知道不是所有的大样都值得去表达！”

2. 构造课知识架构的改变

2.1 当前我国构造课知识架构状况

将建构理念引入构造课，除了产生评价标准的变化之外，必然带来构造课知识架构的改变。所谓构造课知识架构，是指教师将构造知识点以何种组成方式传授给学生，而学生经过课程学习后又将获得怎样的构造知识认知结构。我国高校构造课的知识构架从各种通行教材上可见其貌。构造教材通常分为上下两册，上册一般是建筑构件部分，包括从基础至屋顶的所有建筑构件；下册一般以专题形式或者以特种构造形式呈现，一般包括装修构造、声学构造、大跨构造等等。显然，其知识架构是以具体知识点汇集而成，就单论其中的技术原理，也是分散于章节之中，系统性较弱。比如防水，在屋顶、地下室、外墙都会涉及，那它们之间的共同原理又是什么？有无相关的统一阐述？在此，我们无意评价通行教材之得失，因为教材一旦通行，必然顾及各方，难编也难有特色，但其中知识架构却是一目了然的。

2.2 引入建构的构造课知识架构

当我们引入建构理念之后，构造课的知识架构就由单一的技术线路变成技术结合建构两条线路。毋庸置疑，技术原理是构造课的基本问题，是必须面对和充分掌握的，因此技术原理仍然是主线，建构原理是对技术原理的升华，因此是辅线（图2）。技术原理主线包含了建筑材料、建筑构件、建筑连接三大部分。其中建筑材料包含三个内容：砌筑材料、杆系材料、围护材料；建筑构件包含基础、墙体、洞口、楼板、屋顶、楼梯、坡道、电梯；建筑连接包含结构层面连接和建筑层面连接两个部分。从技术原理架构中可以明显地看出建构思想的影响。知识架构编排打破了以建筑构件分类的单一方式，引入了材料和连接两大知识点，并且在讲授过程中是按照材料、构件、连接的顺序进行。

（1）材料作为相对独立的知识点引入

长期以来，我们对建筑构造的理解都是局限在构件之间的关系而忽略材料。事实上，从建构的角度而言，设计可以通过合理处理建筑材料、构件和细部等，从而塑造空间与形态，最终升华为诗意的表达。而材料如同建筑构件一样，既是一个工程技术问题，又可成为直接的设计表达。所以，构造课知识架构包含材料应是情理之中的事。我们借鉴建构的理念将材料分为了砌筑材料（砖、混凝土）、杆系材料（钢、木）、围护材料（玻璃、保温）三大类别。如此分类，将建筑材料与建造方式、受力原理、建筑物理联系在一起，并在讲授中有意识将材料技术与设计表达相关联，从而将材料纳入建筑学的范围，而非单纯材料学的讲授。

（2）建筑构件技术原理编排受建构思想的影响

虽然沿袭了传统的从基础至屋顶的构件分类方式，但是在具体构件的编排中包含了构件沿革、构件技术、构件与设计表达三个内容。这样的编排是以技术为核心，向上以历史沿革为引导，让学生了解构件之来源，向下以设计为依托，让学生明白技术之用途，其中建构思想的影响是显而易见的。例如在基础的讲解中，从古代的夯土基础一直沿述至今天的桩基础，然后再具体讲解基础的力学、类型、埋深和选型，最后以案例的方式讲解基础与大地的关系，从而让学生形成一个完整的构件知识结构认知（图3）。

（3）将连接原理引入构造课

稍有经验的建筑师都会发现连接是构造的基本原理之一，连接存在于体系之间、构件之间、材料之间，连接不仅是技术性原理而且是具有表现性的方法。然而构造课堂上却鲜有教师将其独立出来专题讲授。究其原因是传统的知识架构限制了连接原理的集中讲授，连接原理散落于各章节之中，难以系统化和条理化。我们将连接原理独立成章，从结构层面入手一直深入到建筑层面的连接原理。结构层面包含梁、板、柱、墙连接以及结构体系的缝（伸缩、沉降和抗震），建筑层面包含连接的基本原则和方式方法，以及连接与设计表达的关系。

（4）建构原理并非构造课的专题讲授内容，而是作为一种认知构造的方法

具体而言，建构原理在以下几个方面影响着南京大学构造课的教学。首先，如前文所述，建构的思想直接影响了构造技术知识架构的编排。其次，在建筑材料和构件的讲解中引入了沿革的视角。沿革分为两个方面，一是技术沿革，也即是技术自身的来龙去脉，例如砖从远古至今的发展历程；二是设计沿革，也即是人们（设计者）如何使用砖的历程。关于沿革讲解所占课堂比例并不多，但不可或缺，其目的一方面是引起学生的兴趣，更重要的是让学生建立构造是设计语言而非纯技术的观念，构造可以被善加运用而成为建筑语言，从而升华成设计的表达。再次，强化案例教学法引入。案例选取不只是构造做法的工地实景照片，而包含基于意图表达的设计案例。这一环节十分重要，它让学生兴趣浓厚，并直观地理解了构造技术如何上升为设计意图的表达，甚至是“诗意的建造”（图4）。

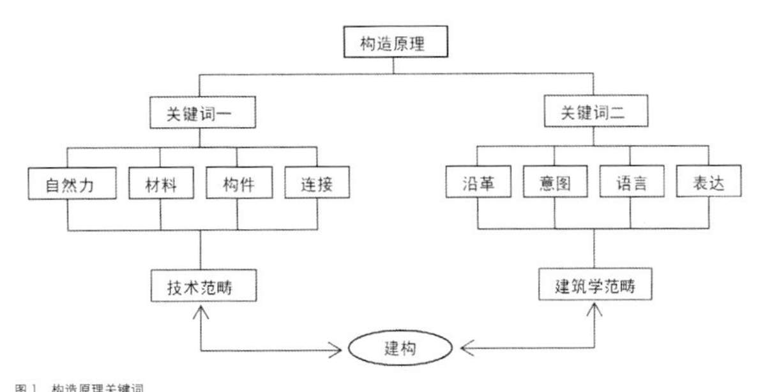

图 1　构造原理关键词

Figure 1 Keywords of Construction Principles

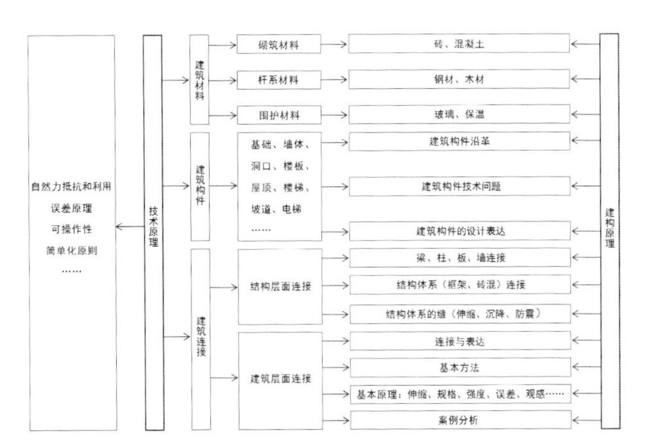

图 2　构造课知识架构

Figure 2 Knowledge Structure of Construction Course

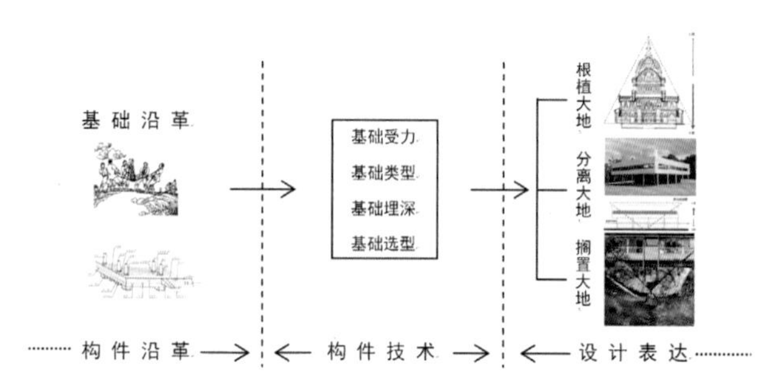

图3 以基础为例的建筑构件技术原理编排
Figure3 Arrangement of Building Component Technologies and Principles for Foundation

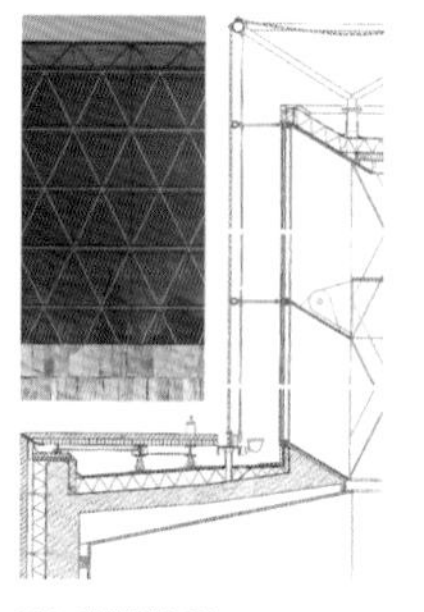

图4 案例教学的引入
Figure 4 Introduction of Case Teaching

2.3 知识架构的整体性把握

所谓知识架构的整体性把握是指在庞杂技术细节中，教师可归纳出一些主要的技术原理来统领技术细节。这些主要技术原理包括“自然力的抵抗和利用、误差原理、可操作性以及简单化原则”等等。例如自然力的抵抗和利用原则是指几乎所有的构造处理都包含了对自然力的反应，对待自然力不仅仅是抵抗，也可以是利用，如抵抗雨水的构造、抵抗侧推力的构造等等（图4）。把握这样的原理将帮助学生读懂构造大样和培养自我判断构造对错的能力。但由于篇幅所限在此不能一一展开论述，这几个整体性技术原理是判断构造设计的基本标尺之一，教师在分析构造时如能将其贯穿进去，将让学生建立起构造的技术理性思维，这较之让学生能够描摹一个复杂的变形缝大样来说，其意义不言而喻。

3. 构造教学体系的改变

构造教学在南京大学建筑学教育中有着十分重要的地位，但是这个重要性不只是由传统的建筑构造课来担纲的，而是依托于一个体系化的构造教学。南京大学的建筑学教育是“2+2+2”的体系[1]，从本科至研究生是一个连贯的教学过程，每个阶段均贯穿了不同形式的构造教学，并在教案和教师安排上具有一定的连贯性和相关性（图5）。

在第一个“2”模式阶段（本科1~2年级），学生将对建筑基本构件进行认知学习，例如在窗构件的学习中，学生首先实际测量1:1的木、铝窗模型，然后将其转换为1:5的徒手大样图纸，这是从物到图的过程（图6）。接下来，学生将根据8个常用的外墙饰面做法大样图纸，制作纸质的1: 2构造模型，这是从图到物的过程（图6）。通过从物到图再从图到物的双向训练，让初涉建筑的学生较为直观地体会到受力、材料、连接、形式之间的关联关系，为进一步的构造学习打下认知基础[2]。在第一个“2”阶段结束，学生将接受严格的古建筑测绘训练，古建测绘训练由经验丰富的教师担任，其目的是通过测绘进一步让学生了解传统建筑构造知识，而非单纯的测绘记录。

在第二个“2”模式阶段（本科3~4年级），学生将学习“构造原理”以及“工地实习”两门课程。南京大学的工地实习并非简单的工地参观，而是将其视做一次重要的构造设计训练，并将其作为构造原理课教学效果的中期检验。工地实习除了要求学生提交传统的实习报告之外，还要求学生根据现场的观察，绘制1:20的工地建筑的外墙构造大样。在绘制过程中教师只给学生1:100的建筑图纸，要求学生根据观察和构造课学到的构造原理自己设计出与建筑立面一致的构造大样。在实习过程中，教师进行设计辅导，并组织实习答辩（图7）。此外，在每次本科设计课中，学生均被要求绘制1:20以上的构造大样，这也加强了构造教学体系化的建立。

在第三个“2”模式阶段（研究生阶段），学生将选修“建造技术研究”课程，该课程包含两门子课程，一是建构设计，二是结构概念设计。建构设计主要是通过设计训练学生对设计概念与构造技术的关联性认识（图7）。结构概念设计主要在学习结构基本知识的基础上，掌握结构与功能，结构与空间、结构与建造的关联关系[3]。研究生建造技术研究课程虽然难度较高，但前面的结构、构造课程训练为其打下了较好的基础，反之建造技术研究课程也是对前面的教学效果的一次有效检验（图8）。

通过体系化的构造教学，可以达到以下几个教学效果。首先，通过连续不断的构造理论、构造实训的学习，加强了学生的构造意识培养，即构造等同于场地、空间、功能、结构、形体，是组成一个建筑必须的基本要素之一，这从本科第一个课程设计题目“材”的训练就可见一斑。其次，每个课程的教学效果都应有其相应的理性评价，体系化的构造教学从根本上解决了构造课的评价标准问题，其教学效果完全可放之于构造教学体系中去检验的，比如，工地实习、本科课程设计以及研究生的建造技术研究等课程既是构造技术实训，同时也是对原理课的有效检验。第三，构造教学体系的建立，促进了不同课程之间的合作与交流，课程之间的协作性和连贯性得到了加强。

4. 结语

虽然南京大学的建筑学教育办学时间不长，但也省去了一些包袱，可以较为自由地思考一些问题。总体而言，南京大学建筑构造课主要从三个方面进行了探索。首先，最为重要的探索是在整体教学框架中强化了构造教学体系的建立，构造将伴随学生整个在校学习过程，其目的让学生通过构造知识的不断学习，增强一种从建造本质上理解设计的意识；其次，在评价标准上划清了原理课与设计课的区别，让构造原理课回归理论认知的范畴，而将绘制构造大样的实训技能放到了工地实习、课程设计、建造技术研究等课程之中；第三，将建构理念引入构造原理课，由此改变了传统构造课的知识架构，由传统的单一的构造技术讲解变成构造技术结合设计的讲解。构造原理课知识架构在一定程度上借鉴了ETH的教学，但在教学方法以及具体内容编排上却有较大的不同，由于篇幅所限难以展开论述。

值得一提的是，当建构引入构造教学之后，任课教师的选择及其知识结构也将发生改变，成熟的建筑职业素养、良好的建筑学理论素养、丰富的教学经验都将成为任课教师的必要知识结构，这将打破构造课只重视教师对于构造技术知识的掌握能力而忽视教师的综合能力的传统标准。事实上，放眼国际，这样的选择标准已是基本的准则。综上，南京大学所做努力，如能为同行提供一定参考，甚是欣慰，或就此能得到同行的关注与批评，从而促进南京大学构造教学的进一步反思与改进，甚是幸事。

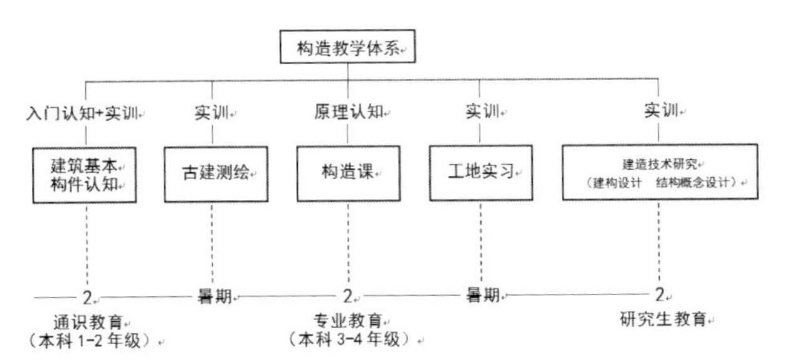

图 5　南京大学构造教学体系
Figure 5　Construction Teaching System at Nanjing University

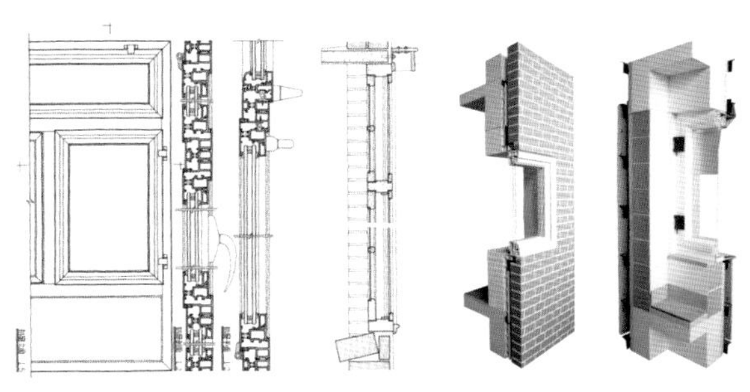

图 6　从物到图的手绘窗大样图与图到物的窗实体模型制作
Figure 6　Freehand Window Detail Drawing from Object to Drawing & Physical Model of Window from Drawing to Object

In the Chinese architectural education system, construction courses are quite like a "shop of traditional Chinese medicines", with numerous and diversified items, extremely difficult to teach and study, but where do the difficulties lie in? How to improve? The teaching group of the School of Architecture and Urban Planning of Nanjing University carried out in-depth thinking and practice on this topic, attempted to explore from the aspects of teaching evaluation criteria, knowledge structure and teaching system of construction courses, achieved good teaching results after several years of teaching practice, and hereby the author would like to take this opportunity to share these efforts with peers in this short paper.

It is necessary to discuss a key issue before introducing the teaching of construction courses at Nanjing University, that is, how should we position construction courses in architectural education, since different positioning would certainly bring different teaching concepts and teaching ways and methods. In order to check positioning of construction courses in Chinese architectural education system, we could examine it from two different angles basically. The first angle is the dividing mode of the system of teaching and research offices. Construction courses are generally placed under teaching and researching office of technologies; the second angle is the edition of construction textbooks. Currently popular construction textbooks in China are all focusing on describing in detail various technological practices. It shows that construction courses are deemed as pure technological courses in architectural education. In comparison with courses about architectural design, pure technological courses are top choice for students skipping class; boring technological principles make students tired, complicated and complex technological details make students not know what to do. How can we make construction courses lively instead of boring, systematic instead of complicated and complex? I think the key of solution lies in reasonable positioning of construction courses. In other words, we must incorporate construction courses into the scope of architecture, other than scope of engineering technology; therefore, we attempted to introduce tectonics into teaching of construction courses. As theories and methods of practical operations linking architecture and engineering, tectonics allows us to view construction teaching from an more overall perspective, and it is derived from technology but beyond technology, enters the perspective of architecture from the perspective of pure technology, which will no doubt lead to change of construction teaching accordingly.

1. Change of Evaluation Criteria of Construction Courses

1.1 Difference between Principle Course and Design Course

In teaching process, teachers of construction courses often were asked, "Why many detail drawings are still impossible tasks for students after finishing your construction course?". Such question shows that evaluation criteria of construction courses are based on practical nature and technological nature. It seems nothing wrong if we do not carefully examine this viewpoint. However, construction courses are principle courses, like principle courses of residential design, they emphasize cognition to theories, while design courses emphasize practice; they are substantially different in terms of teaching content and objective in spite of some connections between them. If we evaluate construction courses with criteria of design courses, teachers may be caught in the trap of professional skill training; they worry about if students already mastered specific detail technology, and in the same time, worry about if he/she missed the imparting of certain knowledge point. In contrast, if we evaluate construction courses with criteria of principle courses, the teaching would focus on screening technological details from a complex system consciously, linking them reasonably, so as to enhance students' cognition to overall principles of construction.

1.2 Technological Principles and Tectonic Principles

In fact, fulfilling objective of practical training with principle courses is a mistake of teaching construction courses at institutes of higher education, let alone that mastering professional skills of construction is by no means a mission that can be achieved in dozens of class hours. Therefore, construction courses are explicitly defined as "principle courses" at Nanjing University, but the comprehension on "principles" is different from that in the past. In construction courses of Nanjing University, there are two types of keywords describing principles; keywords of type 1 are natural force, material, component and connection; keywords of type 2 are evolution, intent, language and presentation. The former belongs to technological scope purely, and the latter belongs to the scope of architecture. For distinguishing principles, the theoretical basis is just the "tectonics". Technological principles included in keywords 1 are basic issues must be faced in construction courses, while content included in keywords 2 is the advancement of basic issues. This concept is

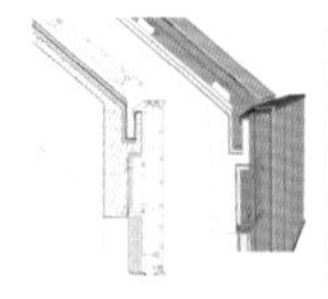
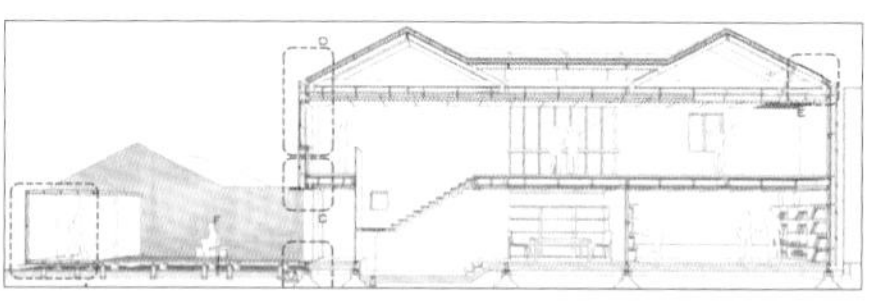
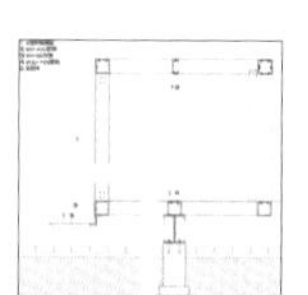
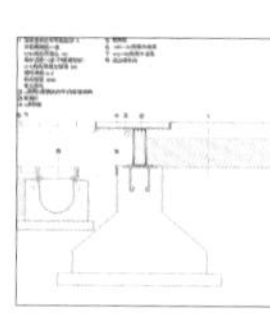
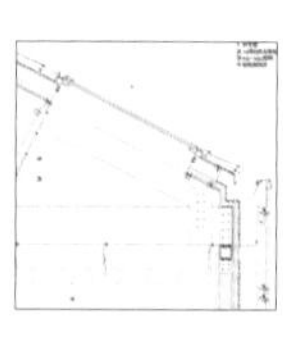
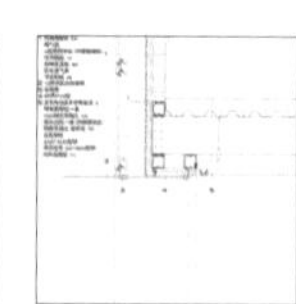

图 7 工地实习与建构设计学生作业
Figure 7 Students' Assignment for Construction Site Practice & Tectonic Design

close to the meaning of "poetics of construction" in tectonics, corresponding to the pure "technological principles", and for convenience of description in this paper, we can also tentatively call them "tectonic principles" (Figure 1).

By distinguishing principle courses and design courses, technological principles and tectonic principles, we clarified elevation criteria of construction courses from tangles between technological training and theoretical cognition, and teachers can release energy from practical training of professional skills to teach students with analysis ability on technological principles within limited class hours. And introduction of tectonics allows students to understand the relationship between technological principles of construction and design, and how to apply constructions as presentations in design. Therefore, when asked with the above question, my answer could be completely like this, "although students cannot draw drawings well, they know that not all details are worth to be presented!"

2. Change of Knowledge Structure of Construction Courses

2.1 Status Quo of Knowledge Structure of Construction Courses in China

In addition to the change of evaluation criteria, introduction of tectonics into construction courses would also cause change of knowledge structure of construction courses undoubtedly. The knowledge structure of construction courses refer to that what composition pattern is adopted by teachers to impart construction knowledge to students, and what cognitive structure of construction knowledge would be acquired by students after studying the course. We can see knowledge structure of construction courses taught at institutes of higher education in China through various popular textbooks. Generally textbooks of construction consist of two volumes; Volume I is about building components, including all building components from foundation to roof; Volume II is generally presented in the form of special topics or special construction forms, including finishing construction, acoustic construction, large-span construction, and so on. Obviously, their knowledge structure is the aggregation of specific knowledge points, of which technological principles are dispersed in different chapters, with weak systematic nature. For example, waterproofing is involved in roof, basement and exterior walls, then what are the common principles among them? Is there any unified explanation? In this paper, we are not intended to evaluate success and failure in popular textbooks, since for a popular textbook, all aspects must be considered certainly, its compilation is difficult and hard to be distinctive, but its knowledge structure would be obvious just at a glance.

2.2 Knowledge Structure of Construction Course Introduced with Tectonics

After tectonics is introduced, knowledge structure of construction course turned from single technological line to double lines integrating technology and tectonics. Undoubtedly, technological principles are basic issues of construction course, which must be faced and fully mastered, so technological principles are still the main line, tectonic principles are advancement of technological principles, hence the secondary line (Figure 2). The main line of technological principles includes three parts: building materials, building components and building connections. Of which building materials include three items: masonry material, framing material and enclosure material; building components include foundation, walls, openings, floors, roofs, stairs, ramps, and elevators; building connections include structural connections and architectural connections. Influence of tectonics can be seen obviously from the structure of technological principles. Arrangement of knowledge structure broke the single approach of classification based on building components, two knowledge points of materials and connections are introduced, and the teaching process is carried out in the order of materials, components and connections.

(1)Introducing Materials as Independent Knowledge Points

For a long time, our understanding on building construction was limited to relations between components, and neglected materials. In fact, from the perspective of tectonics, architects can build space and form by reasonably handling building materials, components and details, and advance it to expression of poetics finally. Like building components, materials not only are issues of engineering technology, but also can become direct design expressions. Therefore, inclusion of materials in knowledge structure of construction courses does make sense. By borrowing the idea of tectonics, we classified materials into three categories: masonry materials (brick, concrete), framing materials (steel, timber), and enclosure materials (glass, insulation). In this way, we linked building materials to construction modes, mechanical principles and architectural physics, and consciously connected material technologies to design expressions in course teaching, thus included materials into the scope of architecture, instead of teaching materials purely.

(2) Influence of Tectonic Ideology on Arrangement of Building Component Technologies and Principles

Although we followed the traditional mode of classifying components from foundation to roof, component evolution, component technology, component and design expression are also included in arrangement of specific components. Such arrangement is around the center of technologies, guiding upward with historical evolution to help students get to know origin of components, and downward relying on design to make students understand purposes of technologies, of which the influence of tectonic ideology is very obvious. For example, when foundation is taught, contents

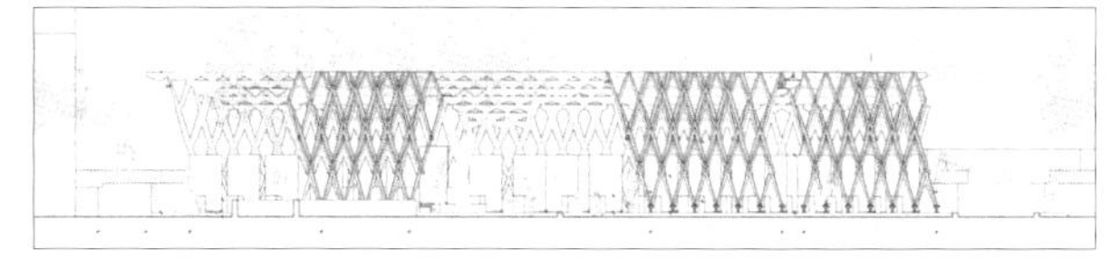
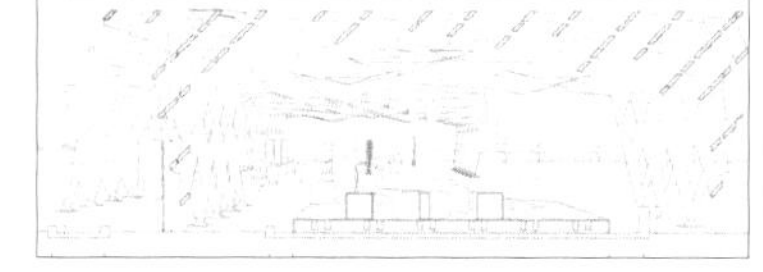
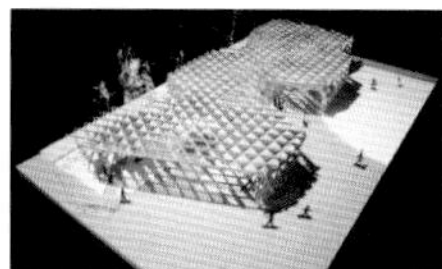

图 8　结构概念设计学生作业
Figure 8　Students' Assignment for Conceptual Structure Design

from ancient rammed soil foundation to today's pile foundation are introduced, and then mechanics, types, placing depth and selection of foundation are elaborated, and relations between foundation and ground are told in the form of case analysis in the last, so as to allow students to achieve entire cognition to knowledge structure of components (Figure 3).

(3)Introducing Connection Principles into Construction Course

Architects with some experience would know that connection is one of the basic principles of construction, which exists between systems, components and materials, and that connection is not only a technological principle, but also a method of presentation. However, few teachers have taught it at construction course as a separated topic. It is because that traditional knowledge structure limited the possibility of teaching connection principles in a concentrated manner, so connection principles were scattered in different chapters, it is hard to be systematic and organized. We compiled connection principles as one separated chapter, starting from connection principles of construction to connection principles of architecture. Content of structure includes beams, slabs, columns, wall connections as well as joints in structural system (expansion, settlement, seismic), and content of architecture includes basic principles and ways and methods of connections, as well as relations between connections and design expressions.

(4)Tectonic Principles are not Teaching Content of Topics in Construction Course, but a Method to Cognize Construction

To be specific, tectonic principles had influence on teaching of construction courses at Nanjing University in the following aspects. First, as mentioned above, tectonic ideology had direct influence on arrangement of knowledge structure of construction technologies; secondly, the perspective of evolution was introduced into the teaching of building materials and components. The evolution consists of two parts: the first one is the evolution of technologies, i.e. the origin and development of technologies, for example, the development course of brick from ancient to today; the second one is evolution of design, i.e. the course of how did people (designers) use bricks. Content of evolution occupies a small proportion of class hours, but it is indispensable. It aims to arouse students' interest on the one hand, but more importantly, to make students establish the idea that construction is design language other than pure technologies, construction can be better applied and becomes architectural language, and be upgraded to expressions of design; on the other hand, intensify introduction of case teaching methodology. Selected cases are not just photos of construction methods at real construction site, and include design cases based on intent expressions. This link is very important, it can trigger students with strong interest, and allows them to understand, in an intuitive manner, how construction technologies are upgraded to expressions of design intent, and even the "poetic construction" (Figure 4).

2.3 Overall Grasp of Knowledge Structure

Overall grasp of knowledge structure means that teachers can summarize some main technological principles from the complex and complicated technological details to guide technological details. Such technological principles include "resistance to and utilization of natural forces, error principles, operability, and simplification principles", and so on. For example, the principle of resistance to and utilization of natural forces means that almost all construction treatments include response to natural forces, and treatment of natural forces is not just to resist, but also could be utilization, for instance, the construction resisting storm water, the construction resisting lateral force, and the like (Figure 4). Grasp of such principles can help students understand construction detail drawings and cultivate their ability of assessing properties of construction. Due to limit of space, it cannot be elaborated one by one in this paper. These overall technological principles are one of the basic benchmarks for assessing construction design, if teachers can integrate them in construction analysis, it will enable students to establish rational thinking on construction technologies, so the significance is obvious in comparison with enabling students to complete detail drawings of a complicated expansion joint.

3. Change of Construction Teaching System

Construction teaching plays a very important role in architectural education at Nanjing University, but this importance is not embodied with traditional construction courses, but a systematic construction teaching program. Architectural teaching at Nanjing University is a "2+2+2" system [1], it is a continuous teaching process from undergraduate to postgraduate, various types of construction teaching courses are incorporated in each stage, and these courses are consistent and related in certain degree in terms of teaching plan and teacher arrangement (Figure 5).

In the first "2" stage (1st~2nd year of undergraduate), students will complete cognitive learning of basic building components. For example, in the process of studying window components, it requires students first to measure a 1:1 wooden-aluminum window model, and then transform it into a 1:5 freehand detail drawing, and this is the process from object to drawing (Figure 6). Next it requires students to complete a 1:2 carton board construction model based on detail drawings of 8 common exterior wall finishing practices, and this is the process from drawing to object (Figure 6). Through the two-way training of from object to drawing and then from drawing to object, it allows fresh students of architecture to understand the relations between forces, materials, connections and types in an intuitive way, and lay down cognitive foundation for subsequent construction study [2]:88-91. When the first "2" stage is finished, students will receive strict ancient building survey training, which is taught by experienced teachers, aiming to allow students to further understand knowledge about traditional architectural construction through surveying, other than pure surveying record.

In the second "2" stage (3rd~4th year of undergraduate), students will study two courses – "Construction Principles" and "Construction Site Practice". Construction site practice of Nanjing University is not just a simple site visit, but an important training of construction design, and is used as mid-term examination on teaching results of construction principle course. For construction site practice, in addition to traditional practice report to be submitted by students, it also requires students to complete a 1:20 detail drawing of exterior wall construction of the building at construction site based on on-site observation. In the process of drawing, the teacher only provides students with 1:100 architectural drawings, and requires students to complete constructional detail drawings in line with building façade based on observation and construction principles learned at construction courses. Teachers coach the design and organize practice defense in the process of practice(Figure 7) . In addition, in each design course of undergraduate education, students are required to complete constructional detail drawings with scale above 1:20, which also enhanced the establishment of systematic teaching of construction.

In the third "2" stage (postgraduate), students will take the course of "Construction Technology Research", which consists of two sub-courses – Tectonic Design and Conceptual Structure Design. Tectonic Design aims to train students with cognition to correlation between design concept and construction technologies through design (Figure 7). Conceptual Structure Design aims to make students grasp relations between structure and function, structure and space, as well as structure and building based on studying basic knowledge of structure [2]. Construction technology research course for postgraduate students is quite difficult, but a good foundation has been laid down through previous training of structure, construction courses, and in return, construction technology research course is also an effective examination on teaching results of previous courses(Figure 8).

Through systematic construction teaching program, we can achieve the following teaching results. First, through continuous study on construction theories, practical training on construction, awareness of students on construction is strengthened, that is, construction equals to site, space, function, structure, and form, and is one of the basic elements required for architecture, which can be seen from the training of first course design topic "Materials" in undergraduate stage; second, each course shall have its rational evaluation on teaching results, systematic teaching on construction provided a solution for evaluation criteria of construction courses, the teaching results can fully examined in the teaching system of construction. For example, construction site practice, undergraduate course design, and construction technology research for postgraduate students and other courses are not only practical training on construction technologies, but also effective examination on principle courses; third, the establishment of construction teaching system facilitated cooperation and exchange among difference courses, so synergy and consistency among courses were increased.

4. Conclusion

Although architectural education at Nanjing University only has a short history, it allows us to get rid of some burdens, so we can think freely on some questions. In general, we made explorations in three aspects for construction courses at Nanjing University. First, the most important exploration is strengthening the establishment of teaching system of construction courses in the entire teaching framework, making construction run through the whole study process of students in school, which aims to make students enhance awareness on understanding design from the nature of architecture through continuous study on construction knowledge; second, evaluation criteria for principle course and design course are distinguished clearly, construction principle course is classified back to the scope of theoretical cognition, while practical training on constructional detail drawings is placed in courses such as construction site practice, course design, and construction technology research; third, tectonic ideology was introduced into course of construction principles, thus changed the traditional knowledge structure of construction courses, from traditional pure construction technology teaching to teaching combining construction technologies with design. In certain degree, the knowledge structure of construction principle course borrowed some ideas from ETH teaching, but it is quite different from ETH teaching in terms of teaching methodology and content arrangement, which will not be elaborated in detail due to limit of space in the paper.

It is worth mentioning that after tectonics is introduced into construction teaching, selection of course teachers and their knowledge structure will also be different, established professional quality of architecture, good theoretical attainment of architecture, rich teaching experience will all become necessary parts of knowledge structure of course teachers, which will break the traditional criteria of only emphasizing mastery of knowledge about construction technologies while neglecting comprehensive competency of teachers. As a matter of fact, at international schools, such selection criteria have become basic criteria. In conclusion, we will be very pleased if the efforts of Nanjing University can provide some reference for peers of architectural education, and we will be privileged if it can cause attention and comments from those peers, thus leading to further reflection and improvement of construction teaching at Nanjing University.

注释

1 从2007年起，南京大学建筑学在原有研究生教育基础上，探索与国际接轨的整体化教学体系，开始实行“2+2+2”的本硕贯通的建筑学教育模式，以探索一条宽基础（通识教育）、强主干（专业教育）、多分支（就业出口多元化）的树形教学之路。

2 结构概念设计课程由毕业于东京工业大学的郭屹民老师指导，同时郭屹民老师还配合该课程主讲结构理论认知。

参考文献

[1]杨维菊.建筑构造设计（上下册）［M］.北京：中国建筑工业出版社，2008.
[2]丁沃沃，刘铨，冷天.建筑设计基础［M］. 北京：中国建筑工业出版社，2014.

作为空间教学的《电影建筑学》课程

CINEMATIC ARCHITECTURE COURSE AS EDUCATION OF SPACE

鲁安东

自现代建筑将空间确立为建筑学的人文原点以来[1, 2]，尽管对空间的知觉和想象能力被普遍认为是建筑师的核心能力之一，针对它的培养方法却较为有限，尤其是对于具体、诗意的空间。一方面摄影、文字描述、拼贴、物理模型等均表现了空间的部分真实特性，另一方面日益普及的计算机工具使得空间成为更加抽象的、分析性的设计对象。在这一背景下，南京大学建筑与城市规划学院自2012年起开设了一门针对空间教学的实验性课程——《电影建筑学》。它试图将一种特殊的空间实践形式——运动影像的拍摄——与建筑学的理论教学结合起来，让学生通过对运动影像的分析、设计和表达提高空间知觉和想象能力，进而对建筑空间的基本问题进行理论性的思考。本文将系统地介绍这一课程的理论基础、工作形式和教学内容。

1.《电影建筑学》课程概要

1.1 电影建筑学的发展

建筑和电影这两种艺术形式都具备空间和时间结构，都带有明显的公共性，都表达着存在空间。在1920年代的现代主义运动中，电影有力地支持了现代主义建筑对于"时空连续体"意识的建构。正如瑞士艺术史学家西格弗里德•吉迪翁（Sigfried Giedion）在论及柯布西耶时指出的，"静态摄影没有清晰地捕捉到它们。人不得不随着眼睛一起运动：只有电影才能让新建筑被人理解"[3]。而德国艺术家汉斯•里希特（Hans Richter）对电影的陈述"电影的独特领域是运动的空间……这个空间既非建构的，也非雕塑的，而是基于时间的，即通过不同属性（光、暗、色彩）的交替创造出的一种光的形式"[4]，则像是对密斯的建筑空间的注解[1]（图1）。在1970年代建筑学自我反思的浪潮中，电影作为"生活空间和生活叙事的动态轨迹"[5]呈现了建筑空间在现实世界中的真实状态，支持了建筑学对混杂城市、异质空间和日常经验的重新定向。1990年代以来，随着数字视频的普及，电影作为建筑媒介的主要障碍不复存在[2]。电影和建筑不再限于彼此启发的两种艺术形式，而是共同构成了一种研究建筑经验、空间使用和城市条件的独立方法。由于它在设计研究上起到的独特作用，"电影建筑学"成为当代建筑重要的试验点之一。

这体现在两个方面。首先，实验性的建筑师更积极地尝试在设计研究中运用影像，例如在荷兰建筑师雷姆•库哈斯（Rem Koolhaas）和委内瑞拉都市智囊团工作室（Urban-Think Tank）那里，影像对城市空间状态（Conditions）的呈现和空间机制（Mechanisms）的揭示有力地支持了一种本地化的、发生型的城市建筑研究[3]。另一方面，部分由于实践的推动，在"电影建筑"研究领域出现了新的热潮[4]。建筑史学家如迪特里希•诺伊曼（Dietrich Neumann）回顾了在不同历史时期电影如何使用建筑来组织和表现空间并反过来构成了一种对建筑的特殊实践[6]，而建筑理论家如尤哈尼•帕拉斯玛（Juhani Pallasmaa）和安东尼•维德勒（Anthony Vidler）则通过对电影媒介的讨论展开了对建筑理论的反思[7, 8]。"电影建筑"这一命题也激发了新的跨学科研究，特别是对建筑和城市空间具有的"电影性"（Cinematics）以及在此基础上对影像空间的新应用领域（例如博物馆、记忆场所等）的探索[9, 10, 11, 12]。

1.2 电影作为空间表现形式的特点

正如戏剧之于新古典建筑、摄影之于现代建筑，我们表现建筑的方式反过来影响了我们思考建筑的方式。建筑设计是在表现媒介的基础上进行的，平面、立面、剖面这类抽象的、分析性的表现媒介在很大程度上将空间"去真实化"了，而我们更加依赖建筑师的知觉和想象能力将被抽象分解的空间整合还原。透视图、轴测图以及实体或数字三维模型可以整体地表现空间的形态、元素和关系，但依然无法捕捉一些最基本的空间特征。与传统的建筑表现媒介相比，电影大大地延伸了建筑学对真实空间的操作范畴，这主要体现在几个方面[1]。电影表现了视觉以外的其他知觉：视觉使人远离空间，而听觉让人回到空间的原点。对身体知觉的表达强化了空间的现场性和亲密性，正如在苏州园林中，大量暗示听觉、嗅觉的题名凸显了空间的现场性（图2 左）[2]。电影表现了运动与时间中的空间。现代空间特有的连续和差异特性使得吉迪翁提出"只有电影才能让新建筑被人理解"（图2 右）[3]。电影表现了身体与空间之间的延伸和互动关系。正如地理学家大卫•西蒙（David Seamon）强调的，日常的身体空间表演是场所感的基础[13] [4]。电影表现了主体的空间经验，并因此打开了空间的情感、记忆和叙事的维度，这些现代建筑学缺失的内容。

1.3 作为空间教学的电影建筑学

自1990年代起，欧美的一些建筑院校开始在常规建筑教学中引入"电影建筑"课程。其中带有一定延续性的有英国建筑联盟学院（AA）帕斯考•舒宁（Pascal Schöning）开设的以心理体验为核心的设计课程[14]、英国剑桥大学弗朗索瓦•潘斯（François Penz）创立的强调影像语言和空间叙事的建筑与动态影像硕士学位课程[15]、瑞士苏黎世联邦高等工业学院（ETH）克里斯托弗•吉罗特（Christopher Girot）建立的关注感知体验的媒体实验室（Media Lab）和相关课程[16]、英国威斯敏斯特大学威廉•费尔布雷斯（William Firebrace）和加比•肖克罗斯（Gabby Shawcross）开设的关注建筑变化的设计课程[17]，以及瑞士门德里西奥建筑学院（Mendrisio）埃里克•拉皮埃尔（Éric Lapierre）开设的围绕建筑氛围的设计课程[18]（表1）。

与上述课程相比，南京大学的《电影建筑学》课程更加注重与建筑教学体系的整合，强调作为基本建筑的空间自身具有的电影性。本课程的前身是2009—2011年开设的短期工作坊课程《电影辅助设计》，并在2012年转为常规课程，而且围绕着建筑空间的基本问题进行了重新设置。主要特点是强调对空间本身特征（包括氛围、时间等维度）和可能性的分析，学生在拍摄影像的过程中利用人物的行动、关系和感受对空间特征进行注解。整个课程由一系列递进的作业构成，每个作业分别对应着特定的理论问题。学生需要在评图环节对自己制作的影像进行分析，并对作业针对的理论问题进行回应。因此，本课程也被定位为一个特殊形式的理论教学，学生需要在拍摄过程中将对空间的感受和分析与对特定理论问题的思考结合起来，用影像语言来寻找和表达自己对空间的独特理解。

1.4 《电影建筑学》课程的几个核心概念

"电影建筑"作为教学命题至始至终谈论的是建筑，"电影"表明的是一种建筑观念，它关注于建筑空间中叙事性的、非物质化的、感知的、诗意的等等内涵，同时强调它们在表面的呈现[19]。为了激发理论性的思考，本课程提出了几个关键性的概念，用于引导学生将注意力从电影本身集中到建筑问题上。

（1）空间的电影性（Spatial Cinematics）。空间本身具有的对人的活动的结构性、场景性和辅助性的功能。一个透视空间因为一览无余而没有变化和悬念，因此是没有时间感的空间。而从电影建筑学的角度，一个好的空间是提供机会和可能性的空间，是人和人的关系发生变化的空间。举例来说，“桥”是一个错失的空间，桥上和桥下彼此能够看见却无法相遇，而“走廊的拐角”是一个遭遇的空间，两端的人彼此看不见却注定相遇。因此“桥”和“L形走廊”在空间本身的电影性上是相反的。通过引导学生对空间本身电影性的发现，这门课试图建立一种摆脱构成美学、注重叙事可能性的空间观念。

（2）叙事表达空间（Narrative Expressive Space）。基于对空间本身电影性的分析，学生需要设计身体的空间表演来演绎空间的可能性。正如伯纳德•屈米（Bernard Tschumi）在《曼哈顿记录》（*Manhattan Transcripts*)中提出的，空间的组织、身体的运动以及空间事件三者的叠加实现了建筑[20]。在这个过程中，身体的运动成为物质空间和空间叙事之间的中介物，它构成了对物质空间的一种特殊的注记（Notation）。

（3）知觉的空间（Perceptual Space）。由于相机有运动的能力，它可以将空间距离的远和近转化为接近或者远离，影像可以带着观众一起运动，并带给她/他强烈的身体感，仿佛相机是我们的身体在电影空间中的替身[5]。在这个密切的“相机—身体”空间之外，影像将真实的空间转译为一个带有情感和记忆的氛围空间（Ambience）。这种对身体空间和氛围空间的区分使学生能够以更加感性的方式去解读和注记空间。例如在《停》这个短片中，女主角被安排在幽暗的亭子内，而男主角被安排在明亮的室外，前者是被限制被窥视的对象，而后者是行动自由的窥视者。这两个角色既受空间限定，又反过来注释了园林空间中内与外、明与暗、动与静的对话（图3）。

2. 课程设置和教学形式

《电影建筑学》教学内容的核心是将拍电影作为一种特殊的空间实践和思考形式。在拍摄过程中，学生既要演又要拍：在演的时候她/他们用身体去感受，同时需要思考自己的身体放置在特定环境中会带给观众怎样的空间感受；而在拍的时候，他们又要用眼睛去观察，感应空间的氛围，捕捉空间的特征，并且根据空间的可能性调度人的表演。在这样往复的过程中，学生逐步养成一种以参与的方式分析空间的思维，进而鼓励她/他们对建筑空间进行理论性的思考。

2.1 课程设置和空间问题

本课程在教学形式上采用了理论教学+作业拍摄+评图研讨的结构，以提出理论问题开始，随后在拍摄的基础上进行思考，最后回归理论讨论。理论教学包括三个讲座：《表达的空间》《影像的逻辑》《存在的直觉》，分别讨论上文介绍的三个核心概念——空间的电影性、叙事表达空间、知觉的空间。课程的主体部分是一系列拍摄作业，而核心难点是如何将一种特殊的空间实践形式——拍摄影像——与建筑学的理论教学进行对接。由于影像媒介有自己的形式语言和空间规律，因此本课程按照空间分析、场所分析和叙事分析三个阶段来设置教学内容，引导学生循序渐进地掌握影像媒介，并在各个阶段探讨相应的空间问题。

（1）空间分析训练的目的是通过对镜头和视角的使用来分析空间的电影性。相关作业有：空间动静分析、空间轨迹分析、空间事件分析、建筑漫步等。

（2）场所分析训练的目的是通过对身体的调度、事件的编排、空间细节和知觉经验的呈现来表达场所的特征。相关作业有：空间交响乐、园林空间等。

（3）叙事分析训练的目的是运用影像语言来分析建筑涉及的心理记忆、日常状态、城市条件等因素。相关作业有：诗意叙事、影像论文等。

本课程为每个拍摄作业设置了对应的空间理论问题（表2）。作业之间既在技术上由易到难，同时在理论思考上也逐渐引导学生从空间自身特性和空间体验这类相对容易理解的因素转向更为抽象和复杂的空间问题。

2.2 工作形式与空间分析

由于影像拍摄工作本身对器材[7]和协作有一定的要求，经过多次试验发现最为理想的合作规模是4人一组，在不需要表演的情况下可减少至2~3人。为了使学生在拍摄过程中将注意力放在对空间的感受和思考上，本课程为小组成员设置了不同的角色，每个角色要对空间进行不同层面的思考和处理。此外学生需要在每个作业中更换新的角色以接受不同的空间分析训练（表3）。

2.3 在工作中思考

《电影建筑学》课程的一个主要目的是让学生通过“拍电影”对建筑空间的基本问题进行理论性的思考。一方面，学生在作业的拍摄过程中需要思考对应的理论问题，另一方面，本课程引入了多种形式的反思过程。首先，在影片完成后，学生需要用图示或模型分析自己拍摄的影像空间——为什么选择某一个或某一组空间？利用了空间的哪些特点或可能性？影像呈现了一种什么样的空间？（图4）其次，每次作业之后都有与设计课程相似的评图环节，学生需要在评图时结合影片、平面图、分析图等陈述自己对理论问题的理解以及影片针对理论问题的设计构思。此外，在最终评图时邀请建筑师、理论家、导演、哲学家等对电影建筑及其空间理论进行研讨，让学生能够从建筑学乃至人文学科的整体视角再一次进行反思。

3.《电影建筑学》教学成果与启示

3.1 在“寻常空间”中发现电影性

本课程第一阶段训练的核心是通过重新审视“寻常空间”，发现空间自身隐藏的电影性。以《空间轨迹》作业为例，虽然作业只要求拍摄一个运动镜头，但难点在于让学生从动态的角度理解空间，例如一个长方形的房间，如果沿着横墙、纵墙或者对角线进行拍摄，空间会呈现为完全不同的面貌。另一方面，镜头运动的空间轨迹不是为了追求新奇的视觉效果，而是基于对空间“电影性”的理性分析，用镜头运动来揭示空间自身结构性的或场景性的可能。此外，学生也需要在这个作业中掌握基本的影像语言。因此在2014—2015学年的课程中，让学生从经典影片中挑选一个巧妙运用空间的镜头，分析它相机的空间轨迹。接着在校园内选择一个具备电影性的场所，设计一个与经典影片镜头相似的空间轨迹对该场所的电影性进行表达。在《囿》这个短片中，教学楼内复杂的走廊系统使得演员可以快速到达空间的各个位置，而相机则被限制在其中一条走廊里运动。由于观众只能看到一条狭长笔直的走廊，演员的出没因此带有了戏剧性，这个戏剧性是由空间自身的结构性特征支持的(图5)。在*Silver*这个短片中，相机的运动轨迹似乎给观众呈现了空间的剖面，在起始位置左侧房间和楼梯间是并置关系，而随着相机的平移，空间关系变为模型室和走廊之间的内外关系，窥视这

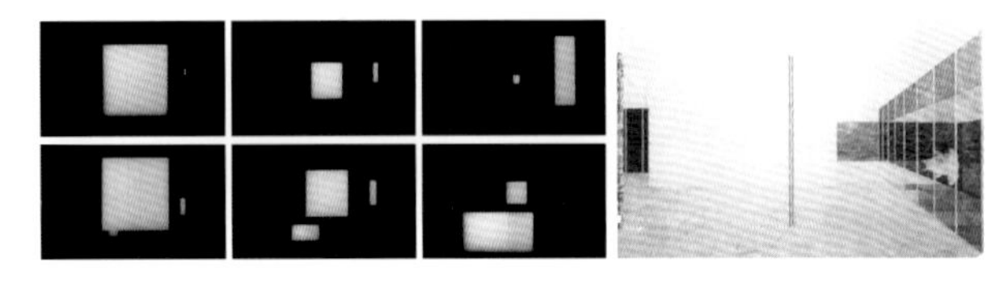

图 1　左：里希特创作的运动影像 Rhythmus 21（1921 年）；右：密斯绘制的巴塞罗那德国馆室内透视（约 1928—1929 年）
Figure 1　Left: Moving Image Rhythmus 21 Created by Richter (1921); Right: Interior Perspective of Barcelona German Pavilion by Mies (roughly 1928-1929)

图 2　左：拙政园听雨轩内景；右：柯布西耶萨伏伊别墅
Figure 2　Left: Interior of Tingyuxuan at Humble Administrator's Garden; Right: Villa Savoye by Corbusier

个动作则成为对这种内外关系的注解（图6）。

作为空间分析训练的第二个作业，《建筑漫步》不仅要求学生基于空间特征对运动进行调度，同时要求从不同镜头角度对空间中的漫步进行拍摄和表现。“建筑漫步”（Promenade Architecturale）这一概念来源于柯布西耶，他认为“建筑通过漫游其间而被体验”以及建筑可以通过是否支持漫游体验而被区分为“死的建筑”和“活的建筑”[21]。同样值得注意的是在柯布西耶的建筑中，通过对楼梯、坡道等空间的表现，“漫游”又成为被看的对象。漫游过程中既看又被看，空间因而具有了表演性（图2）。这个作业要求学生利用空间结构性、场景性或辅助性特征设计一次建筑漫步，并使用连续性剪辑进行表达。其难点在于交替使用主观镜头（看）和客观镜头（被看）来塑造一个带有起承转合的游历体验。在《埃舍尔的楼梯》这个短片中，复杂的旋转楼梯通过镜头画面的捕捉逐渐由正常变得失重和超现实，并最终在一个带有纪念性的画面回复正常（图7）。

3.2 通过“知觉空间”认识场所

本课程第二阶段训练的核心是体验和认知场所。人在身体知觉的基础上感受空间氛围进而建立场所感。彼得•卒姆托（Peter Zumthor）这样描述童年的经历：“我记得脚下砾石的声音，上了蜡的橡木楼梯上闪着微光，当我走过黑暗的走廊进入厨房——这座住宅里唯一真正明亮的房间时，我能听到厚重的前门在我背后关上的声音。”[22] 对空间氛围的感知是迅速而真实的。《场所分析训练》试图引导学生理解场所—知觉—氛围三者之间的关系，并选择了园林空间作为练习的对象。园林在中国建筑空间理论的建构中起到了独特作用[23]，也因此是衔接影像拍摄和理论思考的难得对象。这个作业强调身体知觉体验和空间氛围的统一，要求学生设计的剧情体现园林的某种空间特征，并且在影片完成后用图示表现一个符合该空间特征的理想园林。在《遇》这个短片中，每一个镜头均能看到彼此交错的男女主角，而他们又彼此无法看到。演员用身体诠释了每一个画面中并存着的不同空间，仿佛他们始终游走于以拓扑方式交织着的两个独立空间体系中。这种兼具绵延和并置特征的园林空间使建筑更像是场景调度的媒介。在图示中，学生用墙、廊、亭等建筑元素虚构了一个强化交错体验的园林（图8）。

3.3 用“叙事空间”表达个人的情感和理解。

本课程第三阶段训练的核心是在影像媒介的辅助下对空间提出个人的、思辨性的解读。以《诗意叙事》作业为例，它要求学生在4~5小时内完成对陌生现场的全部体验和拍摄工作，同时也要求学生能够运用影像语言表达身体体验、空间氛围、情感记忆，再现一个带有个人意义的现场，因此它实际上既是对前面所有作业的综合应用，也相当于一个现场考试。作业的难点在于如何建立一个个人的解读，而不仅仅是对现场的记录或者对既有解读（例如建筑师的设计理念）的演绎，这对于拍摄背后的思想性有着更高的要求。另一方面由于不能补拍，学生在现场拍摄时需要留意可能有用的镜头，例如场所的视觉和听觉细节，从而使后期剪辑时有更多的塑造和调整电影空间的余地。

2013—2014学年的《诗意叙事》作业拍摄了绩溪博物馆。由于李兴钢已经用“胜景几何”的概念有力地解读了这个作品，相当于为这座建筑提供了一个标准答案，因此对于学生而言难度最大的是如何在体验这个建筑时找到自己的立场。《听》这部短片用身体直接对建筑进行了注释，对于一个视觉体验如此丰富的场所来说，用身体“聆听”为个人和建筑之间建立了独特的亲密感（图9 左）。短片《漂浮》的拍摄者们敏感地发现了绩溪博物馆在用当代建筑语言转译传统时呈现出的既熟悉又陌生的空间氛围，并用一把透明雨伞对这种超现实的心理空间进行了暗喻（图9 中）。短片《大小》利用儿童作为演员来激发这个建筑中隐藏的微小空间，不同于成人对视觉美学的感应，儿童对空间的尺度以及身体参与的可能性更为敏感。这部影片通过对儿童现场游戏行为的观察和记录呈现了一个充满了身体体验的绩溪博物馆（图9 右）。一个有趣的事实是，在现场拍摄时恰逢绩溪博物馆在清理水池、修缮瓦面、抢救树木，但上面三部短片仍然给我们再现了一个带有强烈美学特征的建筑。而另一部短片《移山》则记录了劳动的场面和施工的细节，通过精心的蒙太奇剪辑表达了在建筑与自然之间的美学关系之下的更为基本的生产性的关系，并为绩溪博物馆中的水池、漏窗、铺地、抹灰等精美细部赋予了更为深刻和动人的含义（图10）。

3.4 跋：启示

本课程的教学实验显示了电影作为空间教学法的巨大潜力——它以更加具体、诗意的方式让我们直面空间本身而不是它的再现。本课程同样显示了影像的拍摄可以很好地结合对空间的思考，而理论性的思考反过来也促进了对空间的知觉和想象能力的培养。本课程通过一系列的任务——在“寻常空间”中发现电影性，通过“知觉空间”认识场所，用“叙事空间”表达个人的情感和理解——引导学生从对空间本身的认识到对空间经验的分析到对复杂空间问题的思辨性的表达。对空间知觉和想象能力的培养是建筑教育最基本的任务之一，其目的不仅仅是为了提高设计能力，更是为了给建筑学一个人文的原点。从“寻常空间”到“知觉空间”到“叙事空间”的过程引导着学生由外向内逐步深化对空间本质的认识。作为一门面向空间教学的课程，《电影建筑学》尝试将体验与思考、分析与表达结合起来。

本课程的理论基础和工作形式在很大程度上基于之前的教学经历，包括：2009-2011年南京大学《电影辅助设计》硕士课程；2010—2011年英国剑桥大学《影像城市》硕士课程；2011—2012年德国德绍建筑研究所《微观城市研究》《电影空间研究》硕士课程。在此感谢丁沃沃教授、Francois Penz教授、Alfred Jacoby教授对上述课程的支持。

图3　短片《停》
Figure 3　Short Film *Stop*

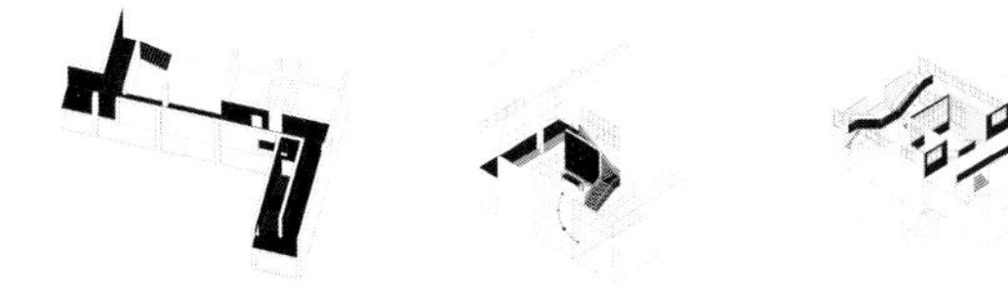

图4　左：短片《对一个楼梯的空间注记》；中：短片《格网》；右：短片《层叠》
Figure 4　Left: Short Film *Spatial Notation to a Stair*; Mid: Short Film *Grid*; Right: Short Film *Overlapping*

表1　代表性的“电影建筑”常规课程

学校	时间	主要导师	课程类型	侧重点
英国建筑联盟学院	1993—2005	Pascal Schöning	硕士设计型课程	情感：心理体验
英国剑桥大学	1998—2005	François Penz Maureen Thomas	硕士研究型课程	叙事：空间叙事
瑞士苏黎世联邦高等工业学院	2003—	Christopher Girot	硕士研究型课程	感知：感知体验
英国威斯敏斯特大学	2008—	William Firebrace Gabby Shawcross	硕士设计型课程	时间：建筑变化
瑞士门德里西奥建筑学院	2012—	Éric Lapierre	硕士设计型课程	氛围：建筑氛围
南京大学	2012—	鲁安东	硕士研究型课程	空间：空间自身的电影性

（注：不含非常规课程或未进入建筑学教学体系的通识课程）

表2　《电影建筑学》课程设置方式及对应理论问题

<table>
<tr><th>阶段划分</th><th colspan="2">作业设置方式</th><th>对应的空间理论问题</th></tr>
<tr><td rowspan="4">第一阶段：空间分析训练</td><td colspan="2">空间分析练习I：本练习的目的是认识空间自身的电影性，同时掌握基本的影像语言</td><td rowspan="2">时间—空间
从“普遍语言”到“时间图像”，影像呈现了空间特有的一种动态美学</td></tr>
<tr><td>I-a：运动——空间动静分析
在寻常空间中发现一个带有电影性的片段，拍摄一个运动镜头和一个静止镜头对它进行表达</td><td>I-b：运动——空间轨迹分析
在寻常空间中发现一个带有电影性的片段，设计一个相机运动路径对它进行表达</td></tr>
<tr><td colspan="2">空间分析练习II：本练习的目的是理解空间—行动—事件三者之间的关系，并用影像叙事来表达空间本身的电影性</td><td rowspan="2">表演空间
电影将身体在空间中的活动再现为一种“表演”，它在实践着空间的同时也反过来塑造着空间</td></tr>
<tr><td>II-a：表演——空间事件分析
在寻常空间中发现一个带有电影性的片段，设计一个利用空间结构性、场景性或辅助性特征的空间事件，并使用连续性剪辑加以表达</td><td>II-b：表演——建筑漫步
在寻常空间中发现一个带有电影性的片段，设计一次游历性的建筑漫步，并使用连续性剪辑从不同镜头角度对建筑漫步进行观察和表达</td></tr>
<tr><td rowspan="2">第二阶段：场所分析训练</td><td colspan="2">场所分析练习：本练习的目的是进一步体验和认知场所，并用影像语言客观地表现场所特征。与第一阶段相比，不再强调身体运动对空间电影性的注记，而更注重身体对空间的知觉体验</td><td rowspan="2">存在空间
电影帮助我们理解经验、感觉和意义在物质空间和精神空间之间的交流</td></tr>
<tr><td>a：场所精神——空间交响乐
挑选一个带有强烈场所感的环境（如历史场所和记忆场所），对该场所进行深度观察，把握它的关键特征并运用蒙太奇手法加以表达</td><td>b：场所氛围——园林空间
挑选一个带有强烈空间氛围的环境（如园林），根据空间氛围设计和编排剧情，主要的行为或事件应符合场所特征</td></tr>
<tr><td rowspan="4">第三阶段：叙事分析训练</td><td colspan="2">叙事分析练习I：本练习的目的是进一步理解情感、记忆、想象创造的空间，要求综合运用影像语言再现一个带有个人意义的现场</td><td rowspan="2">心理空间
电影揭示了空间的情感维度。我们通过情感、记忆和想象使空间适应我们内心世界的形式。空间因而不是物体和身体的容器，而是一种主体投射的产物</td></tr>
<tr><td colspan="2">I：情感空间——诗意叙事
体验和拍摄一个陌生现场。运用影像语言将身体感受、空间氛围、情感表达结合起来。在对陌生场所的再现中表达个体的“诗意的经验”[6]</td></tr>
<tr><td colspan="2">叙事分析练习II：本练习的目的是发现建筑和城市条件在日常空间中的冲突，同时学会运用影像语言表达自己的观点</td><td rowspan="2">日常空间
影像提供的表达性的、参与式的和客观的多种观察视角帮助我们分析真实的城市情境。它特别善于呈现无名的空间实践和现象并揭示它们的意义</td></tr>
<tr><td colspan="2">II：空间实践——影像论文
在城市中寻找一个异化的建筑，通过影像语言以该建筑为例表述自己对“建筑异化”这一命题的理解</td></tr>
</table>

表3　《电影建筑学》的拍摄分工及相应任务

角色	空间分析任务	说明
导演	对叙事进行空间布局	导演将叙事分解为不同场景和动作并在“平面图”上进行调度；此外导演负责控制整个拍摄进程。导演以一种空间“编舞”的方式思考空间和动作之间的关系
编剧	建构场景	编剧需要对一个特定空间的形态和视觉特征进行详细分析，并在空间内将场景分解为一组镜头序列，即通过不同空间视角的组合对场景进行描述
摄像	镜头的视觉构成	摄像需要设计相机的位置、运动路径以及相机与演员或空间内元素（例如洞口或者家具）的相互关系，同时需要确立影像画面的视觉美学
剪辑	塑造电影空间	剪辑将镜头组合成叙事，并对图像和声音的不同效果进行试验。剪辑对电影空间的塑造不需要符合被拍摄的真实空间

图5 短片《回》
Figure 5 Short Film *Hui*

图6 短片 *Silver*
Figure 6 Short Film *Silver*

Since space was determined as the humanistic origin of architecture in modern architecture[1, 2], although perception and imagination of space are universally considered as one of the core competencies of architects, education methods targeted at it was relatively limited, especially for specific, poetic space. On the one hand, photos, text description, collages and physical models all present part of the real characteristics of space, and on the other hand, the increasingly popular computer tools make space become a more abstract, analytic design object. In this context, the School of Architecture and Urban Planning of Nanjing University started an experimental course for space education in 2012 – *Cinematic Architecture*. It tries to combine a special form of spatial practice – film-making of moving images – with the teaching of architectural theory, so as to allow students to improve spatial perception and imagination through analysis, design and expression of moving images, and to conduct theoretical thinking on some basic issues of architectural space. Theoretical basis, working mode and teaching content of the course are introduced systematically in this paper.

1. Summary of the Course of *Cinematic Architecture*

1.1 Development of Cinematic Architecture

Both art forms of architecture and cinema have structure of space and time, have apparent public nature, and express existing space. In the movement of modernism in 1920s, cinema provided strong support to the construction of the awareness of "space-time continuum" of modern architecture. Just as the art historian Sigfried Giedion once pointed out when he talked about Corbusier, "Static images did not capture them clearly. We have to move along with eyes: only cinema made new building understood by us." [3] And description of cinema by German artist Hans Richter, "the unique domain of cinema is the space of movement......this space is not tectonic, nor sculptural, but is based on time, i.e. creating a light form through alternation of different attributes (light, darkness, color)"[4], looks like an annotation to architectural space of Mies (Figure 1)[1]. In the tide of self-reflection of architecture in the 1970s, as "dynamic trajectory of life space and life narratives"[5], cinema presented the real state of architectural space in real world, and supported the reorientation of sophistiated cities, heterogeneous space and daily experience by architecture. Since 1990s, with popularity of digital video, the main obstacle for cinema as an arcitectural medium does not exist anymore[2]. Cinema and architecture are no longer limited to two artistic forms that inspires each other, but jointly become an independent approach of studying architectural experience, space use and urban conditions. Given its unique role in design research, Cenematic Architecture becomes one of the important experimental points of modern archiecture.

This is reflected in two aspects. First, experimental architects were more actively trying to apply images in design research, for example, presentation of urban space conditions and revelation of space mechanisms with images by Dutch architect Rem Koolhaas and Venezuelan Urban-Think Tank provided powerful support to localized, occuring urban architecture research[3]. On the other hand, partially driven by practice, new upsurge appeared in the field of Cinematic Architecture research[4]. Architectural historians like Dietrich Neumann looked back how did cinemas in different period use architecture to organize and present space, and in return, established a special practice of architecture[6], while architectural theorists like Juhani Pallasmaa and Anthony Vidler carried out reflection on architectural theory through discussion on the medium of cinema[7, 8]. The proposition of Cinematic Architecture also inspired new interdisciplinary research, especially the "cinematics" of architecture and urban space, as well as exploration in new application field (e.g. museum, memory place) of image space based on the research[9, 10, 11, 12].

1.2 Characteristics of Cinema as Space Expression Form

Just as drama to new classical architecture, photography to modern architecture, the way expressing architecture in return affected the way of thinking architecture by us. Architectural design is carried out on basis of media of expression, such abstract, analytic media of expression as plane, elevation and profile "de-actualized" space to a great extent, while the perception and imagination that we are more relying on architects are consolidated and restored by abstract, disintegrated space. Perspective drawing, axonometric drawing and physical or digital models can represent forms, elements and relations of space integrally, but still cannot capture some most basic features of space. In comparison with traditional expression media of architecture, cinema substantially expanded the operation scope of architecture to real space, which is reflected in the following aspects[1]. Cinema expresses other perceptions beyond visual sense: visual sense makes people away from space, while auditory sense makes people back to origin of space. Expression of body perception intensifies live sense and proximity of space, for example, a large number of topics implying auditory sense and olfactory sense in Suzhou gardens highlights the live sense of space (Figure 2 Left) [2]. Cinema expresses space in movement and time. The distinctive continuous and differentiated feature of modern space allowed Giedion to point out, "Only cinema can make new architecture understood by people" (Figure 2 Right) [3]. Cinema expresses extended and interactive relations between body and space. As geographer David Seamon emphasized, daily body space performance is the basis of the sense of place[13][4]. Cinema expresses spatial experience of the subject, and hence opens the dimensions of emotion, memory and narratives of space, which are missed in modern architecture.

1.3 Cinematic Architecture as Education of Space

Since 1990s, some European and American schools of architecture began to introduce Cinematic Architecture course into regular architectural education. Those with certain continuity include the design course centring psychological experience taught by Pascal Schöning from British AA School[14], the MPhil program of architecture and dynamic images emphasizing cinematic language and space narrative created by François Penz from University of Cambridge[15], the MediaLab

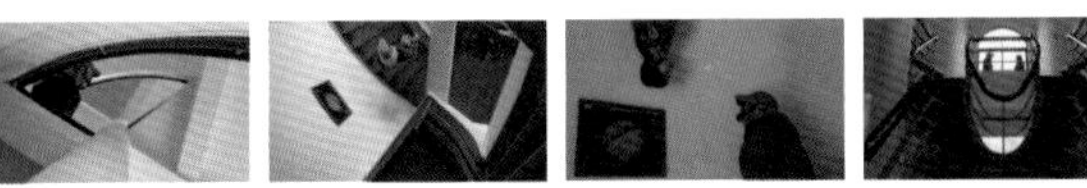

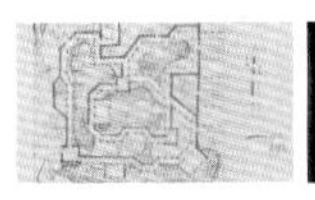

图7 短片《埃舍尔的楼梯》
Figure 7 Short Film *Escher's Stairs*

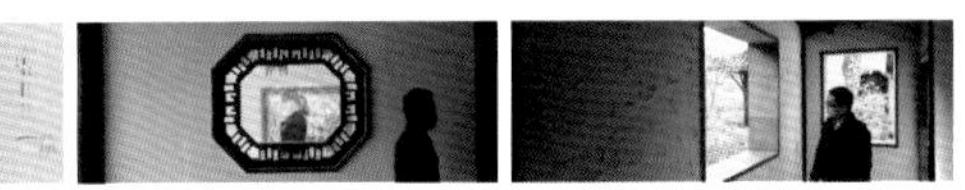

图8 短片《遇》
Figure8 Short Film *Meeting*

and relevant course focusing on perceptive experience set up by Christopher Girot from ETH Zürich[16], the design course focusing on architectural change established by William Firebrace and Gabby Shawcross from University of Westminster[17], as well as the design course centring architectural ambience taught by Mendrisio and Éric Lapierre from Accademia di Architettura di Mendrisio [18] (Table 1).

In comparison with above courses, the *Cinematic Architecture* course at Nanjing University pays more attention to combination with the system of architectural education, and emphasizes the cinematics of space itself of basic architecture. The former of the course is the short-term workshop course *Cinema Aided Design* taught in 2009-2011, which was transformed as a regular course in 2012 and re-arranged centring fundamental issues of architectural space. The main feature is emphasizing analysis on inherent features (including ambience, time and other dimensions) and possibility of space, which requires students to annotate spatial features with actions, relations and feelings of characters in the process of film-making. The whole course consists of a series of progressive assignments, and each of them corresponds to certain theoretical issue. Students are required to carry out analysis on images made by them in the link of drawing review, and respond to theoretical issue in assignments. Therefore, this course is positioned as a special form of theoretical teaching, students are required to combine feelings and analysis of space with thinking of theoretical issues in the process of film-making, and use cinematic language to search and express their peculiar comprehension on space.

1.4 Several Core Concepts of the Course of *Cinematic Architecture*

As a teaching topic, Cinematic Architecture discusses architecture throughout the process, "cinema" refers to a kind of architectural idea, it focuses on the narrative, non-material, perceptual, poetic and other implications, while emphasizes their superficial presentation[19]. To inspire theoretical thinking, this course comes up with several key concepts, for the purpose of guiding students to pay attention from cinema to architectural issues.

(1) Spatial Cinematics. Space itself has structural, scenic and auxiliary functions to human activities. A perspective space has no change and suspended interest because everything is conspicuous, so it is the space with no sense of time. From the angle of cinematic architecture, A good space should be one providing opportunities and possibilities, and the one with change of person-and-person relations. For example, "bridge" is a space of missing, those on and under bridge can see each other but unable to meet with each other, while the "corner of corridor" is a space of meeting, those at two ends cannot see each other but are meant to meet with each other. Therefore, "bridge" and "L-shaped corridor" are contrary in term of cinematics of space. By guiding students to find cinematics of space, this course attempts to establish a kind of spatial concept getting rid of formation aesthetics while emphasizing narrative possibilities.

(2) Narrative Expressive Space. On basis of analysis on spatial cinematics, students need to design spatial performance of body to deduce possibilities of space. As Bernard Tschumi pointed out in *Manhattan Transcripts*, overlapping of spatial organization, body movement and spatial event realizes architecture. [20] In this process, body movement becomes the medium between material space and space narrative, and it constitutes a special notation of material space(Figure 3).

(3) Perceptual Space. Since camera has the ability of movement, it can transform far and close spatial distances into approaching or departing away, images can take audience to move together, and bring her/him with strong body sensation, as if that camera is the substitute of our body in cinematic space. [5] Beyond the close "camera-body" space, real space is translated into a space of ambience with feelings and memories by images. This differentiation of body space and ambience space enable students to interpret and notate space in a more perceptual way. For example, in the short film "*Stop*", the heroine is arranged in a dark pavilion, and the hero is arranged in bright outdoor space, the former is the object being confined and peeped at, while the latter is the peeper with freedom of action. These two characters are confined by space, and in return annotated the dialogue between interior and exterior, brightness and darkness, movement and stillness in the garden space(Figure 3).

2. Course Arrangement and Teaching Mode

The core of teaching content of *Cinematic Architecture* is making film-making as special spatial practice and thinking mode. In the process of film-making, students need to make the film and paly roles in it; they have to feel with body during performance, while think about what spatial feeling it will bring to audience when their bodies are placed in specific environments; when they are making films, they have to observe with eyes, to feel ambience of space, to capture features of space, and to arrange performer's performance according to possibilities of space. By repeating the process, students can cultivate the thinking mode of space analysis in the way of participation gradually, and encourage them to carry out theoretical thinking on architectural space.

2.1 Course Arrangement and Spatial Issues

In term of teaching mode, this course adopted the structure of theoretical teaching + assignment film-making + drawing review and discussion, starting from theoretical questions, and then thinking based on film-making, and getting back to theoretical discussion in the end. Theoretical teaching includes three seminars: *Expressive Space, Cinematic Logic, Intuition of Existence,* which discusses the three core concepts mentioned above - Spatial Cinematics, Narrative Expressive Space, and Perceptual Space. Main part of the course includes a series of film-making assignments, while the core difficulty is how to match the special form of spatial practice – film making – with theoretical teaching of architecture. The cinematic medium has it form language and spatial rules, so teaching content in this course is arranged based on three stages – spatial analysis, place analysis and narrative analysis, so as to guide students to master cinematic medium gradually, and discuss associated spatial questions in various stages.

图9 左：短片《听》；中：短片《漂浮》；右：短片《大小》
Figure 9 Left: Short Film *Listening*; Mid: Short Film *Floating*; Right: Short film *Size*

图10 短片《移山》
Figure 10 Short Film *Removing the Mountain*

(1)Spatial analysis training aims to analyze cinematics of space through the use of scenes and angles. Related assignments include: dynamic and static analysis of space, space trajectory analysis, spatial event analysis, promenade architecture, etc.
(2)Place analysis training aims to express characteristics of the place with presentation of body dispatching, event arrangement, spatial details and perceptual experience. Related assignments include: spatial symphony, garden space, etc.
(3)Narrative analysis training aims to analyze psychological memory, daily state, urban conditions and other factors involved in architecture with cinematic language. Related assignments include: poetic native, cinematic essay, etc.

Corresponding issues of spatial theory are arranged for each assignment of film making in this course (Table 2) . Assignments are arranged from easy to difficult in term of technology, and in term of theoretical thinking, students are guided gradually from factors that are easy to understand such as spatial features and spatial experience to spatial issues that are more abstract and complicated.

2.2 Working Mode and Spatial Analysis

Since film-making work requires certain equipment[7] and cooperation, we found that the most optimal cooperative scale is a group with 4 members after many experiments, which can be reduced to 2~3 members in case of no performance needed.In order to allow students to pay their attention to the feeling and thinking of space, this course defines different roles for group members, and each role should carry out thinking and treatment of different levels of space. In addition, students are required to change to new roles in each assignment, so as to receive different spatial analysis trainings (Table 3).

2.3 Thinking in the Work

One of the primary goals of the course *Cinematic Architecture* is making students to carry out theoretical thinking on basic issues of architectural space through "film-making". On the one hand, students are required to think corresponding theoretical issues in the process of assignment film-making, and on the other hand, many types of reflection process are introduced in this course. First, after the film is completed, students are required to analyze the cinematic space filmed by them with graphics or models – why select one or a group of space? What features or possibilities of space are utilized? What space does the cinema present? (Figure 4) Secondly, there has a drawing review link similar to that of design course after each assignment, students are required to present their understanding on theoretical issues and design idea of the film for theoretical issues by combining film, plans, and analysis graphics in the link of drawing review. In addition, in the final drawing review link, architects, theorists, directors, and philosophers will be invited to discuss cinematic architecture and its spatial theory, so as to allow students to reflect again in an overall perspective from architecture to humanities.

3. Teaching Results and Inspiration of *Cinematic Architecture*

3.1 Discovering Cinematics in Ordinary Space

The core of training in stage I of the course is to review "ordinary space", and discover the cinematics concealed in the space. In assignment *Space Trajectory* for example, the assignment only requires completing a moving shot, but the difficulty is asking students to understand space from a dynamic angle, for example, for a rectangular room, if we shoot along with cross wall, longitudinal wall or diagonal, the space will present different appearance. On the other hand, the space trajectory of camera movement is not to pursue novel visual effect, but to carry out rational analysis on Cinematics of space, and use camera movement to reveal possibilities of structure or scene of space. Moreover, students need to master basic cinematic language in this assignment. So in the course of academic year 2014-2015, students were asked to select a shot of ingenious utilization of space from classic movies, and to analyze its space trajectory of camera. And then select a place with cinematics in the campus, and design a space trajectory similar to that of the shot in classic movie to express the cinematics of the place. In the short film "*Hui*", complicated corridor system in teaching building allows performers to reach specific locations in the space rapidly, and the camera was confined to move in one of the corridors. Since the audience can only see a long, narrow, straight corridor, emergence of performers bears dramatic nature, which is supported by the structural feature of the space (Figure 5). In the short film "Silver", moving trajectory of camera seemingly presents the profile space to audience, left room and staircase at starting point are of the juxtaposition relationship, while along with translation of the camera, the spatial relationship is turned into internal-external relationship between modeling room and corridor, and watching this movement becomes annotation to this internal-external relationship (Figure 6).

As the second assignment of spatial analysis training, *Promenade Architecture* requires students not only to allocate movement based on spatial features, but also to shoot and express promenade in space from different camera angles. The concept "Promenade Architecture" was created by Corbusier, who believes that "buildings can be experienced by promenading among them" and that buildings can be classified as "dead buildings" and "live buildings" based on if they support the promenade experience[21]. It is also worth noting that in buildings of Corbusier, Promenade becomes the object being watched through the presentation to stairs, ramps and other spaces. The promenade process watches and is watched, so performance nature is rendered to space (Figure 2). This assignment requires students to design an architectural promenade with structural, scenic or auxiliary features of space, and use continuous editing to express it. The difficulty lies in the alternative use of subjective shot (watching) and objective shot (being watched) to create a travel experience with beginning, developing, changing and concluding. In the short film "*Escher's Stairs*", complicated spiral stairs changes from normal to agravic and surreal gradually by capturing with shot scenes, and finally returns to normal with a memorial frame (Figure 7).

3.2 Perceiving Place through "Perceptual Space"

The core of training in stage II of the course is experiencing and perceiving place.

Perceive spatial ambience and build the sense of place on basis of physical sensation. Peter Zumthor described his childhood like this, "I remember the sound of the gravel under my feet, the soft gleam of the waxed oak staircase. I can hear the heavy front door closing behind me as I walk along the dark corridor and enter the kitchen, the only really brightly lit room in the house." [22] His perception to spatial ambience was prompt and real. *Place Analysis Training* attempts to guide students to understand relations among place, perception and ambience, and selected garden space as object of the training. Garden plays a unique role in the construction of spatial theory of architecture in China[23], so it is also a rare object to link film-making and theoretical thinking. This assignment emphasizes the unification of physical perceptual experience and spatial ambience, and requires students to design a story presenting certain spatial feature of garden, and express an ideal garden matching the spatial feature with graphics after the film is completed. In the short film "Meeting", hero and heroine staggered with each other can be seen in every frame, but they cannot see each other. Performers interpreted different spaces that coexist in each frame with their bodies, as if they are always walking in two separated spatial systems interwoven in topological pattern. This garden space with both stretching and juxtaposition features makes the building more like the media of scene allocation. In graphics, students made up a garden stressing intertwined experience with walls, corridors, pavilions and other architectural elements (Figure 8).

3.3 Expressing Personal Feelings and Understanding with "Narrative Space"

The core of training in Stage III of the course is coming up with personal, dialectical interpretation on space with assistance of cinematic media. By taking assignment *Poetic Narrative* as an example, it requires students to complete all experiencing and shooting work of a strange site within 4~5 hours, and in the meantime, to represent a site with personal meaning by expressing physical experience, spatial ambience and emotional memory with cinematic language. Therefore, it actually is not only a comprehensive application of all previous assignments, but also equivalent to an on-site examination. Difficulty of the assignment lies in how to build personal interpretation, other than just recording the site or deducing based on existing interpretation (e.g. design concept of the architect), this has higher requirement on the thought behind film-making. Moreover, since reshooting is not permitted, students must notice possibly useful shots in the process of film-making at site, for example, visual and auditory details of the place, so that subsequent film editing may have larger room for creating and adjusting cinematic space.

The assignment "*Poetic Narrative*" of the academic year 2013-2014 filmed the Jixi Museum. Since Li Xinggang has powerfully interpreted this work with the concept of "Scenic Geometry", like provided a standard answer for this building, the most difficult point for students is how to find their own standpoint when experience the building. The short film "*Listening*" interpreted the building with human body directly, for a place so rich in visual experience, unique intimacy between person and building was built with "listening" of human body (Figure 9 Left). Filmmakers of the short film "*Floating*" sensitively discovered the familiar but unacquainted spatial ambience presented by Jixi Museum when they were translating tradition with modern architectural language, and used a transparent umbrella to metaphorize this surreal psychological space (Figure 9 Mid). The short film "*Size*" used children as performers to inspire the micro-space concealed in the building, which is different from adults' sensation of visual aesthetics, and the children's spatial dimension and possibility of body participation are more sensitive. This film presented a Jixi Museum full of physical experience through observation and record on children's on-site behaviors of playing games (Figure 9 Right). What interesting is that pool cleaning, tiles renovation and tree rescuing were carried out when the film was made, but the above three short films still brought us a building with strong aesthetic features. Another short film "*Removing the Mountain*" recorded the laboring scene and details of construction work, expressed a more fundamental relationship of production under the aesthetic relationship between building and nature with elaborate montage editing, and rendered the pool, ornamental perforated windows, pavement, plastering and other fine details of Jixi Museum with more profound and touching implications (Figure 10).

3.4 Postscript: Inspiration

Teaching experiment of the course shows great potential of cinema as teaching method of space – it allows us to face space directly in a more specific, poetic way other than its representation. This course also shows that film-making can be combined with thinking of space in a better way, and in return theoretical thinking can promote cultivation of perception and imagination ability to space. Through a series of tasks – discovering cinematics in "ordinary space", recognizing place through "perceptual space", and expressing personal feelings and understanding through "narrative space" – this course guided students to complete tasks from recognition to space itself, to analysis on spatial experience and then to dialectical expression on complicated spatial issues. Cultivation of perception and imagination to space is one of the most fundamental tasks of architectural education, which aims not only to improve design capability, but also to provide a humanistic origin for architecture. The process from "ordinary space" to "perceptual space" and to "narrative space" guided students to deepen cognition to nature of space gradually from outside to inside. As a course facing space education, *Cinematic Architecture* attempts to combine experience with thinking, analysis and expression.

Theoretical basis and working mode of the course are based on previous teaching experience to a large extent, including: master's program "*Cinema Aided Design*" at Nanjing University in 2009-2011; master's program "*Cinematic City*" at University of Cambridge in 2010-2011; master's program "*Research on Micro Cities*", "*Research on Cinematic Space*" at the Bauhaus and its sites in Weimar and Dessau in 2011-2012. Hereby I would like to extend my appreciation to Professor Ding Wowo, Professor Francois Penz, and Professor Alfred Jacoby for their support to the course.

Table 1 Representative Regular Course of "*Cinematic Architecture*"

School	Time	Tutor	Type of Course	Focus
AA School	1993-2005	Pascal Schöning	Diploma unit 3	Emotion: psychological experience
University of Cambridge	1998-2005	François Penz Maureen Thomas	MPhil program	Narrative spatial narrative
ETH Zürich	2003-	Christopher Girot	MAS LA MediaLab	Perception: perceptive experience
University of Westminster	2008-	William Firebrace Gabby Shawcross	Diploma studio 17	Time: architectural change
Accademia di Architettura di Mendrisio	2012-	Éric Lapierre	Atelier	Ambience: architectural ambience
Nanjing University	2012-	LU Andong	M.Arch	Space: cinematics of space

(Note: not include non-regular courses or general courses beyond the system of architectural education)

Table 3 Roles and Assignments of Film-Making for *Cinematic Architecture*

Role	Spatial analysis task	Description
Director	Spatial arrangement for narratives	The director breaks up narratives into different scenes and actions, and allocates them on the "plan"; in addition, the director takes in charge of the whole film-making process. Director thinks over relations between space and actions in the form of spatial "choreography"
Scriptwriter	Scene construction	The scriptwriter carries out detailed analysis on form and visual characteristics of a specific space, and breaks up the scene in the space into a group of shot series, i.e. describing the scene with different groups of spatial angles
Cameraman	Visual construction of camera	The cameraman designs position, moving path of camera, as well as mutual relations between camera and performers or elements in space (e.g. openings or furniture), and determines visual aesthetics of video clips
Film editor	Building film space	The film editor combines shots into narratives, and tests different effect of image and sound. The film space created by film editor does not have to match the real space filmed

Table 2 Course Arrangement and Corresponding Theoretical Issues of *Cinematic Architecture*

<table>
<tr><th>Stage</th><th colspan="2">Assignment arrangement</th><th>Corresponding issues of spatial theory</th></tr>
<tr><td rowspan="4">Stage I: Spatial Analysis Training</td><td colspan="2">Spatial analysis training I: This training aims to recognize the cinematics of space, and master basic cinematic language</td><td rowspan="2">Time - Space
From "universal language" to "time image", images present the peculiar dynamic aesthetics of space</td></tr>
<tr><td>I-a : Movement – dynamic spatial analysis
Find a fragment with cinematics in ordinary space, and film a dynamic scene and a static scene to express it</td><td>I-b: Movement – space trajectory analysis
Find a fragment with cinematics in ordinary space, and design a camera movement trajectory to express it</td></tr>
<tr><td colspan="2">Spatial analysis training II: This training aims to understand relations among space, action and event, and use cinematic narrative to express cinematics of space</td><td rowspan="2">Performing Space
Cinema represent movement of body in space as "performance", it practices the space and in return builds the space</td></tr>
<tr><td>II-a: Performance – spatial event analysis
Find a fragment with cinematics in ordinary space, design a spatial event using structural, scenic or auxiliary feature of space, and use continuous film editing to express it</td><td>II-b: Performance - promenade architecture
Find a fragment with cinematics in ordinary space, design an architectural promenade with travel nature, and use continuous film editing to observe and express promenade architecture from different camera angles</td></tr>
<tr><td rowspan="2">Stage II: Place Analysis Training</td><td colspan="2">Place analysis training: This training aims to further experience and recognize place, and use cinematic language to express features of place objectively. In comparison with Stage I, it never emphasizes notation of bode movement to spatial cinematics, but pays more attention to the perceptual experience of body to space</td><td rowspan="2">Existential Space
Cinema helps us understand exchanges among experience, feeling and significance in material space and spiritual space</td></tr>
<tr><td>a: Place spirit – spatial symphony: Select a setting with strong place sense (e.g. historical place and memorial place), observe the place deeply, grasp its key features, and express it with montage approach</td><td>b: Place ambience – garden space: Select a setting with strong place sense (e.g. garden), design and arrange story according to spatial ambience, and main actions or events shall match features of the place</td></tr>
<tr><td rowspan="4">Stage III: Narrative Analysis Training</td><td colspan="2">Narrative analysis training I: This training aims to further space created by emotion, memory and imagination, and requires presenting a site with personal meaning by using cinematic language comprehensively</td><td rowspan="2">Psychological Space
Cinema reveals emotional dimension of space. We make space to accommodate the form of our inside world with emotion, memory and imagination. So space is not an object or organ of human body, but the product of subjective projection</td></tr>
<tr><td colspan="2">I: Emotional Space – poetic narrative: Experience and film a strange site. Combine body sensation, spatial ambience and emotion with cinematic language. Express individual "poetic experience" in presentation of the strange place[6]</td></tr>
<tr><td colspan="2">Narrative analysis training II: This training aims to discover conflict between building and urban conditions in daily space, and learn how to use cinematic language to express our opinions</td><td rowspan="2">Daily Space
The multiple expressive, participatory and objective observation angles offered by cinema help us analyse real urban settings. In particular, it is good at presenting unknown spatial practice and phenomena, and revealing their meanings</td></tr>
<tr><td colspan="2">II. Spatial practice - cinematic essay: Find an alien building in the city, and express your understanding on the topic of "building alienation" with cinematic language by taking this building as an example</td></tr>
</table>

注释

1 里希特和密斯结识于1921年，随后共同创办了G杂志。里希特对于“普遍视觉语言”的认识对密斯有很大影响。在巴塞罗那德国馆中，不同材质的墙面随着人的运动呈现不断变化的对比，这是一种带有强烈时间感的形式。

2 艺术史学家安德列斯·扬赛（Andres Janser）指出：“电影并没有代替摄影成为建筑视觉表现的主要媒介。这主要由于结构性原因，例如电影制作较大的资金需求以及播放它们所涉及的复杂因素。”JANSER A. Only Film Can Make The New Architecture Intelligible [M]//PENZ F, Thomas M. Cinema & Architecture. BFI Publishing, 1997: 34-46.

3 库哈斯邀请导演Bregtje van der Haak参加了对尼日利亚城市拉各斯（Lagos）的研究并完成了纪录片《拉各斯/库哈斯》（2002）、《拉各斯：宽镜头与近镜头》（2005）。都市智囊团工作室与导演Rob Schröder合作对委内瑞拉首都加拉加斯进行考察并完成了纪录片《加拉加斯，非正式城市》（2007）。

4 英国建筑期刊*Architectural Design*（《建筑设计》）在1994和2000推出了两期专辑：Maggie Toy, ed. *Architecture & Film. AD, Vol. 64, 1994*; Bob Fear, ed. *Architecture & Film II*. AD, Vol. 70, 2000. 美国*Journal of the Society of Architectural Historians*（《建筑史学家学会会刊》）自2005年起每期开设《多媒体》专栏，先后由电影理论家爱德化·蒂门伯格（Edward Dimendberg）和建筑理论家比特瑞兹•克罗米娜（Beatriz Colomina）等主持，迄今已讨论了近百部建筑电影。

5 华中科技大学汪原指出了另一种“被相机拍摄的身体”的空间：“在教学中强调学生要用自己的身体置入空间：当学生面对镜头的时候，他对周围空间的感知能力会被激发，从而形成身体对镜头的感知和对空间要素的感知，进而叠加产生身体对空间感知的圆。”（汪原在《影像空间教学研讨会》上的发言，南京大学，2014年5月10日。）

6 绩溪博物馆的设计师李兴钢认为：“电影也有建筑无法表达的，而建筑也有电影无法表达的。但是所有这些艺术所追求的有一个共同点，就是使人有一种诗意的体验。这种诗意不是狭义的，而是更加深刻动人的。”（李兴钢在《影像空间教学研讨会》上的发言，南京大学，2014年5月10日。）

7 在本课程教学中需要提供的设备有：摄像机三脚架及匹配的滑轮底座、摄像滑轨。

参考文献

[1] SCOTT G. The Architecture of Humanism: A Study in the History of Taste [M]. Boston: Houghton Mifflin Company, 1914.

[2] GIEDION S. Space, Time and Architecture: The Growth of a New Tradition [M]. Cambridge: Harvard University Press, 1941.

[3] GIEDION S. Building in France, Building in Iron, Building in Ferroconcrete [M]. Trans. BERRY J. Getty Research Institute, 1995.

[4] RICHTER, H. De Stijl 6.5 [J]. 1923(3).

[5] BRUNO G. Bodily architectures [J]. Assemblage, 19: 106-111.

[6] NEUMANN, D. Film Architecture: From Metropolis to Blade Runner [M]. New York: Prestel Publishing, 1996.

[7] PALLASMAA J. The Architecture of Image: Existential Space in Cinema [M]. Helsinki: Rakennustieto Publishing, 2001.

[8] VIDLER A. Warped Space: Art, Architecture, and Anxiety in Modern Culture [M]. Cambridge, : MIT Press, 2000.

[9] SCHWARZER M. Zoomscape: Architecture in Motion and Media [M]. New York: Princeton Architectural Press, 2004.

[10] PENZ F, LU A. Urban Cinematics: Understanding Urban Phenomena through the Moving Image [M]. Chicago: University of Chicago Press, 2012.

[11] BRUNO G. Public Intimacy: Architecture and the Visual Arts [M]. Cambridge: MIT Press, 2007.

[12] KOECK R. Cine-scapes: Cinematic Spaces in Architecture and Cities [M]. London & New York: Routledge, 2012.

[13] SEAMON D. Body-Subject, Time-Space Routine, and Place-Ballets[M]// BUTTIMER A, Seamon D. The Human Experience of Space and Place [M]. London Croom Helm: 1980, 148-165.

[14] SCHÖNING P. Manifesto for a Cinematic Architecture [M]. London: Architectural Association, 2006.

[15] THOMAS M, PENZ F. Architectures of Illusion: From Motion Pictures to Navigable Interactive Environments [M]. Bristol: Intellect Press, 2003.

[16] GIROT C. Landscape Video: Landscape in Movement [M]. Zurich: Gta Verlag, 2009.

[17] FIREBRACE, W, SHAWCROSS G.[2015-3-24]. https://designstudio17.wordpress.com/ accessed.

[18] 刘泉，郭廖辉. “想象之空间” ——门德里西奥建筑学院Atelier Lapierre “建筑—电影”教学述评[J]，Der Zug, 2015, 2(1): 25-36.

[19] 李华.一个关于“电影建筑”的建筑文本[J].新建筑，2008(1): 4-8.

[20] TSCHUMI B. Manhattan Transcripts, 1977-1981 [M]//TSCHUMI B. Architecture and Disjunction. Cambridge: MIT Press, 1994.

[21] SAMUEL F. Le Corbusier and the Architectural Promenade [M]. Basel: Birkhäuser, 2010.

[22] 彼得·卒姆托.思考建筑[M]. 张宇, 译. 北京：中国建筑工业出版社，2010: 7.

[23] 鲁安东. 迷失翻译间：现代话语中的中国园林[M]. 建筑研究01：词语、建筑图、图. 北京：中国建筑工业出版社，2011: 47-80.

图片来源

图1 左：RICHTER H. Rhythmus 21. Online video, accompanied by new soundtrack. 右：Envisioning Architecture: Drawings from the Museum of Modern Art. The Museum of Modern Art, 2002: 71.

图 9 建筑摄影由李兴钢工作室提供。

其他图片均由作者提供。

建筑技术课程中能耗模拟软件Ecotect教学探讨

BUILDING SIMULATION TOOLS IN ARCHITECTURAL TECHNOLOGY COURSES：AN EXPERIENCE TEACHING ECOTECT

吴蔚 董姝婧

近几年来，计算机能耗模拟技术在我国建筑节能设计领域占有越来越重要的地位，将能耗模拟教学作为建筑节能设计的一部分引入建筑学必修课程中已不可避免。然而相较于传统的建筑学教学，如何教授学生正确地使用建筑能耗模拟软件，以及如何能让学生利用它们更好地进行建筑节能设计，还面临着许多新的问题。

本文讲述笔者在南京大学建筑与城市规划学院建筑学三年级学生的建筑技术课中，教授计算机能耗模拟软件Ecotect的经历。通过观察分析法和问卷调查法，总结和分析学生在学习建筑能耗模拟软件时所常遇到的问题，以及应用到节能设计的过程中所提出的疑问和困惑。本文旨在帮助我国的建筑学院、系在引入计算机能耗模拟训练的时候，让建筑学学生更快、更好地掌握和使用能耗模拟软件，并运用到建筑节能设计上面。

1.课程设置介绍

南京大学建筑与城市规划学院的前身是2000年建立的南京大学建筑研究所，以前主要从事研究生教育，从2007年起，开始招收建筑学本科学生，并在2010年与南京大学城市规划系合并。南京大学建筑与城市规划学院一直积极探索和尝试建筑学的教育改革，致力于培养具有坚实基础理论和宽广专业知识的高层次实用型人才。南京大学建筑学本科教育的建构是与国际建筑教育界接轨的2+2+2模式。即充分利用南京大学基础学科的优势，头两年进行通识教育，后两年进行建筑学专业基础培训。本科毕业后，对建筑学有兴趣及能力较好的学生可以进入研究生教育，进行建筑学专业高级培养。相较于我国其他五年制建筑学教育而言，这种模式的优点是南京大学建筑学本科学生具有较坚实宽广的基础知识，但缺点是建筑专业基础理论课不得不压缩在较短的时间内，这给教师教学与学生学习都带来较大的压力

笔者教授的建筑学本科三年级下半学期的建筑技术II：建筑物理和建筑技术III：建筑设备，这两门都是建筑技术的核心课程。每门建筑技术课程的跨度为16周，共38个课时。建筑技术课程一般分为基础和应用理论大部分，此外还有实验室训练和参观实习。由于时间紧，要掌握的知识较多，还要配合建筑设计课题，所以计算机能耗软件的学习穿插在两门技术课程中，一般会占用一共3周课程。

在能耗软件上，笔者选择介绍Ecotect。选择Ecotect的主要原因是由于Ecotect是一款可视化程度较高的计算机能耗模拟软件，且操作简便，功能全面且结果表达直观，很受广大的建筑设计人员的喜爱[1,2]。目前很多国外大学的建筑系，都将Ecotect的学习纳入他们必修的课程内[3,4]，我国一些大学的建筑学专业也开始将该软件列为建筑物理课程的教学中。此外，经过协商，AutoDesk公司答应为南京大学建筑学学生提供免费的软件以及一些相应的培训资料。

南京大学建筑学三年级下半学期的建筑设计课题是住宅设计训练。为配合设计课，建筑技术的课程报告也是以住宅建筑为主。基本上包括两部分：一是小组作业，主要是选择一个真实的住宅单元进行现场调研，实地测量建筑声、光、热环境，该训练是为了让学生学习巩固建筑物理的声、光、热基础知识。此外，该作业也要求学生进行计算机模拟与物理实测的相互验证训练。第二部分是单人作业，利用他们所测量的现有住宅建筑进行节能改造设计，并要求运用所学的计算机能耗模拟技术做节能改造设计的定量分析。

2.教学过程解析

因为教学内容多，时间紧，Ecotect培训的安排为3周课（6个学时）。笔者通过授课期间的观察以及与学生的沟通互动，分析学生常遇到问题及出现的原因。

第一周课首先认识Ecotect，包括认识一些用户界面和基本的操作命令，然后带领学生一起学习建立一个简单的三维数字模型，设置分析网格，以及插入其他CAD文件等。在引领学生建立数字模型时，笔者发现教授计算机软件完全不同于理论教学。在教授计算机软件时，因为学生要跟着老师一同操作，有些计算机基础好的同学很容易跟上学习进程，但有相当一部分同学跟不上节奏，这时候就需要放慢教学速度，重新演示操作来引领较慢的学生。同时，最好有一个助教或者已会Ecotect的研究生在旁协助，因为总有个别学生会出各种各样的问题，如果只为个别学生停下来的话会影响总体进度，这时有人在旁边单独辅导的话，会稳定学习较慢学生的情绪，帮助他们赶上学习进度。

在第二周课就开始学习热工能耗分析。利用学生在第一次课上所建立的三维数字模型，学习认识、下载、导入和分析气象数据；笔者发现学生较为喜欢的Ecotect功能是寻找本地的相对最佳建筑朝向，由于这一功能适合于建筑平面设计和规划设计，基本上所有学生都在他们的设计课题中采用这个功能。在使用Weather Tool进行气候数据分析时，最大的问题就是关于焓湿图的分析。尽管焓湿图在专业建筑热工学中一个较为重要的概念，但笔者采用的是刘加平主编的普通高等教育土建学科专业“十一五”规划教材《建筑物理》[5]中并没有涵盖这一概念，笔者只是在介绍软件功能时对焓湿图做了大概的解说，但后面还是发现很多学生对此概念产生很大疑问并乱用这一概念。

Ecotect在教学中的另一难点是定义材质。在教授时就发现部分学生在一些基础知识上极为薄弱，如不知道基础建筑热工学中的“比热”和“导热系数”这两个概念。此外，学生仅会使用在Ecotect软件库默认的材质，一旦找不到符合自己设计意图的材料时，就不知如何处理了。例如，有的学生的设计里采用了U形玻璃幕墙采光，但是Ecotect材料库没有这种材料的，当被告知U形玻璃的材料材质属性需要自己查询时，学生就不愿坚持最初的设计转而采用软件材质库提供的材料。这一现象与南澳大利亚阿德莱德大学的Soebarto在Ecotect教学中所遇到的问题很相似[3]。

在第三周课上，笔者介绍了Ecotect的光模拟、可视度模拟和声模拟分析法。在引领学生学习使用Ecotect模拟的时候，笔者发现学生会被Ecotect各种各样的分析功能所迷惑，热衷尝试各种模拟分析，但并不是很清楚自己在做什么，需要做什么，很多学生对分析结果不求甚解，或者根本看不懂分析表格。

3.问卷调查

为了更好地了解学生对学习建筑能耗软件Ecotect的感受，在学期结束后笔者请助教做了问卷调查。为确保问卷调查的客观准确性，采用完全自愿原则，调查的对象是参与课程的南京大学2007级建筑学系学生，共发放问卷33份，收回问卷22份。问卷调查的项目主要包括学生对建筑能模拟软件的学习使用所遇到的问题，以及对Ecotect软件的评价和使用感受等。

根据在教学和作业中发现的问题，问卷首先询问了学生在应用软件时遇到的问

题。问卷显示学生在建立模型和模拟结果分析方面遇到的问题最多，这与教学观察过程中的反馈相一致。如建模时参数设置上有32%的学生遇到问题，这主要是当学生在材料设置上出现材质库以外的材料时不会定义新的材质。另外有16%的学生在网格设置上出现问题，这与模型的复杂程度和操作有关。对于模拟结果分析方面，询问了是否能够看懂软件的模拟结果并对结果做出判断。其中96%的学生认为自己能够知道在模拟什么，但有44%的学生认为不能够从中解读出正确的分析信息。此外，有76%的学生认为软件模拟出的物理环境结果与所预期的设计结果有些出入或者不一致。

问卷询问了对建筑能耗软件Ecotect的认知问题。有88%的学生认为Ecotect本身是相对准确的软件。但60%学生认为由于建立数字模型和参数设定等原因，人为因素会造成模拟结果不准确。并且72%的学生认为模拟结果的准确性对设计深化有影响。

总体而言，有98%的学生非常认同或是认同Ecotect是个有用的能耗模拟软件，对辅助建筑节能设计有一定帮助，仅有2位学生对该问题持中立的态度。当被询问在以后的设计工作中应用Ecotect的意愿度，80%的学生表示非常愿意和愿意在以后的设计工作中使用ECOTECT辅助建筑设计，说明了他们对软件的认可以及对课程教学的肯定。

问卷的最后是希望学生对Ecotect教学提出一些建议。学生主要反映的问题有两点。（1）普遍反映课时太少，同时上课所教授的三维数字模型有点简单，不能够满足课程报告和设计课上的需求。（2）希望能够对软件的基本原理做一个介绍，并且能够学习读懂模拟分析数据、不同的模拟分析有什么不同作用、意义何在等。

4.问题与经验教训

根据在授课时的观察和事后作业分析，以及后期的问卷调查，可以总结出教授Ecotect时需要注意以下一些问题：

建立数字模型时的模型简化问题。建筑学学生在学习和使用普通的计算机建筑辅助设计软件时，建立三维模型会被要求越细致越好。但能耗模拟软件则需根据情况进行必要的简化，在不失准确度的情况下减少计算机的计算时间，问题是如何简化数字模型是能耗模拟软件学习的一个难点。尽管在上课时一再强调简化数字模型的重要性和简化模型的要素，但学生在课后做作业时，还是忍不住将数字模型复杂化，或者对于如何简化还存在着很多疑问。例如：楼梯间需不需要表示？室内楼梯要不要建模？屋顶有挑檐怎么处理？对于从没有接触过建筑能耗模拟的同学来说，什么是会影响模拟结果的因素、怎么简化以提高模拟速度但不影响模拟精度等的确是很难判断的问题，需要学生自己逐步摸索。

数字模型的精准性问题。例如有些模型在屋顶和墙面的结合处出现了缝隙，光模拟会出现漏光的现象。造成这个问题的主要原因还是能耗模拟与建筑辅助设计软件的差异引起的，建筑辅助设计在建立数字模型时，对其准确性要求并不严格，因此学生在建模时很容易忽略这方面的问题。由于Ecotect支持AutoCAD导入的的三维模型，一些学生会使用在AutoCAD中建立的不准确数字模型。此外，学生将AutoCAD三维模型导入Ecotect后，往往忘记将门窗和墙面进行绑定，因此会出现在有采光口的情况下反而采光系数为零的现象。因此在教授Ecotect建模时，有必要强调模型的精准性，以及养成在模拟之前检查模型的习惯，预防这些问题的发生。

在能耗模拟时最常出现的问题是学生对软件不求甚解的模拟应用问题，这在作业中表现得更为明显，自以为是地认为摆放分析图表越多越好，或将模拟分析图摆放到错误的地方，很多学生对分析图表没有任何解释说明。其次是学生不明白Ecotect软件本身存在的缺陷也会造成模拟结果异常。如在模拟天然采光时，Ecotect会以窗口作为光源面进行模拟计算。如果在室内添加内窗的话，Ecotect是无法识别出外窗与内窗的差别，而是作为相同光源进行模拟。这就造成靠近内窗的采光系数反而远高于靠近真正窗口位置的采光系数的异常现象。当一些学生发现这个问题时，并没有意识到这是软件缺陷带来的模拟结果异常，反而以为是自己出错了。

建筑学学生这种对计算机能耗软件的盲目信任是一个令人忧心的普遍问题。如建筑物理作业中要求学生进行物理实测和计算机模拟的相互验证，当二者出现误差时，很多学生会认为是实测数据出现问题。当做节能模拟设计的时候，部分学生以为Ecotect能自动生成节能设计，但是当被告知软件只是用来做能耗模拟分析，而不能代替自己作出设计时，学生对此表示出了极大的失望，进而对Ecotect的作用发生怀疑。如有学生会问"Ecotect不能直接自动生成建筑节能设计方案，这个软件还有什么用？"这种问题表明建筑学学生对能耗模拟软件的认知有一定误解。他们以为只须建立一个建筑三维模型，再按一些命令，软件就会跟他们使用的建筑渲染软件一样自动生成节能设计。面对学生的质疑，只能不断重申能耗模拟软件只是建筑节能设计的辅助工具，并不可能替代设计者的决策。正确的节能设计方法还是需要结合在建筑技术课上学到的一些建筑节能设计手段，如通过改变建筑的形体系数、开口大小或者应用各种主动和被动式节能策略等做不同的节能设计方案，然后利用计算机能耗软件帮助分析比较不同设计方案的优劣，逐步优化出可行的建筑设计方案。

5.结论

计算机能耗模拟技术的引入无疑为传统的建筑技术理论教学提出了新的机遇和挑战。通过学习，学生不仅可以更扎实地掌握基础理论知识，也能更好地开展建筑节能设计。但计算机能耗模拟软件与传统的理论教学不同，本文结合笔者在南京大学教授建筑学三年级学生学习计算机能耗模拟软件Ecotect的经历，根据观察分析和问卷调查，总结了在Ecotect教学中可能遇到的的四个问题：（1）在建模阶段，最容易出现的问题是混淆计算机辅助建筑设计软件（CAAD）与建筑能耗模拟软件在使用的异同，如模型简化和模型精准性问题。（2）参数设定时的主要问题是材质设定。（3）盲目追求各种模拟分析图表，但对模拟结果不求甚解。（4）对计算机能耗模拟软件产生认知错误。如对能耗模拟结果产生盲目认同，甚至认为软件可以代替设计者做节能设计，当发现不可能自动生成节能设计时，又对软件的作用产生质疑。

根据以上的经验教训，笔者提出以下几点建议：（1）需要加强建筑技术理论基础知识的学习，以及充分介绍不同的节能设计策略，尤其是反复强调可持续建筑设计的重要性；（2）对建筑能耗模拟技术的认知需要反复说明，既不能让学生对软件产生盲目信任，尽早点出软件本身所有缺陷，可以防止学生做一些不必要的无用功。（3）强调能耗模拟软件与普通建筑辅助设计软件的不同。（4）有必要学习如何分析模拟结果并做出正确的判断，杜绝不求甚解的瞎用乱用现象。总而言之，需要让学生了解到学习软件的最终目的是更好地做建筑节能设计，因此掌握多种节能设计手段和积累节能设计经验要比仅仅学会软件工具的使用更为重要。

In recent years, Building information modeling (BIM) plays an increasingly important role in the field of building energy efficiency design in China, and introducing BIM as part of sustainable architectural design in architectural schools has now become inevitable. However, in comparison with traditional architectural teaching, many new questions need to be faced with. For example, how can teach students to use building energy simulation software properly, and how could students use BIM in their architectural design in a better way.

This paper introduces the author's experiences that she taught building energy simulation software "Ecotect" in the architectural technology course for third-grade students, in the School of Architecture and Urban Planning, Nanjing University. Through observation analysis and questionnaire survey, this paper is to identify problems and lessons from teaching Ecotect. Also, it is to develop strategies to synchronize the building simulation studies and traditional architectural technology education. The significance of this research is in seeking understanding of the strategies required to integrate building simulation techniques into the design process, both professionally and in design studio in architectural schools in China.

1. Course Introduction

The School of Architecture and Urban Planning, Nanjing University, former the Research Institute of Architecture, Nanjing University was founded in 2000, was mainly focusing on postgraduate education in the early years. It started to admit undergraduate students in 2007, and was merged with the Department of Urban Planning in 2010. The Department of Architecture has always been actively exploring and experimenting reform on architectural education, and is dedicated in cultivating high-level practical talents with solid fundamental theories and broad professional knowledge. The undergraduate architectural program in the Department of Architecture uses 2+2+2 mode that is in line with the international architectural education. That is, by using excellent general education of Nanjing University in the first two years, focus architectural professional training in the last two years. After graduated, only outstanding students who are interested in architecture could enroll in the postgraduate study, receiving advanced training of architecture. In comparison with the five-year program in other architectural schools in China, the advantage of this mode is that the undergraduate architectural students in Nanjing University have solid and broad fundamental knowledge, but the disadvantage is that all professional architectural training have to be compressed in a relatively short period, which brings large pressure to both professors and students.

The authors teach Building Technology II (Acoustic, Lighting and Thermal Environment) and Building Technology III (Building Services System) in the second half semester of third undergraduate year. These two courses are both required technology courses in the architectural program. Each building technology course is taught in16 weeks, a total of 38 hours. The lectures were divided into two parts: (1) lectures about the fundamentals of building knowledge, and (2) lectures on application of sustainable design. In addition, laboratory training and field study are involved. Given a tight schedule and a lot of information, as well as coordinating with the design studio, Ecotect training interspersed with two energy courses. It would take 3 weeks in total..

Ecotect is chosen to be introduced to lectures. The main reason is that Ecotect features with high visualization, and its operation is relatively simple. As a result, Ecotect is popular among architectural designers [1, 2]. Currently, Ecotect has been incorporated into required curriculum in many oversea architectural schools [3, 4]. Also, many Chinese universities begin to teach Ecotect in their architectural schools. After negotiation, the AutoDesk company agreed to provide free software and some appropriate training materials.

The design studio in the second semester of third year is a residential building design project. In order to coordinate with the design studio, residential buildings are also used for students' assignments in authors' courses. The assignments includes two parts: (1) The first part is group work. The students are asked to study acoustic, lighting and thermal environment in a real residential unit. And then results of the on-site measurements and Ecotect simulation are required to be compared to understand the difference. (2) The second part is an individual assignment, which requires students to carry out green design for an existing residential building. A quantitative analysis with Ecotect is required in their design.

2. Teaching Arrangement

Because of the tight schedule and a lot of materials to be taught, Ecotect teaching is arranged within 3 weeks (A total of 6 class hours). The following are the major problems that the authors summarized during teaching Ecotect.

Task of the first week is to know Ecotect, including getting familiar with user interfaces and basic operation commands. And then the authors would teach the students to build a simple BIM , lay out analysis grid, and import different kinds of CAD files. The authors found that teaching simulation software is completely different from teaching general lectures. For example, the students need to follow the teacher's demonstration step by step when they are studying simulation software. If one or two steps missed, the whole learning process may not be followed up. Some students with better computer background may follow up easily, while a few of students may have trouble. At this time, the teachers have to slow down for a few unsatisfied students. It would be better if there has a teaching assistant who can help individual students during the teaching process. This is because that there would always be some students who have various kinds of questions or problems. If there is an assistant by the side for individual counseling, it will stabilize the mood of students who learn slowly, and help them catch up with the progress of learning.

The energy simulation started in the second week. Based on the BIM built in the in first week, the students begin to learn climate information and how to apply it to BIM. The authors found that the students like to use "Optimum Orientation" tool in Ecotect to find the best orientation for local climate environment. "Optimum Orientation" is suitable for site planning and floor planning,. Almost all the students applied this function in their design subject. When "Weather Tool" was used, the biggest problem is analysis of psychrometric chart. Although the psychrometric chart is a relatively important concept in energy simulation, this concept is not covered by the textbook *Building Physics* edited by Liu Jiaping [5], which is main textbook suggested by the "11th Five-year Plan" program of construction engineering for higher education. The authors explained this function only roughly when "Weather Tool" was introduced. However, the authors noticed that many students were confused with this concept and applied it mistakenly.

Another difficulty for teaching Ecotect is defining building materials. It was found that some students were poor on some fundamental knowledge. For example, they did not know two concepts of "specific heat" and "thermal conductivity" in fundamental thermal engineering. In addition, students like to apply "default" materials from material library in Ecotect. They did not know what to do once they cannot find certain material in the material library. For example, one student wanted to used "U-shape" glazed curtain wall in his design. However, the student gave up his design when this special "U-shape" glazed curtain wall cannot be found in the material library of Ecotect. This is very similar to the problem encountered by Soebarto from University of Adelaide in South Australia when he taught Ecotect[3].

In the third week, the authors introduced light simulation, visibility simulation and acoustic simulation analysis. The authors found that students were attracted by various analysis tools provided by Ecotect. However, most students were not clear why they do all these simulations and how to apply them to their design. Many students had no in-depth understanding on results of analysis, or did not understand simulation results at all.

3. Questionnaire Survey

In order to better understand the learning process, the authors asked a teaching assistant to carry out a questionnaire survey after the semester was finished. To ensure objective and accurate results, the survey was carried out in voluntariness. The samples were architectural students in year 2007 in Nanjing University, who attended the course. Totally 33 questionnaires were distributed, and 22 questionnaires were taken back. Items in the questionnaire survey include main problems encountered when they were studying and using Ecotect. Also, evaluation and feelings on use of the Ecotect were asked.

Based on problems detected in the teaching process and the students' assignments, the questionnaire first asked what main problems are when they were using Ecotect. Results of questionnaires show that most problems encountered by students are related to modeling and analysis of simulation results, and this is consistent with feedback from teaching observation. For instance, 32% students had problem about parameter setting in the process of modeling, this is mainly because that students did not know how to define new material if their designed material was not included in Ecotect library. And 16% students had problem about grid setting, this is related to the complexity and operation of models. In term of simulation result analysis, the questionnaire asked if the student can understand simulation results and make design judgement accordingly. Of which 96% students thought they could understand what was simulated, but 44% students said they could not read out correct analysis information from results. In addition, 76% thought that simulation results were different from or inconsistent with expected design results.

The questionnaire asked a question about cognition to Ecotect. Of which 88% students said that Ecotect itself is an accurate software tool. But 60% students said that simulation results may not be accurate if digital model, parameter setting, and human-related factors have mistakes. And 72% students thought that accuracy of simulation results will affect future design decisions.

In general, 98% students were positive about or agreed that Ecotect is a useful energy simulation tool, and it is helpful in certain degree to facilitate building energy efficiency design. Only 2 students maintained neutral standpoint to this question. When asked if they are willing to apply Ecotect in design work in the future, 80% students said that they are very willing or willing to apply Ecotect to their design in the future. This question suggested that the students are positive about the courses of teaching BIM.

In the last, the questionnaire asked students to give some suggestions to the teachers. The students gave two main feed back: (1) The most students said that class hours are not sufficient, and that the 3D digital model that they learned during the course is relatively simple in comparison with the model which they had to build in their assignments. (2) The students hope to obtain an introduction about basic principles of Ecotect, and hope that they can learn how to understand different simulation results.

4. Lessons Learned

According to observation of the lectures and analysis of the students' assignments, as well as the subsequent questionnaire survey, some lessons learned from teaching Ecotect are summarized as follows:

How to simplify digital models special for BIM? When architectural students use general CAAD software, the more detailed the better it is for setting up 3D models.

However, BIM should not be too complicate. It is necessary to ensure reduced computing time without compromising accuracyt tine. And the key problem is that a new learner always has difficulty to simplify BIM. Although the importance of simplification of BIM and key factors for model simplification were emphasized repeatedly in class, the students could not help complicating BIM. They always had lots of questions about how to simplify BIM. For example: Does staircase room need to be built in 3D model? Is it necessary for modeling the detail staircase? How should I deal with eaves, if any? For students who never used BIM before, it definitely needs to take a lot of time and efforts to explore how to make a compromised accurate 3D model without increase computing time.

The accuracy of BIM is another problem. For example, if the junction between roof and walls in 3D models has gap, results of lighting simulation will not be accurate. This is because of the differences between BIM and CAAD. Since accuracy requirement is not strict when 3D digital model is built by application of CAAD, the students always ignore accuracy of 3D models. Since Ecotect supports 3D models imported from AutoCAD, some students may use inaccurate 3D models established in AutoCAD. Moreover, after 3D models were imported into Ecotect from AutoCAD, the students often forget to "link" materials of "windows" to walls. As a result, Ecotect would not "see" any windows in the walls, and consequently lighting simulation would make mistakes. It is important for emphasizing accuracy of 3D models and double check 3D models before doing any simulation.

When energy simulation was carried out, the most common problem is that the students did not understand simulation results. This problem can be found everywhere in the students' assignments. The students mistakenly thought that the more analysis charts the better it is, or even simulation results cannot match their design. The authors noticed that many students made no interpretation or introduction for the analysis charts. Secondly, the students did not understand that abnormal results may also be caused by software's own shortage. For example, when daylight simulation is applied in Ecotect, Ecotect will treat "windows" which defined in 3D models as daylighting source. Ecotect cannot identify difference between exterior and interior windows, instead, it will treat all these windows as daylight source equally. As a result, the simulation results may mistakenly show that daylight factor near interior windows is higher than near exterior windows which are real daylighting source. When the students noticed this problem, they did not realize that this is a problem of Ecotect, but thought that they made mistakes somehow.

The blind belief of the architectural students in computer energy simulation software is a common worrying problem. For example, one of the assignments required the students to carry out validation study between on-site measurement and computer simulation, many students would think that their measured data are not correct if two sets of data could not match. Some students expected that Ecotect can generate energy saving design automatically. When they were told that Ecotect cannot do energy-saving design, but only for energy analysis. The students seem disappointed about the answer, and became skeptical about the role of Ecotect. For instance, some students asked, "If Ecotect cannot do energy-saving design automatically, then what is the software for?" These surprising questions show that the architectural students have some misunderstanding of BIM. They thought that what they need to do is to build a 3D model, and pressing some commands, then BIM would generate energy-saving design automatically, just like the computer rendering software they used. Facing such questions from the students, we only reiterate that energy simulation software is just an analysis tool for building energy-saving design, and that it cannot replace decisions of the designer. Computer simulation software can be used to compare different kinds of energy saving design strategies taught in the course, such as orientation, opening size, building coefficient, and passive solar designs strategies. The students need to analyze different simulation results, and work out optimal and feasible architectural design step by step.

5. Conclusion

Undoubtedly, introduction of BIM brought new opportunities and challenges to the traditional teaching of building technology. Through study, students can not only master fundamental theory knowledge more solidly, but also can carry out building energy-saving design in a better way. However, Teaching BIM is different from the traditional lecture. This paper introduced experiences that the authors teach Ecotect for third-grade architectural students in Nanjing University. Based on observation and questionnaire survey, the authors concluded four problems that may be encountered in the process of teaching Ecotect: (1) When 3D model is being built, the students are easy to confuse CAAD with BIM, such as model simplification and model accuracy. (2) During the stage of parameter settings, the main problem is how to define material. (3) The students like to pursue various simulation analysis, and make no in-depth understanding on simulation results. (4) The students have mistaken cognition on BIM. For example, the students believed in simulation results blindly, even thought that it can replace designer to complete energy saving design. When they found out BIM cannot do design for them, the students became skeptical about the software.

According to above experiences and lessons learned from teaching Ecotect, the following suggestion would like to be provided: (1) The fundamental knowledge of building technology is very important. It is necessary to repeatedly emphasize the importance of sustainable architectural design and different energy-saving design strategies; (2) Information and knowledge about BIM should be introduced as early as possible. It could help the architectural students understand abilities and shortages of BIM. (3) The differences between energy simulation software and common computer aided architectural design software should be emphasized again and again during BIM training. (4) It is necessary to make students learn how to analyze simulation results and make correct judgement, so as to eliminate blind use and mistaken use of software with no in-depth understanding. In a word, the main purpose of studying BIM is to do better energy-saving design. Therefore, letting the architectural students to know sustainable design is more important than learning simulation tools.

参考文献:

[1] LI Kun, YU Zhuang. Design and Simulative Evaluation of Architectural Physical Environment with Ecotect[J]. CADDM, 2006, 16(2): 44-50.
[2] 严钧，赵能，梁智尧. Ecotect在建筑方案设计中的应用研究 [J]. 高等建筑教育，2009，18(3): 140-144.
[3] SOEBARTO V I. Teaching an Energy Simulation Program In An Architecture School: Lessons Learned [C]. Proceedings of Building Simulation 2005: 9th International IBPSA Conference, Montréal, Canada. August 15-18, 2005: 1147-1153.
[4] PALME M. What Architects Want? Between BIM And Simulation Tools: An Experience Teaching Ecotect [C/OL]// 12th Conference of International Building Performance Simulation Association. Sydney, 2011: Proceedings of Building Simulation 2011 [2015-12-2]. http://www.ibpsa.org/proceedings/BS2011/P_1689.pdf.
[5] 刘加平.建筑物理[M]. 北京：中国建筑工业出版社，2009.

年度改进课程
WHAT'S NEW

设计基础(一)

BASIC DESIGN 1

季鹏

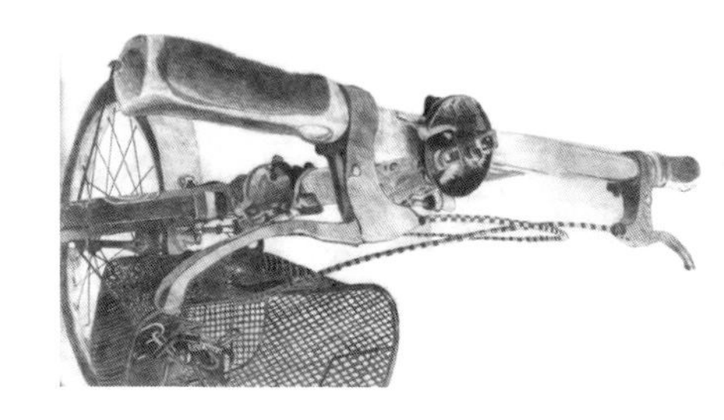

教学目标

提升学生感知美、捕捉美和创造美的能力。

研究主题

1. 看到世界的美。
2. 看到形式的美。
3. 理解空间的规则。
4. 理解使用的规则。

教学内容

1. 研究对象的素描表达、黑白灰归纳、拼贴表现和陶土烧造。
2. 针对 Zoom in 概念的训练与纸构成训练。
3. 建筑摄影。
4. 观念与材料的关系。

Training Objective

Improve ability of students to perceive, capture, and create beauty.

Research Subject

1.See beauty of the world.
2.See beauty of forms.
3.Understand rules of space.
4.Understand rules of use.

Teaching Content

1.Sketch presentation, black-white-grey induction, collage expression and pottery making for research objects.
2.Training on the concept of “Zoom in” and paper construction.
3.Architectural photography.
4.Relations between ideas and materials.

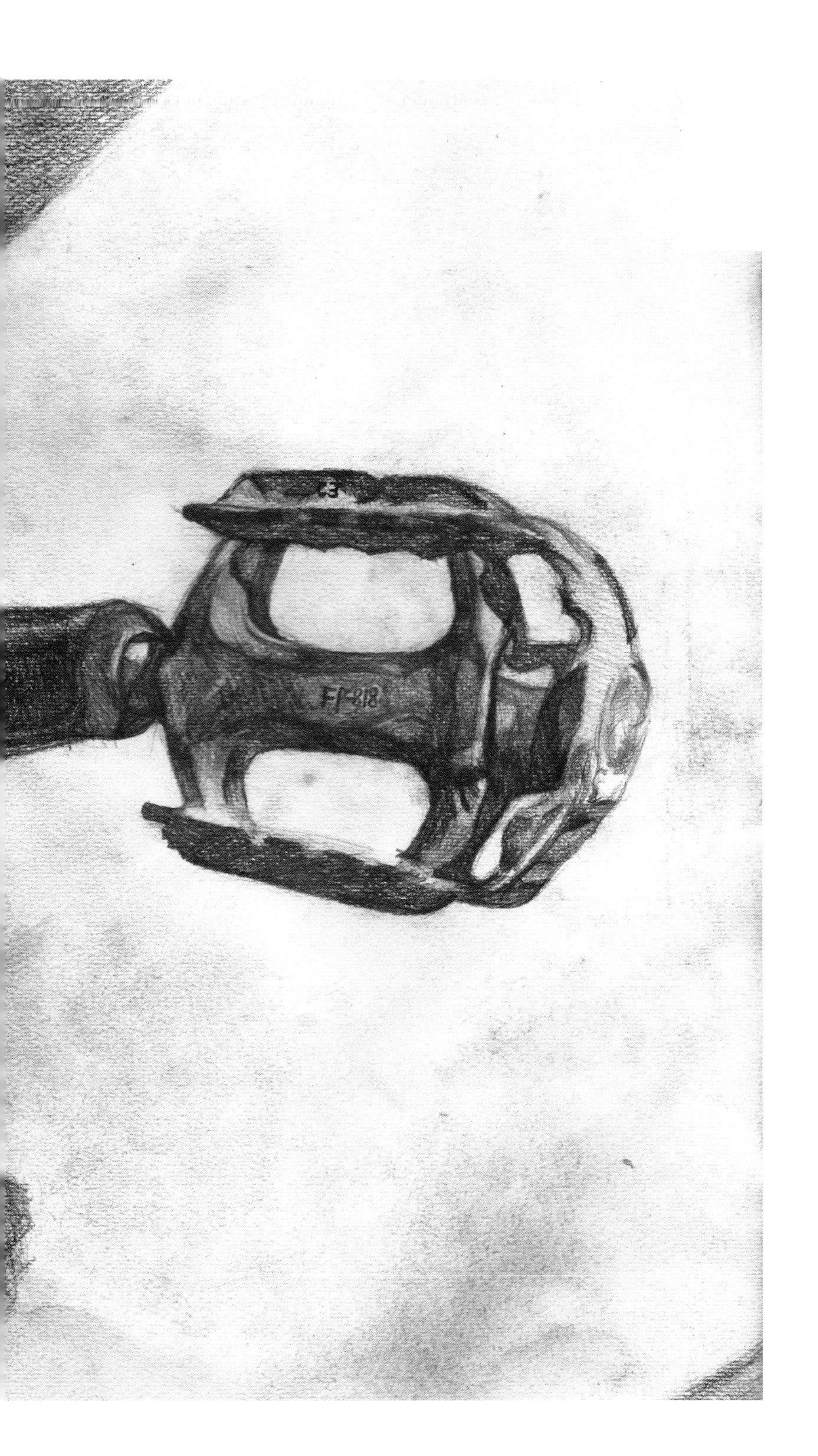

设计基础(二)

BASIC DESIGN 2

鲁安东 丁沃沃 唐莲

动作—空间分析

鲁安东 丁沃沃 唐莲

通过影像记录人在空间中的动作，挑选关键帧进行动作、尺度、几何与感知分析。目的在于：使学生初步认识身体、尺度与环境的相互影响；学会观察并理解场地；初步认识形式与背后规则的关系；学会发现日常使用中的问题并解决问题；学会使用分析图交流构思。

折纸空间

鲁安东 丁沃沃 唐莲

折叠纸板创造空间，用轴测图、拼贴图再现空间。目的在于：使学生初步掌握二维到三维的转化，初步认识图和空间的再现关系；利用单一材料围合复杂空间，初步认识结构与空间的关系；学会用分析图进行表述。

互承的艺术

丁沃沃 鲁安东 唐莲

运用互承结构原理，搭建人能进入的覆盖空间。目的在于：初步理解结构知识对于构筑空间的意义；在搭建过程中初步建立材料、节点、造价等概念；强化场地意识（包括朝向、环境、流线等）；学会用分析图进行表述。

Action-space Analysis

LU Andong, DING Wowo, TANG Lian

Record human's actions in space with video and select key frames to conduct analysis on actions, dimensions, geometry and perception. It aims to enable students to preliminarily understand the mutual influence among body, dimensions and environment; learn how to observe and understand the site; preliminarily understand relations between forms and rules behind them; learn how to find out problems in daily use and how to solve these problems; learn how to use analysis charts to exchange ideas.

Folding Space

LU Andong, DING Wowo, TANG Lian

Fold paperboard to create space, and represent space with axonometric drawing and collage. It aims to enable students to preliminarily grasp the transformation from 2D to 3D, preliminarily understand the reappearance relationship between drawing and space; enclose complex space with single material, and preliminarily understand the relations between structure and space; learn how to express with analysis charts.

Art of Mutually-supporting

DING Wowo, LU Andong, TANG Lian

Apply principles of mutually-supporting structure to erect an accessible covered space. It aims to preliminarily understand meaning of structure knowledge to construction of space; roughly establish the concepts material, joints, and cost during erection; enhance site awareness (including orientation, environment, streamline, etc.); learn expression with analysis charts.

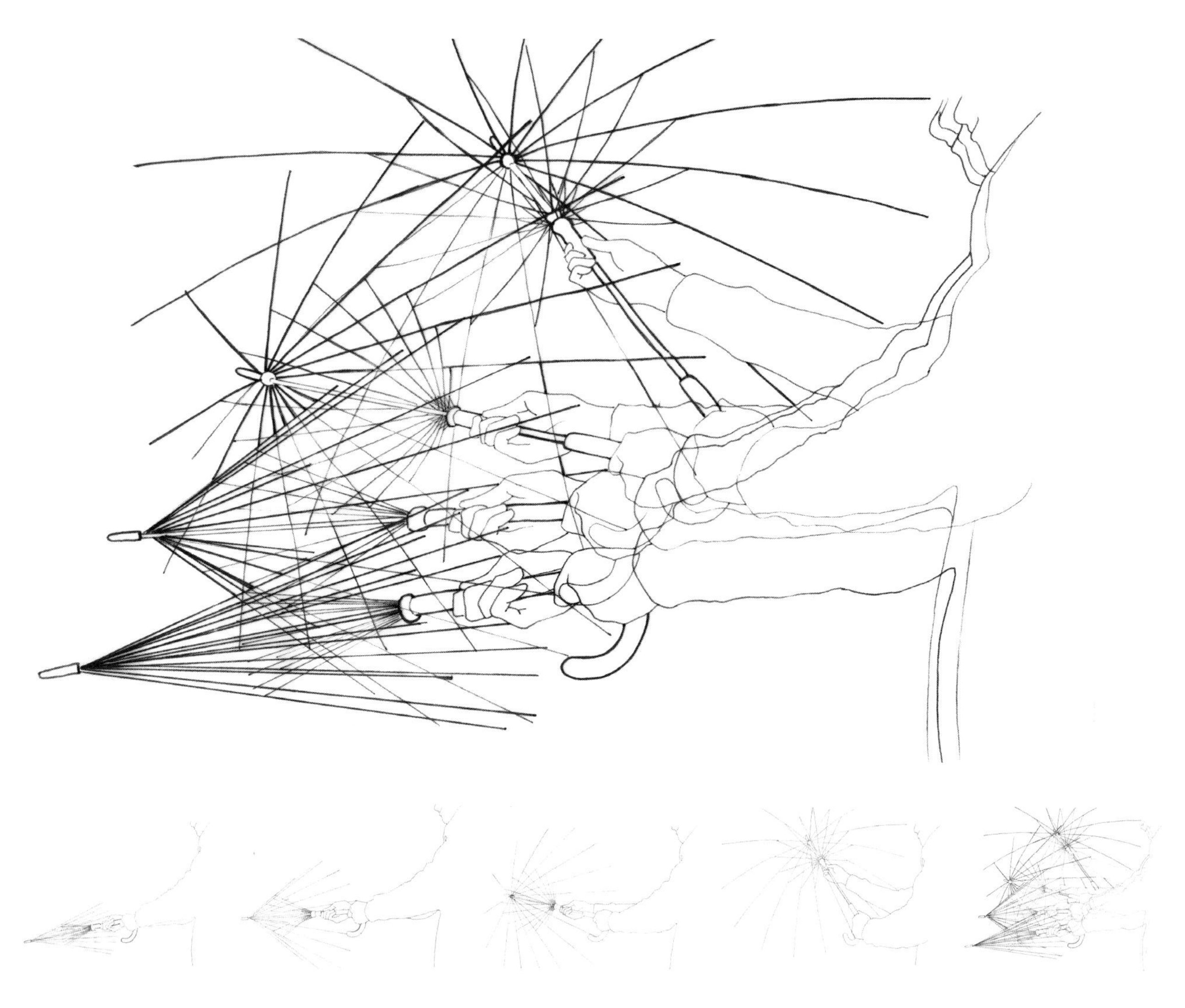

探索与再现动作与人体尺度的关系。

Explore and reappear relations between actions and dimensions of human figures.

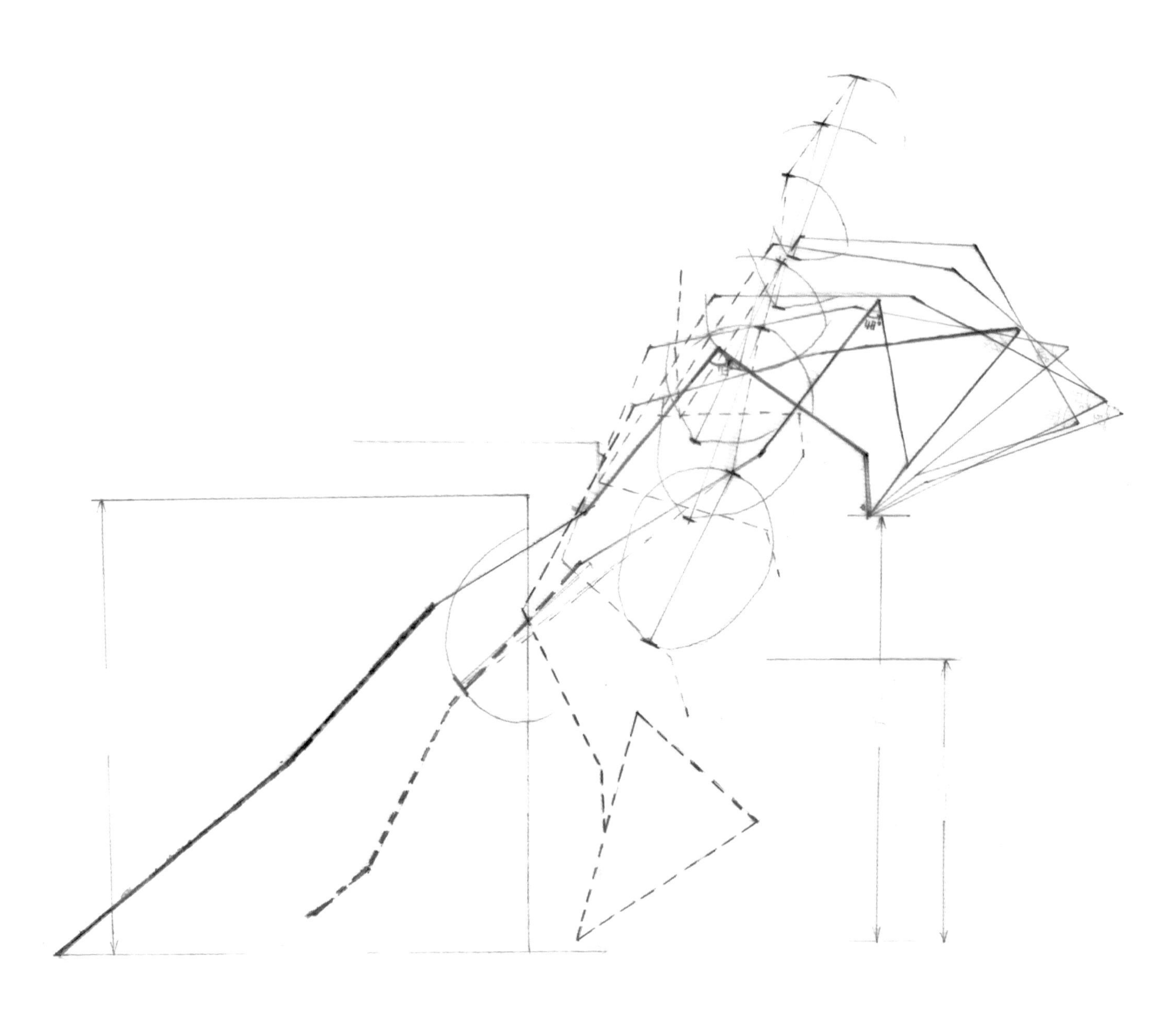

将动作、尺度抽象为几何模数与几何关系。

Abstract actions and dimensions into geometrical modulus and geometrical relations.

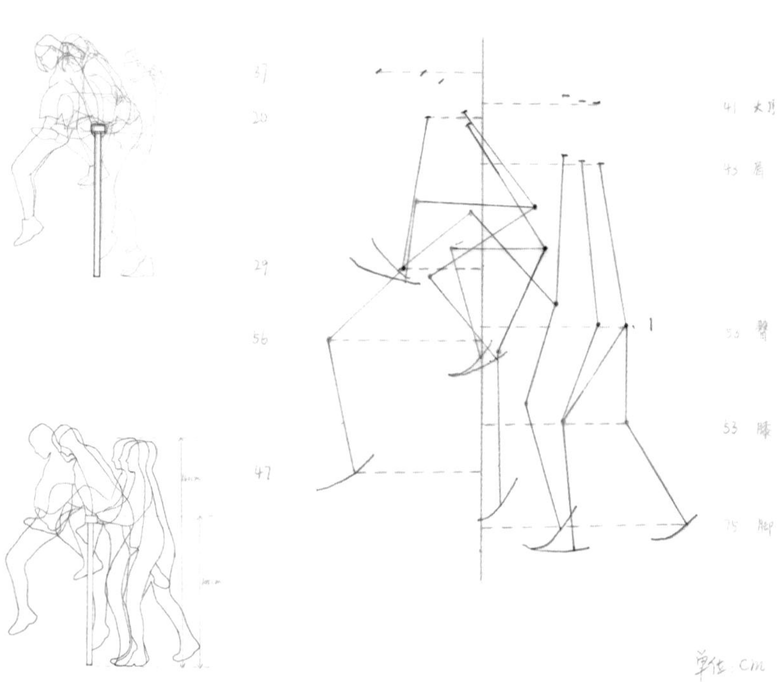

从不同视角观察动作“翻越”；分析手脚的配合与空间的限定关系；针对该动作设计的联动装置；对该装置的作用原理进行分析图示。

Observe “climbing over” of actions from different angles; analyze definite relations between space and coordination of hands and feet; design an interlocked device for the actions; analyze and illustrate action principles of the device.

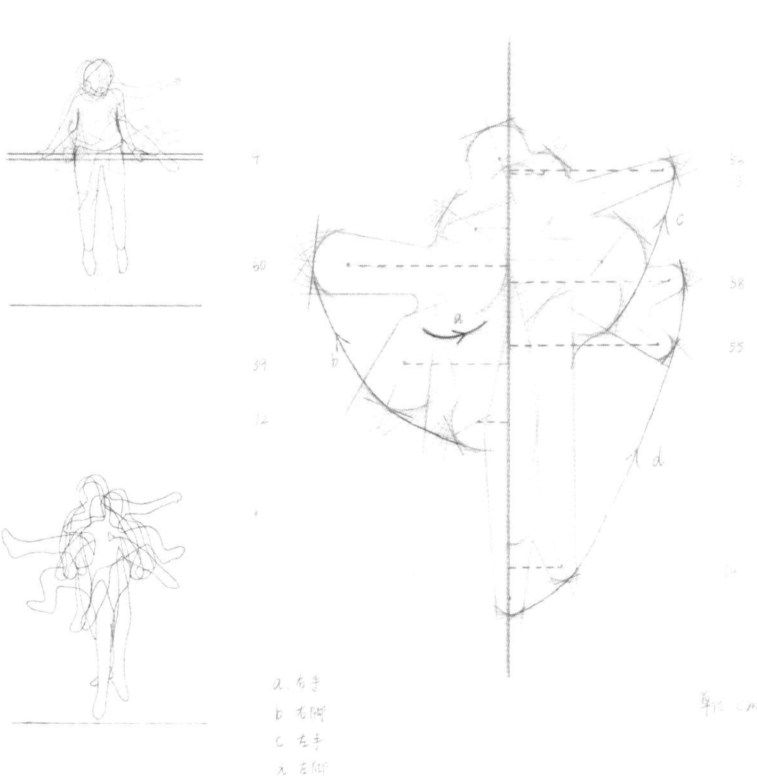

a. 右手
b 右脚
c 左手
d 左脚
单位 cm

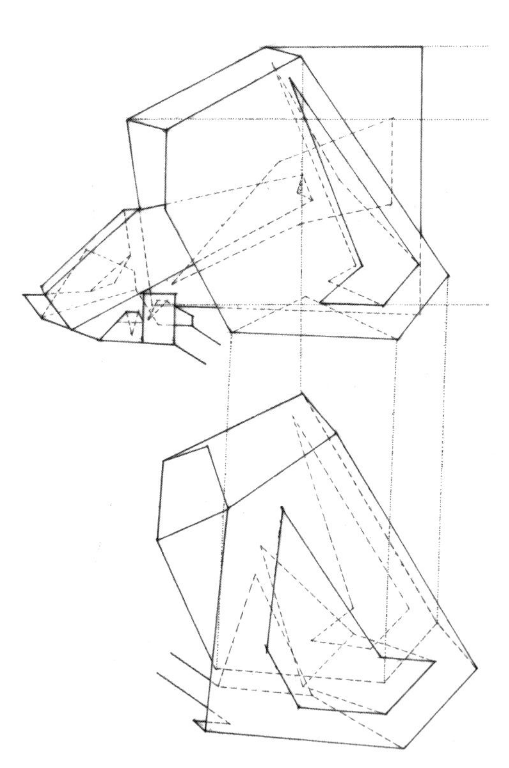

对一张 A2 纸板进行折叠、切割、穿插等产生多个不同大小、不同类型的空间。对折叠成的纸结构内的空间关系进行测量和分析，绘制不同角度的轴测图、空间分解图。设计一条路径将空间串联并用拼贴的方式再现空间序列。

Fold, cut and intersect one A2 paperboard to create several spaces of different sizes and types. Measure and analyze spatial relations within the folded paperboard structure, and complete axonometric drawings and space breakdown drawings from different angles. Design a path to link spaces and represent spatial sequence with collage.

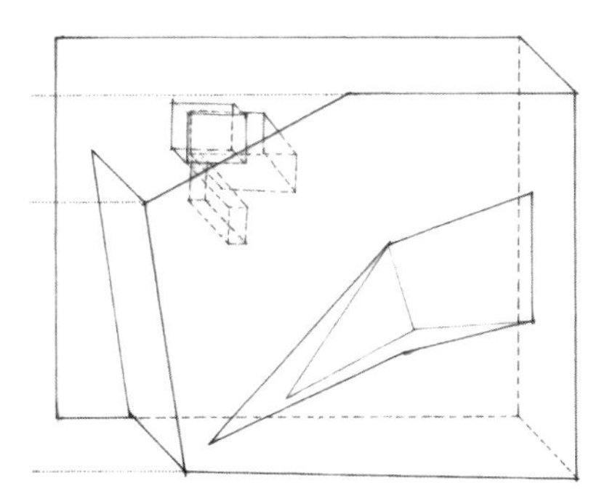

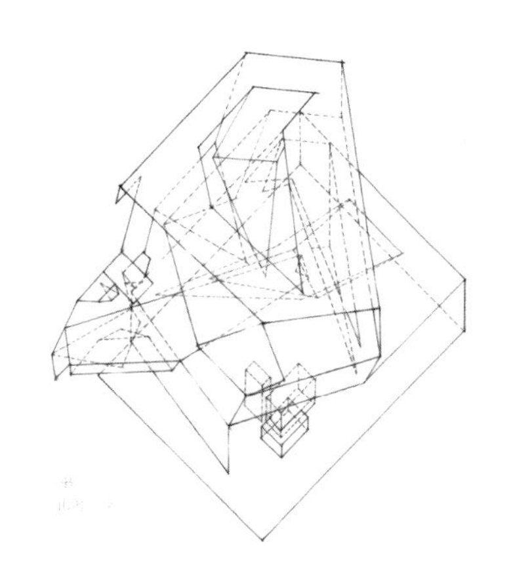

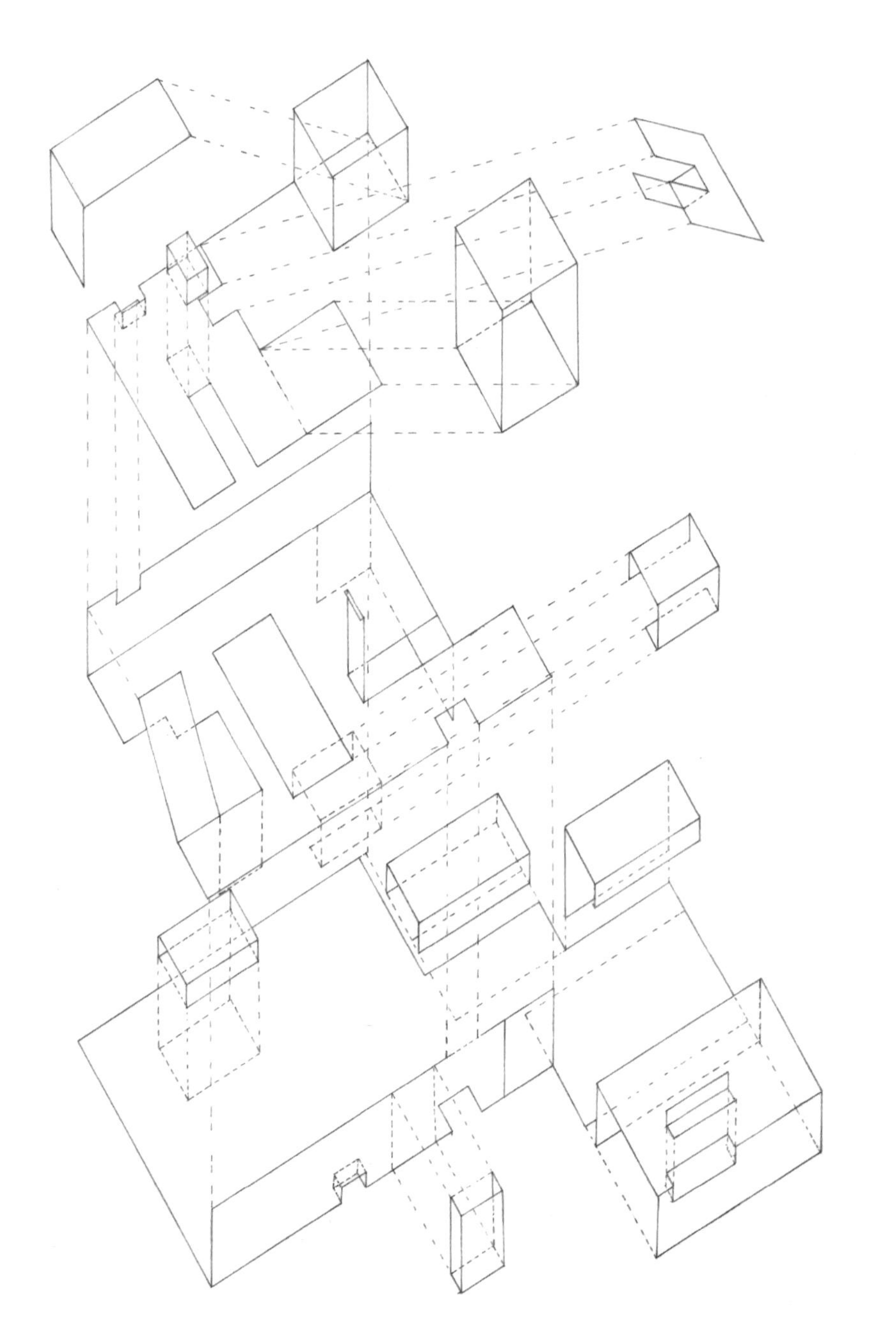

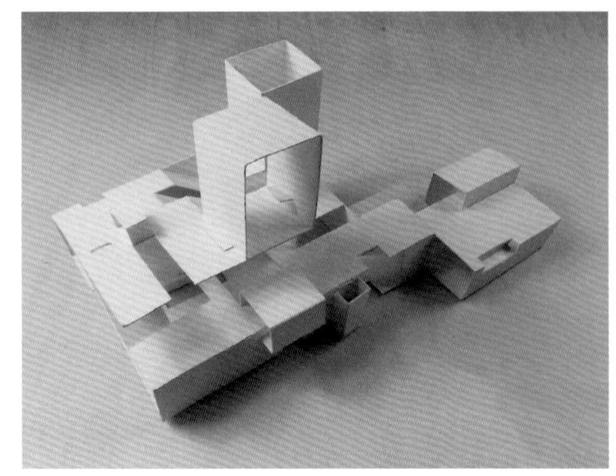

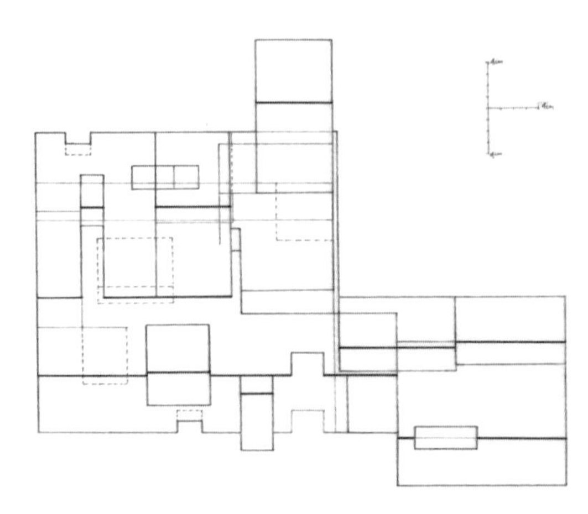

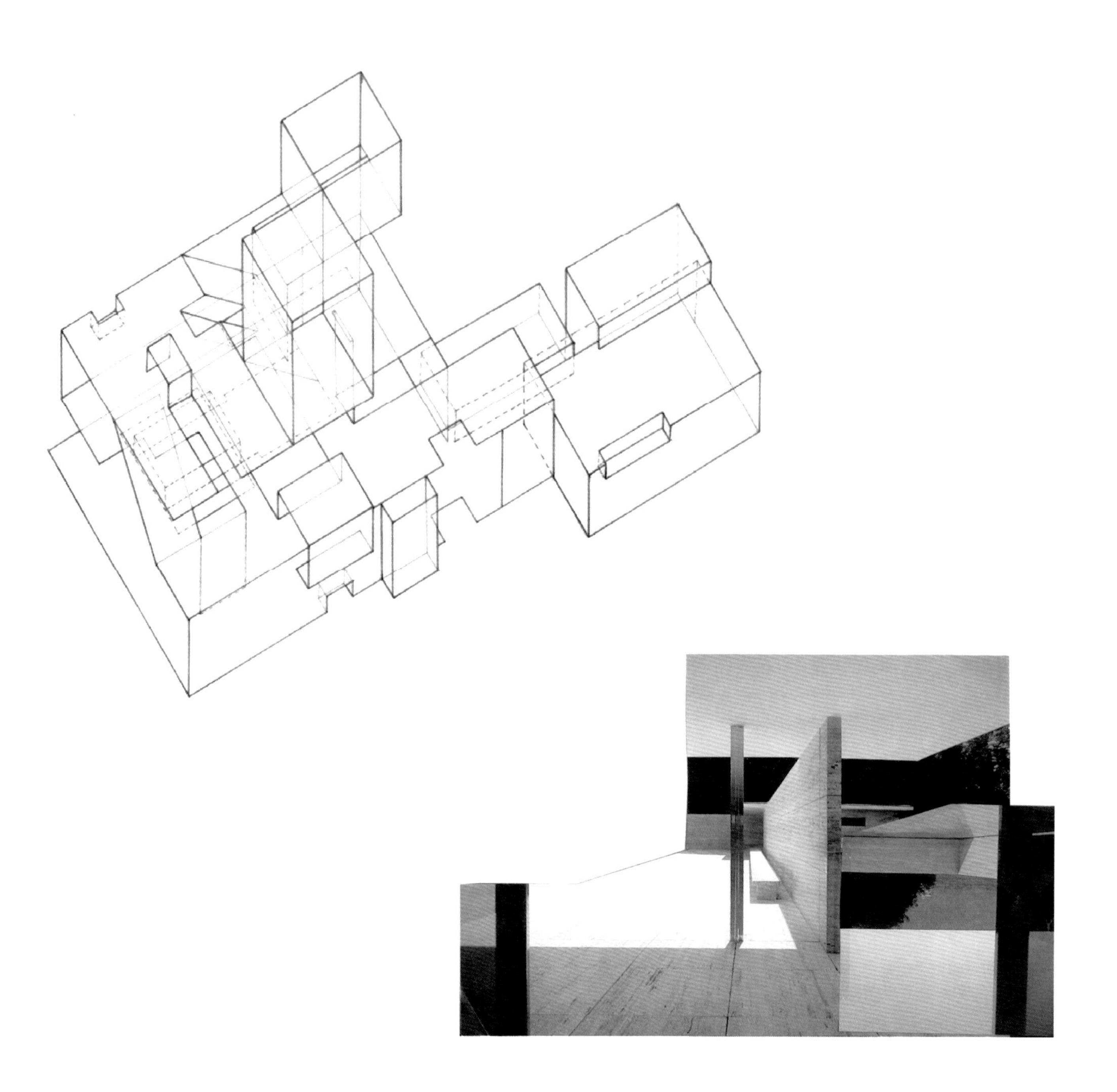

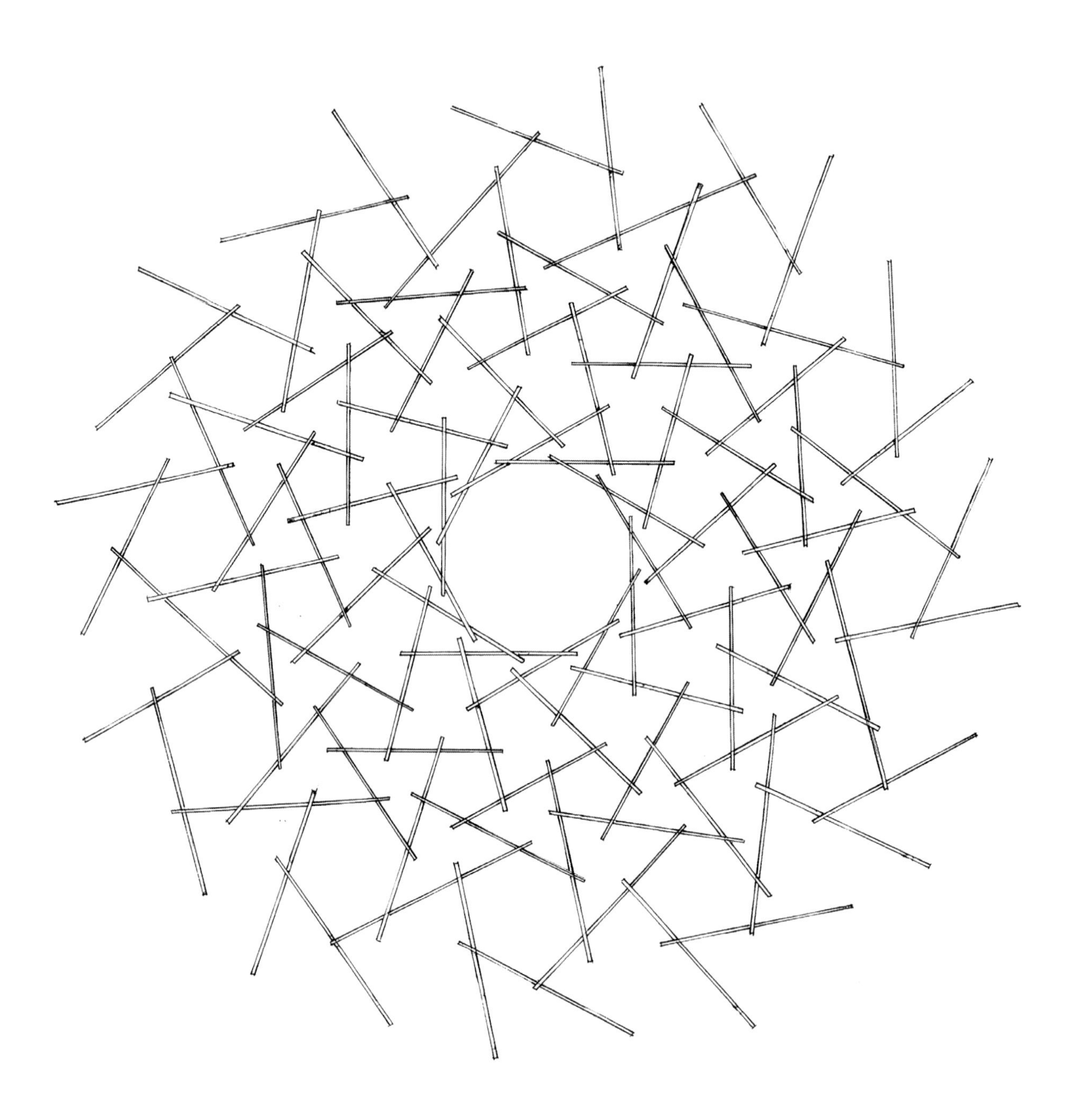

实验两种材料：竹子与 PVC 管，运用互承结构原理用小的杆件完成“大”的覆盖空间。材料长度 1.2~1.5 m，截面 4~5 cm；搭建出的空间高 2 m，宽 3~5 m。尝试通过变化杆件的截面尺寸、搭接窗口的形状与大小、搭接方式等获得多样的空间形式。

Experiment with two types of material: bamboo and PVC tubes, apply principles of mutually-supporting structure to complete a “large” covered space with small tubes. Length of material is 1.2~1.5m, sectional size is 4~5cm; height of completed space is 2m, width is 3~5m. Try to achieve various types of space by changing sectional size of tubes, shapes and sizes of lapped windows, and forms of lapped joints.

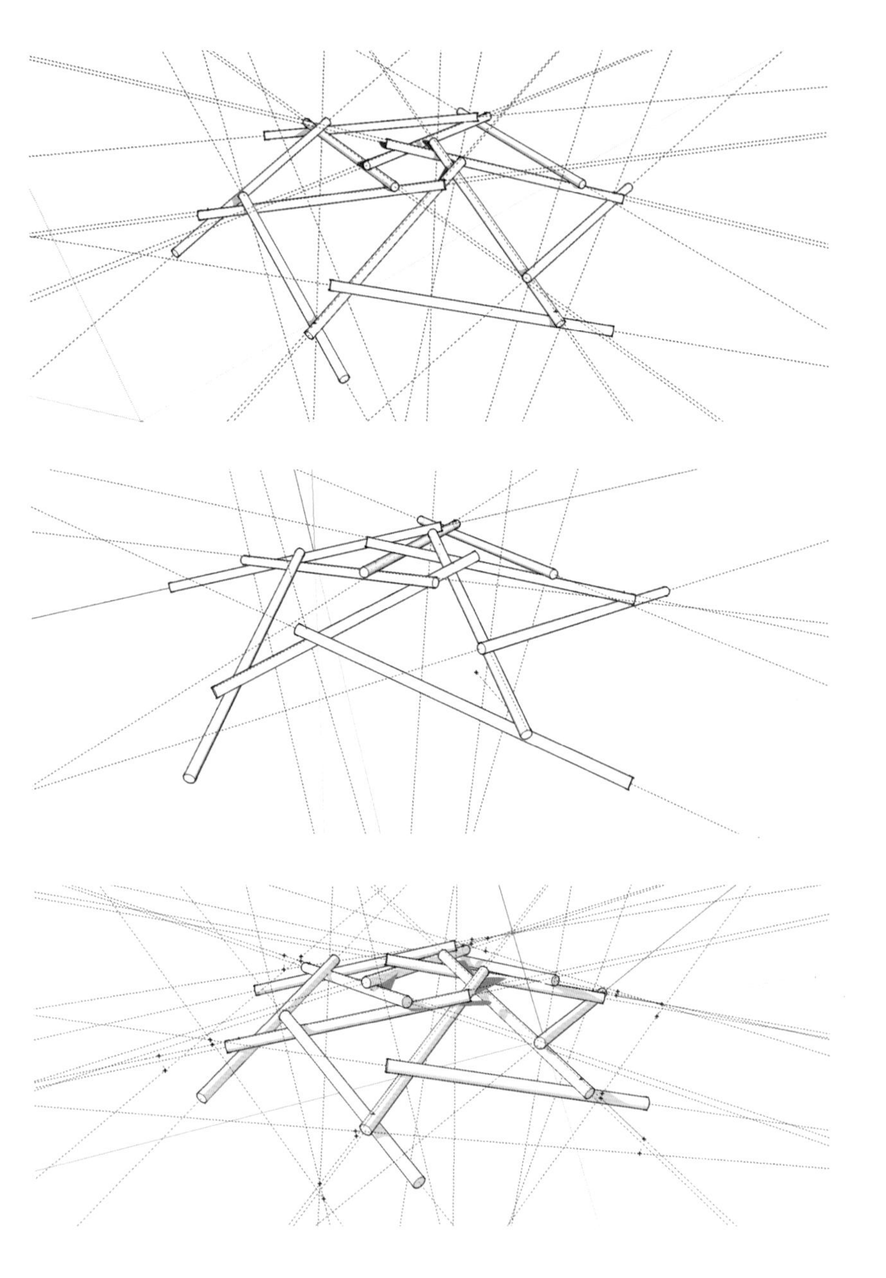

软件学院

风景区茶室设计

TEA HOUSE DESIGN

刘铨 冷天 王丹丹

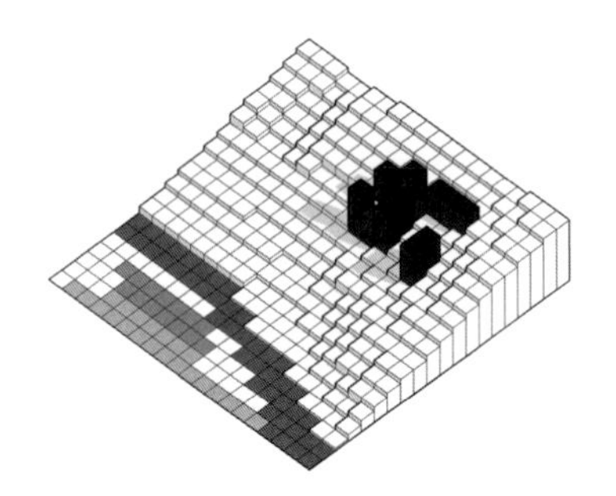

教学目标

本课程训练使用空间形式语言进行设计操作。在设定上，教案希望学生进一步认识到“空间形式语言”作为设计工具，去解读、应对场地环境与功能需求并推动设计进展的作用。

研究主题

场地与界面、功能与空间、空间的组织、尺度与感知。

教学内容

1. 本次设计场地是紫金山风景区内的一处坡地，要求学生从场地水平向界面的限定来考虑设计建筑的形体、布局及其最终的空间视觉感受。

2. 本次设计的建筑功能设定为风景区茶室，建筑面积 300 m^2，建筑层数 2 层，其中包括入口门厅、容纳 60 人的大厅，容纳 30 人的 2~4 人雅座若干，以及必要的辅助空间。同时，必须进行室外场地的设计。

3. 基于地形的形式化表达，寻求适应建筑空间需求的场地改造形式进行高差的处理，把对地形的理解与公共建筑的基本功能空间组织模式及形态塑造联系起来。

4. 在空间形式处理中注意通过图示表达理解空间构成要素与人的空间体验的关系，主要包括尺度感和围合感，注意景观朝向问题。

Training Objective

This course aims to train design operations with the language of spatial forms. In term of course setting, the teaching plan is intended to allow students to further understand the role of “the language of spatial forms”, as a design tool, in interpreting, dealing with field environment and function demand, and promoting design progress.

Research Subject

Site and interface, function and space, organization of space, dimensions and perception.

Teaching Content

1. The design site is a sloped field within the Zijinshan scenic spot, and it requires students to consider the profile, layout and final visual effect of space of the designed building based on limits of horizontal interface of the site.

2. The building to be design will functions as a teahouse at scenic spot, covers an area of 300 m^2, consists of two floors, including an entrance hallway, a public hall with seating capacity of 60 people, several private rooms with seating capacity of 2~4 people which can serve 30 guests in aggregate, as well as other necessary auxiliary spaces. Meanwhile, design for outdoor space must also be conducted.

3. On basis of formalized expression of topography, try to deal with height difference in the form of field modification by accommodating demand of building space, and link understanding of topography to the space organization patterns and form modelling for basic functions of public building.

4. Pay attention to the understanding of relations between factors of space formation and spatial experience of human through graphic expression in the process of handling spatial forms.

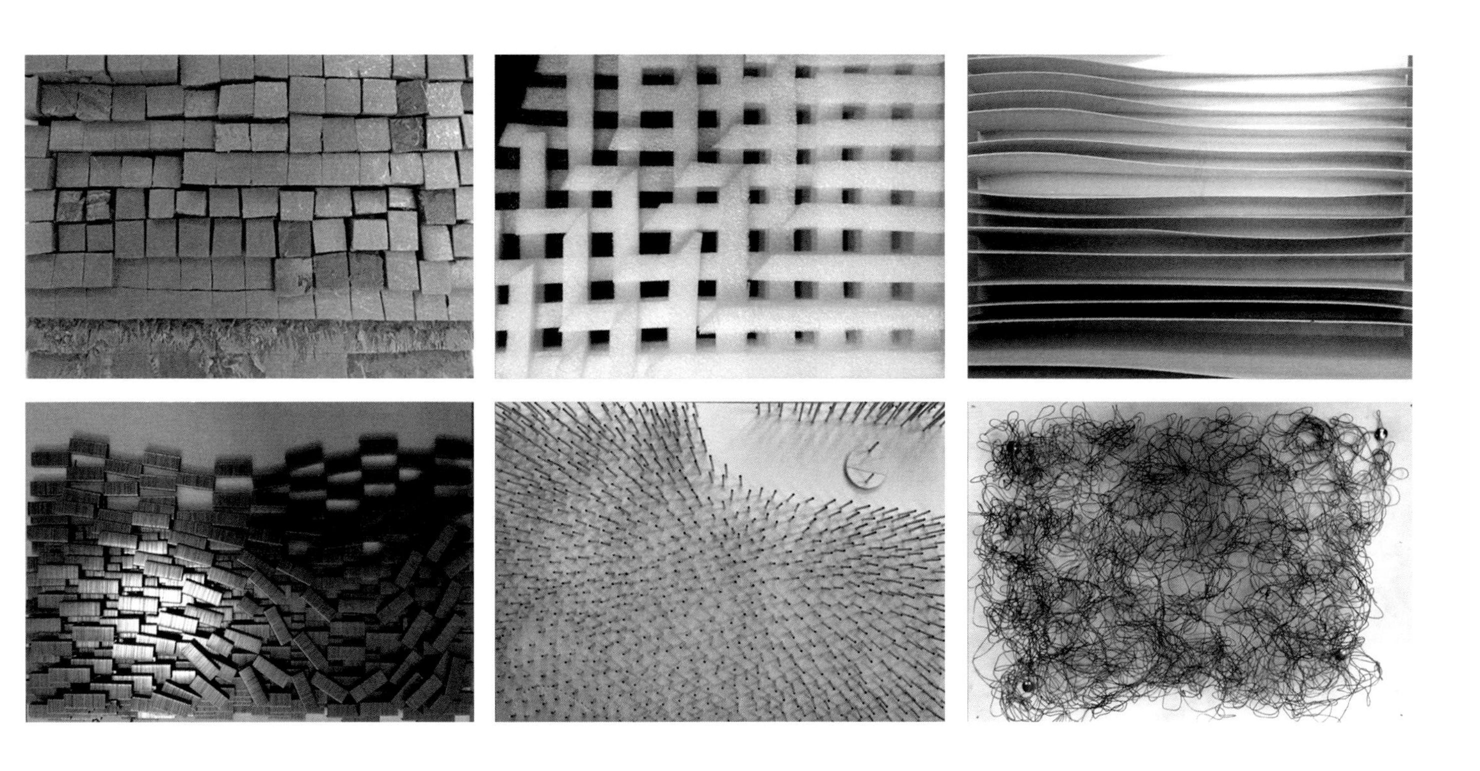

探索地形表达的形式秩序，材料与可操作性，剔除无序的表达。

Explore form order of topographic morphosis, material and operability, and get rid of unordered morphosis.

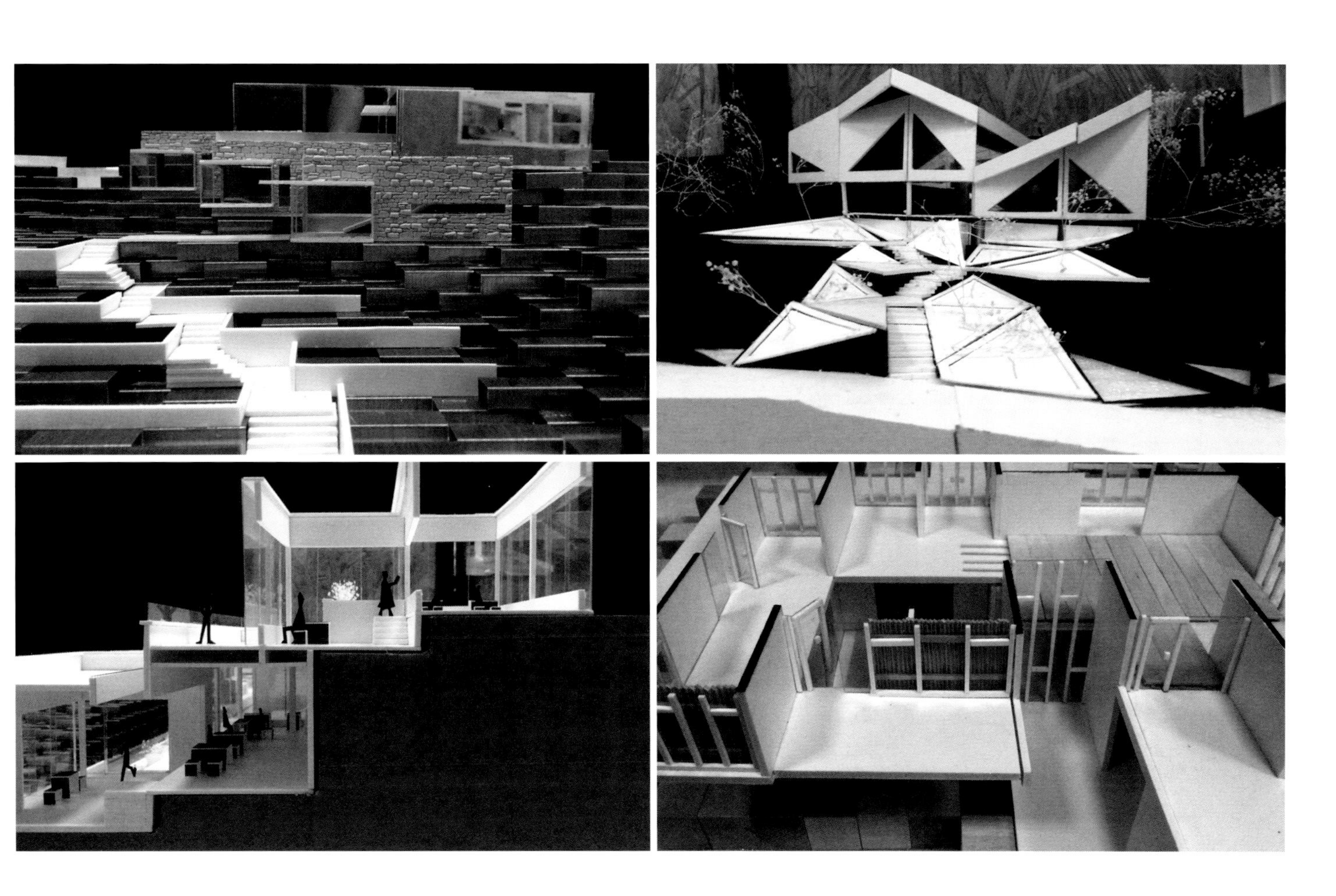

根据景观的朝向、体量尺度要求对形态进行修正，并通过提压、推拉等操作形成建筑空间，进而深化室内微空间的设计。

Correct morphology according to orientation, physical measurements of landscape, and shape building space via lifting and pressing, pushing and pulling and other operation, so as to deepen the design of indoor micro-spaces.

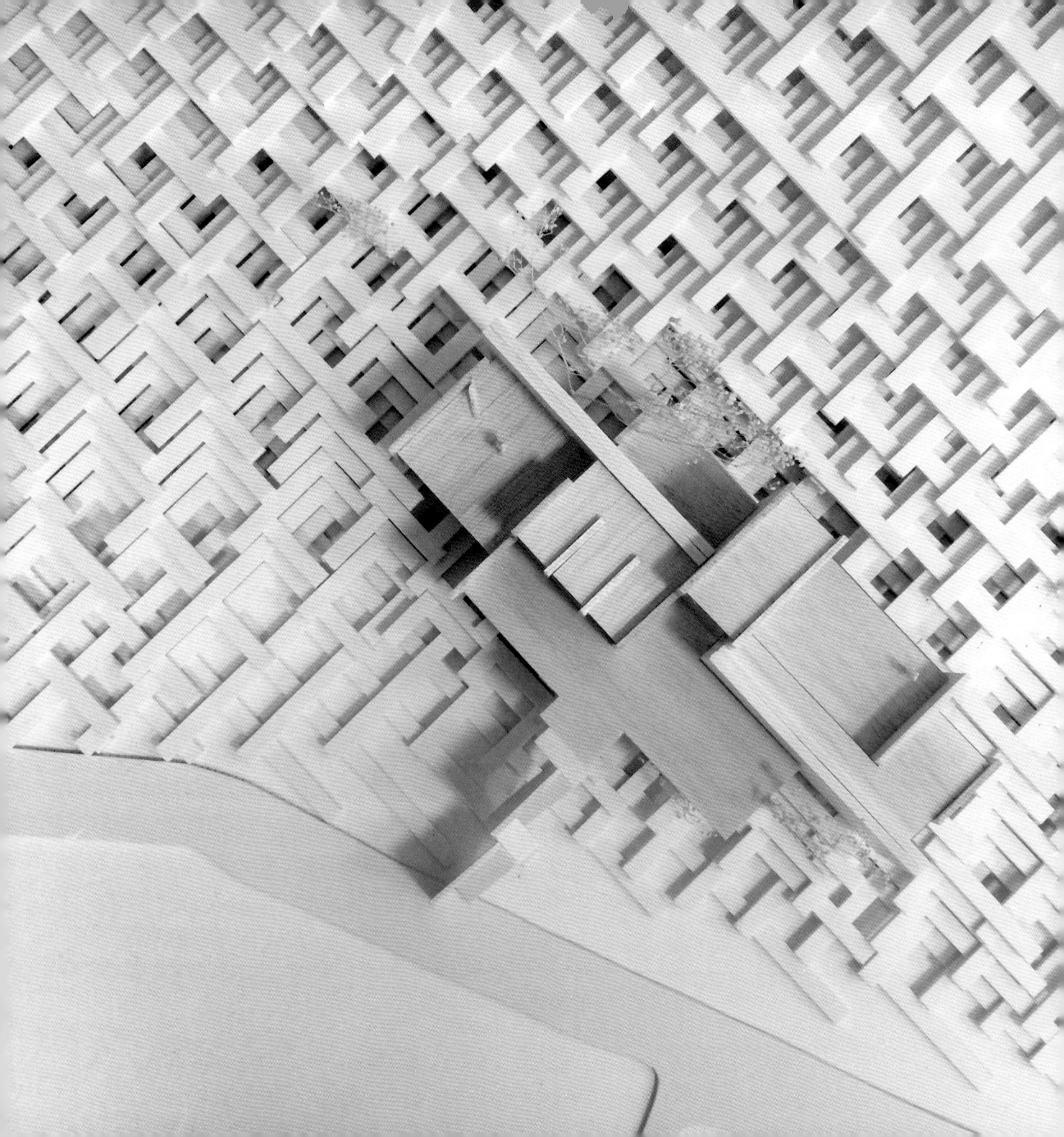

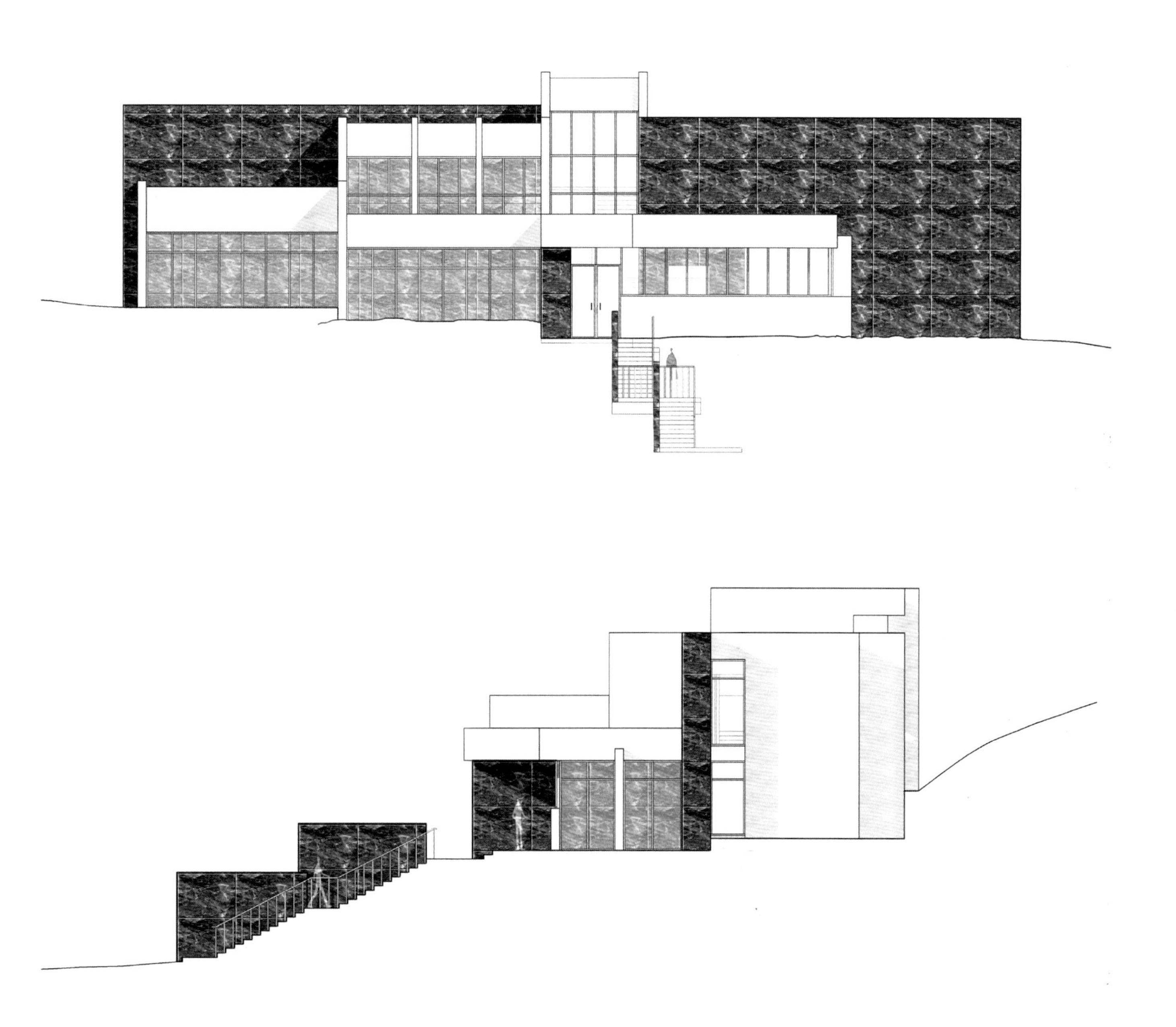

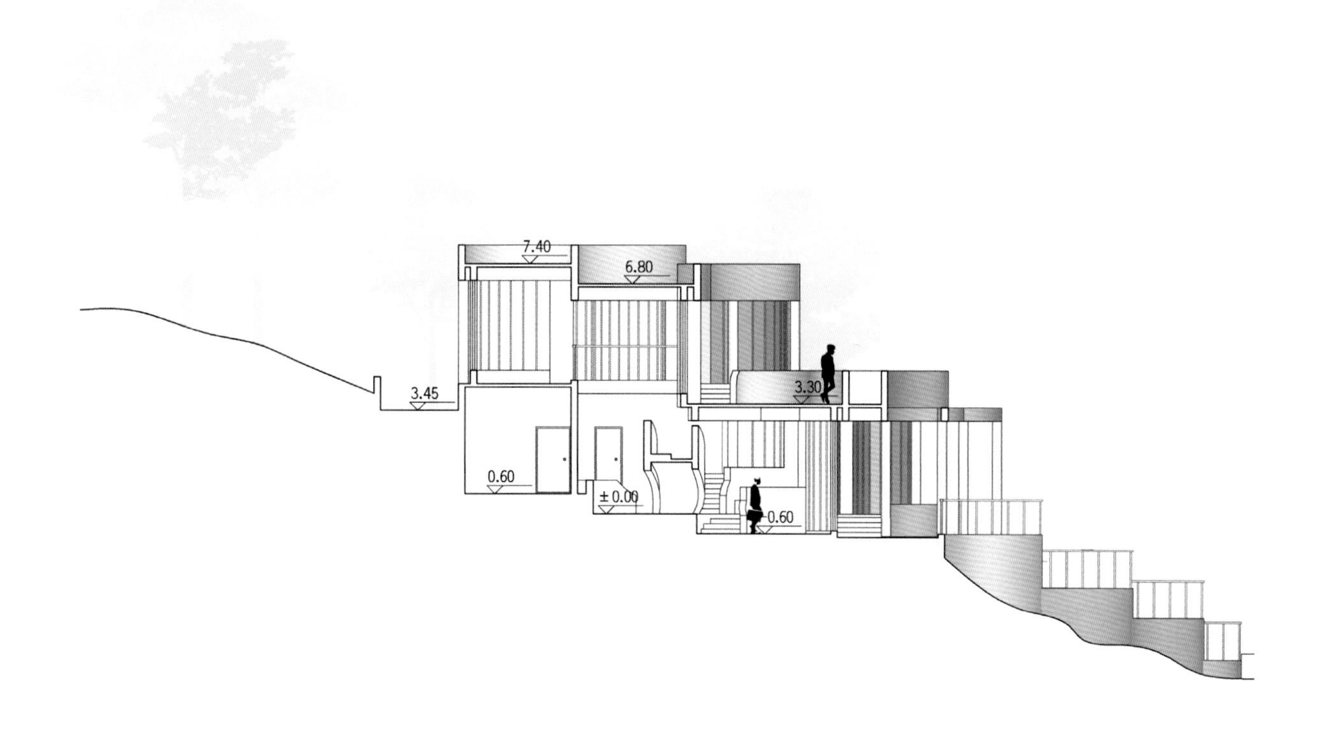
7.40
6.80
3.45
3.30
0.60
± 0.00
0.60

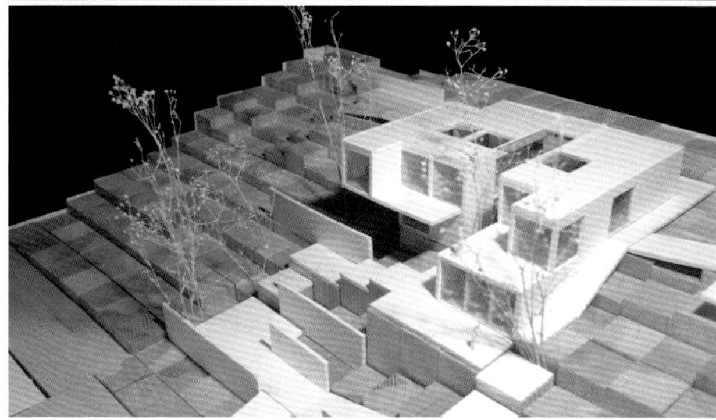
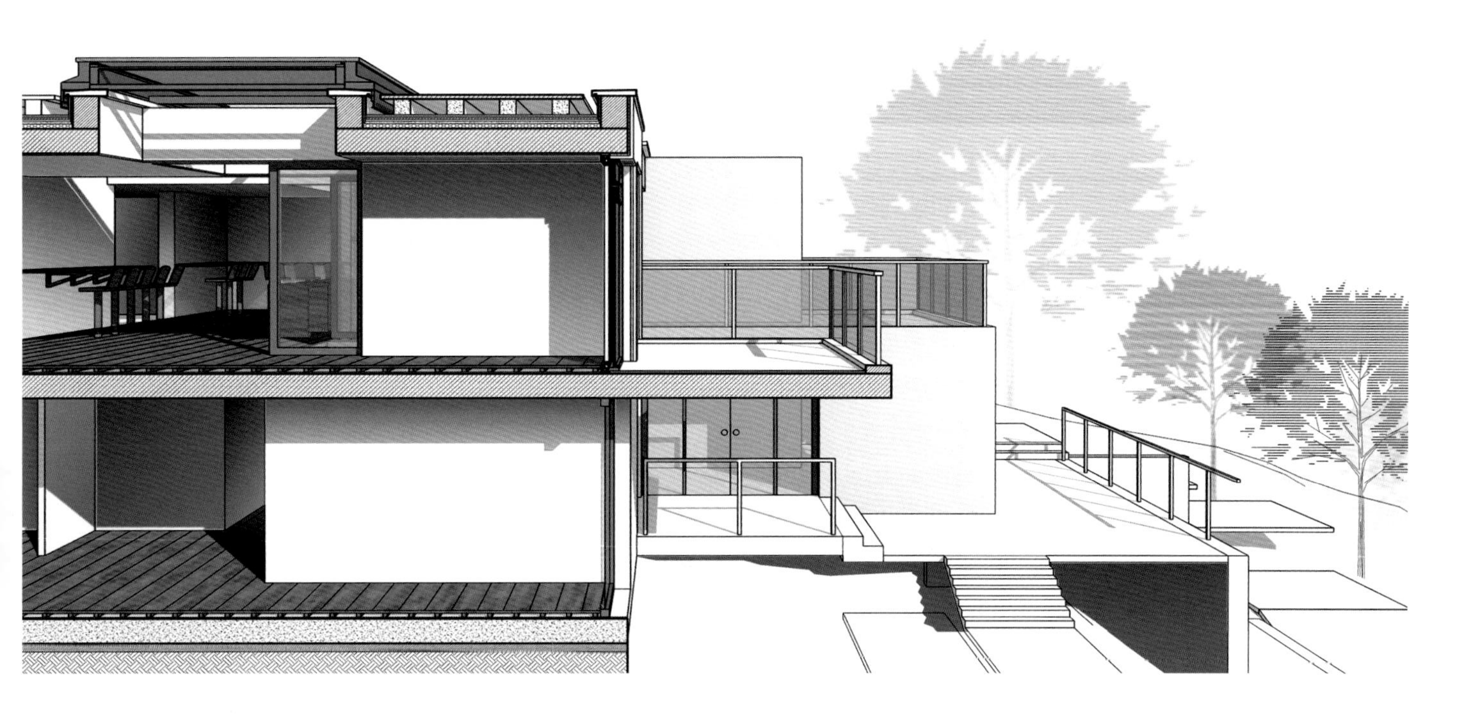

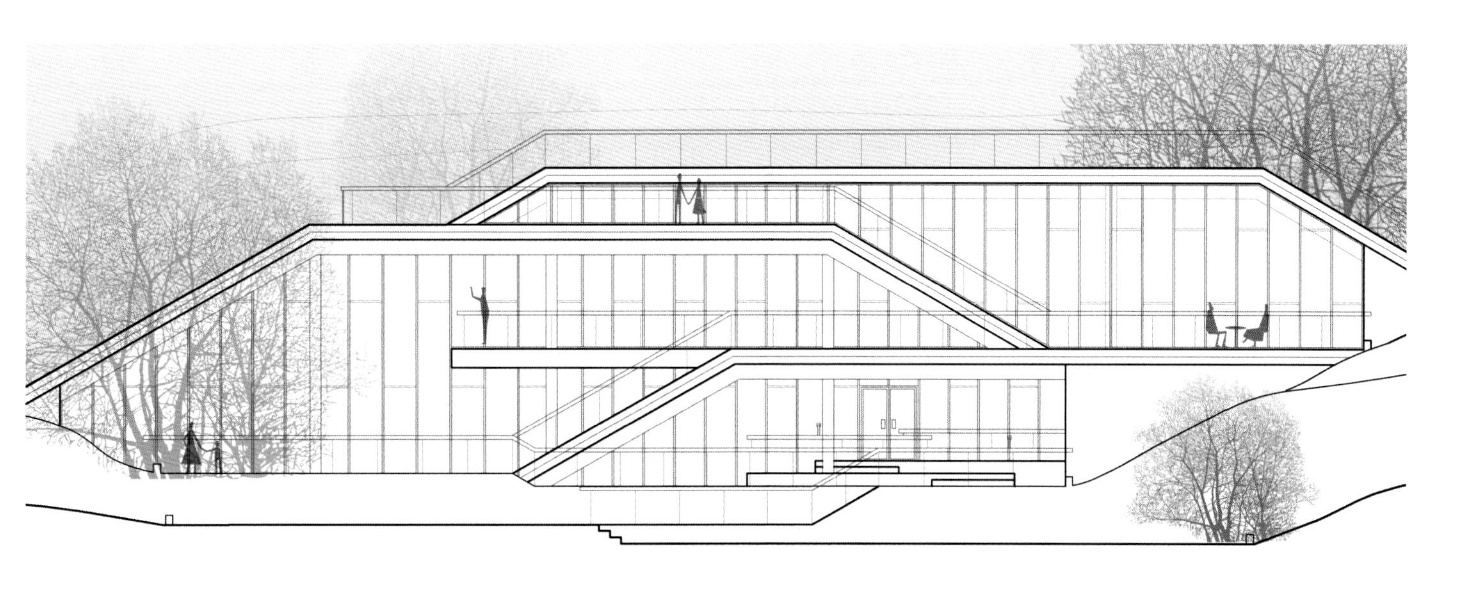

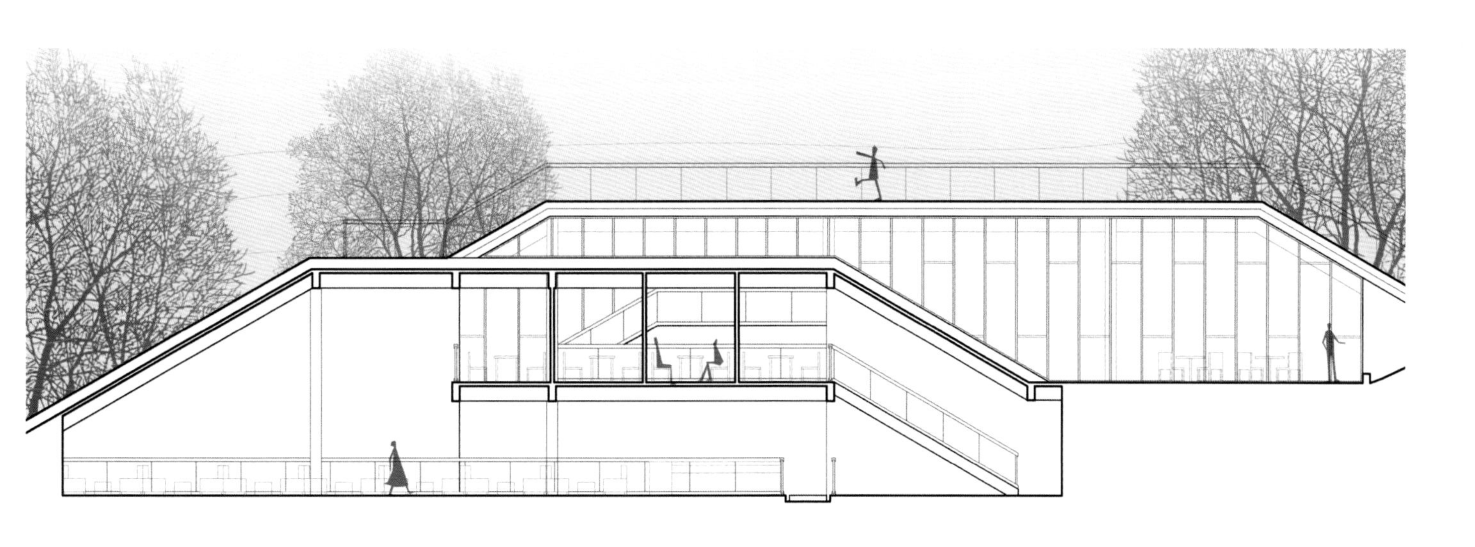

社区商业中心+活动中心设计

COMMUNITY BUSSINESS CENTRE & ACTIVITY CENTRE

华晓宁 钟华颖 王铠

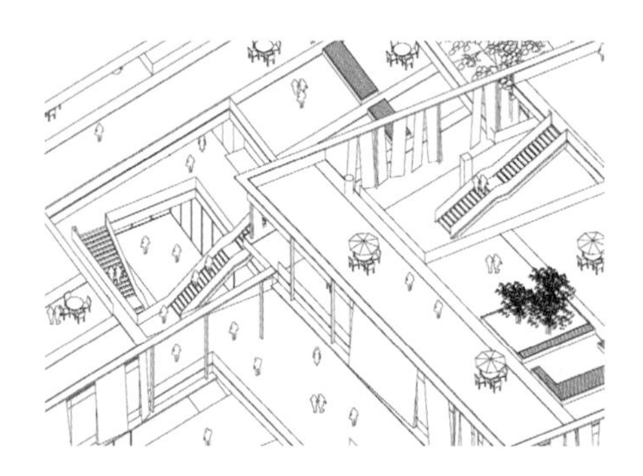

研究主题

实与空：关注城市中建筑实体与空间的相互定义、相互显现，将以往习惯上对于建筑本体的过度关注拓展到对于“之间”的空间的关注。

内与外：进一步突破“自身”与“他者”之间的界限，将个体建筑的空间与城市空间视为一个连续统，建筑空间即城市空间的延续，城市空间亦即建筑空间的拓展，两者时刻在对话、互动和融合。

层与流：不同类型的人和物的行为与流动是所有城市与建筑空间的基本框架，当代大都市中不同的流线在不同的高度上层叠交织，构成一个复杂的多维城市。必须首先关注行为和流线的组织，由此才生发出空间的系统和形态。

轴与界：城市纷繁复杂的形态表象之后隐含着秩序和控制性，并将成为新的形态介入。

教学内容

在用地上布置社区商业中心（约 15000 m^2）、社区文体活动中心（约 8000 m^2），并生成相应的城市外部公共空间。

Research Subject

Entity and space: Pay attention to mutual definition, mutual representation of architectural entity and space in cities, and extend traditional excessive attention to the building itself to the space “among them”.

Interior and exterior: Further break through the boundary between “self” and “others”, and consider space of individual buildings and urban space as a continuum. Architectural space is the continuation of urban spce, While urban space is the expansion of the architectural space. Both keep on dialoguing, interacting and fusing.

Stack and flow: Behaviors and flows of different types of people and objects are the basic framework of all urban and building spaces, and a complex multi-dimensional city is formed by stacking up and interweaving of different flow lines at different altitudes in modern metropolis. We must pay attention to organization of behaviors and flow lines first, and then can generate system and morphology of space.

Axis and boundaries: Order and control are concealed behind the morphologic appearance of complexity of cities, which will be involved as new forms.

Teaching Content

Lay out community commercial center (about 15000 m^2) and community recreational and sports activities center (about 8000 m^2) on the land, and generate associated outdoor urban public spaces.

A-A剖面 1:300
B-B剖面 1:300

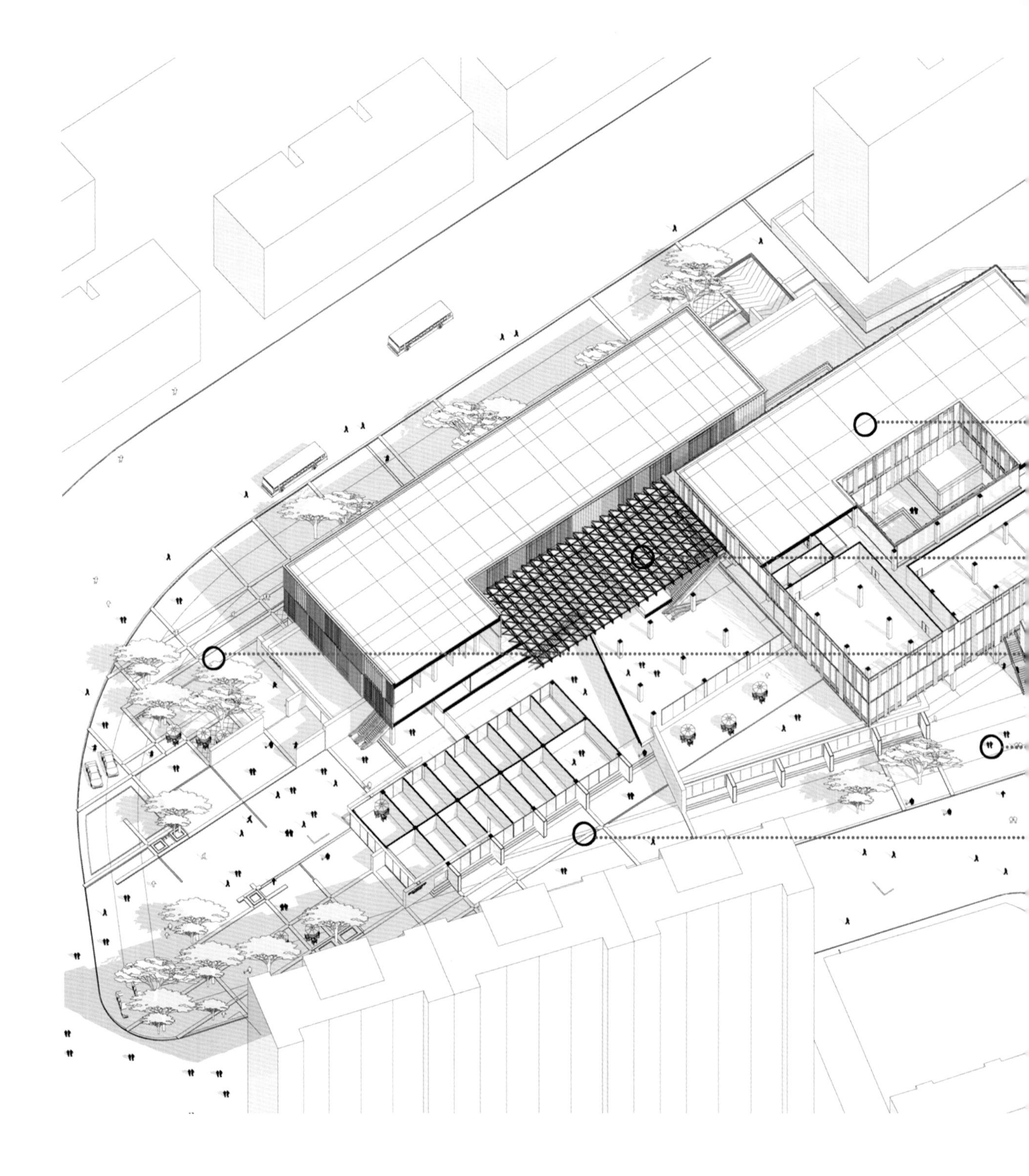

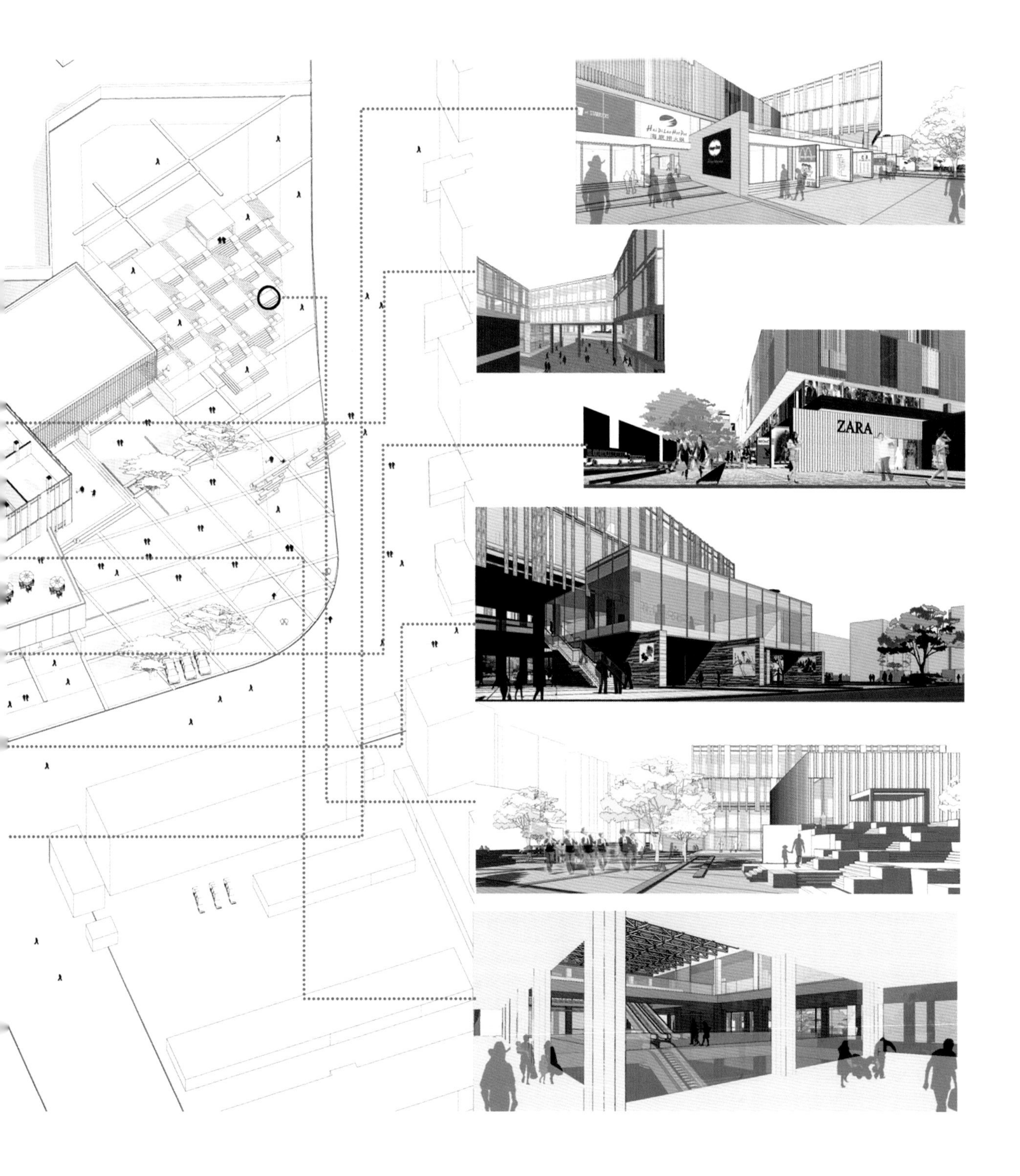
ZARA

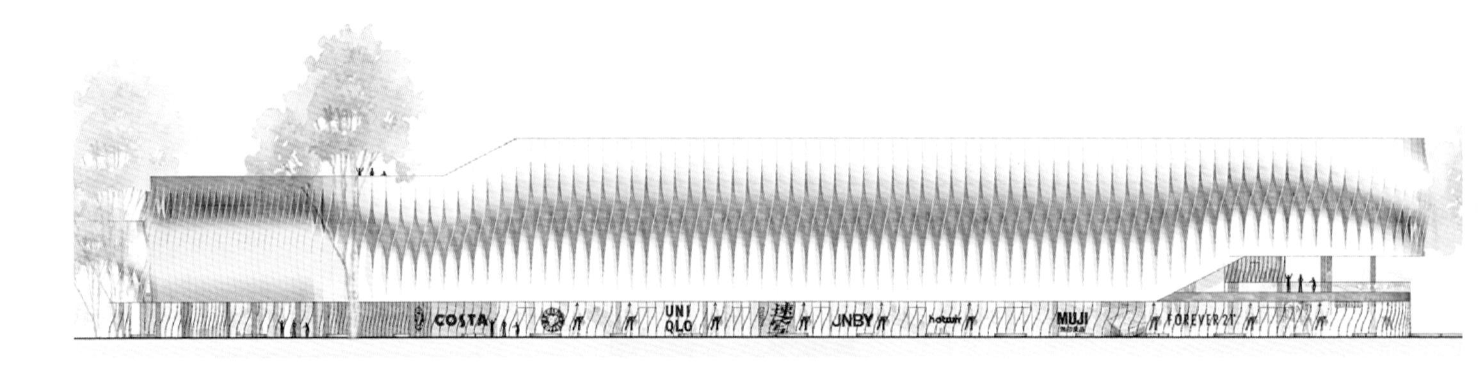
COSTA
UNI
QLO
JNBY
MUJI
FOREVER 21

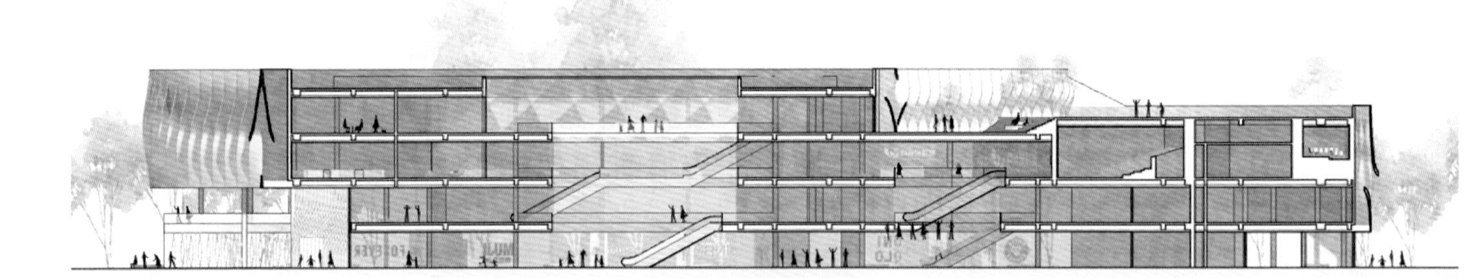

UNI
QLO
FOREVER 21

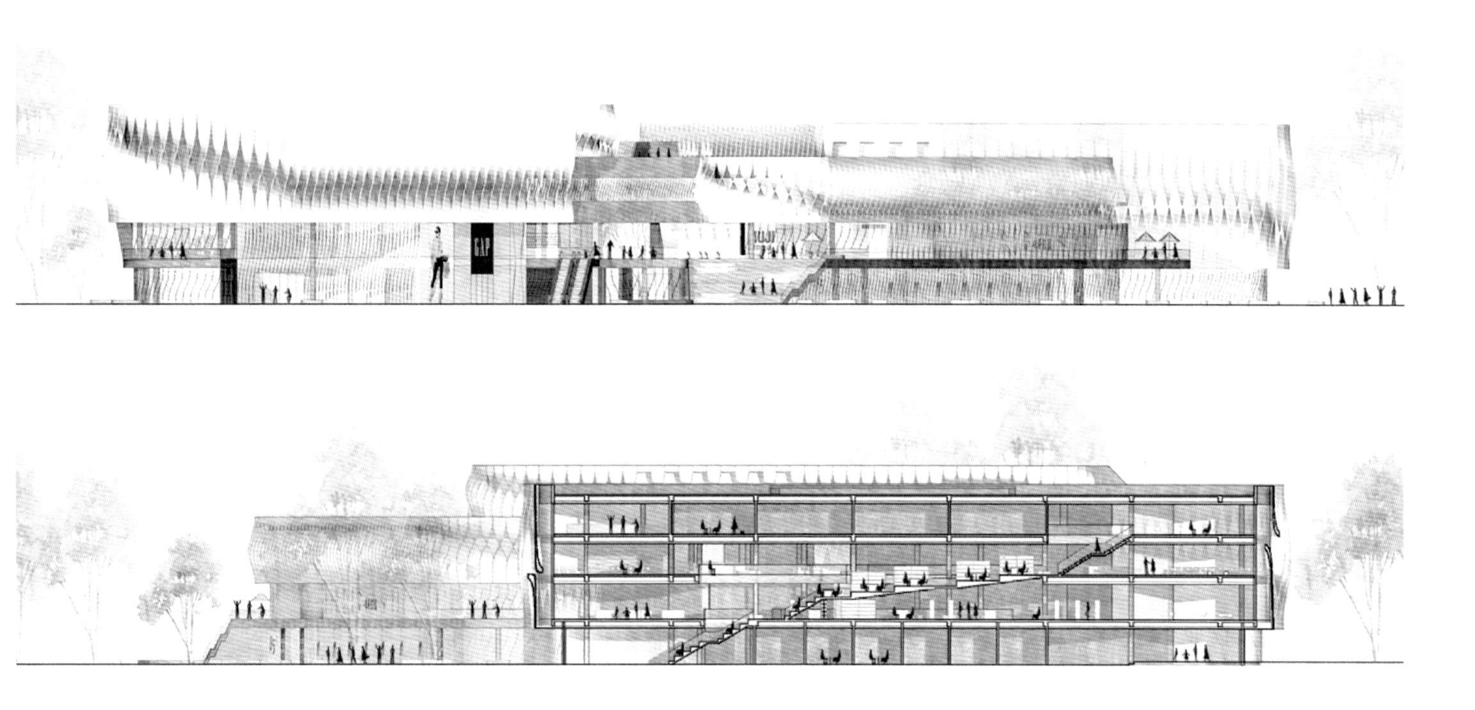
GAP

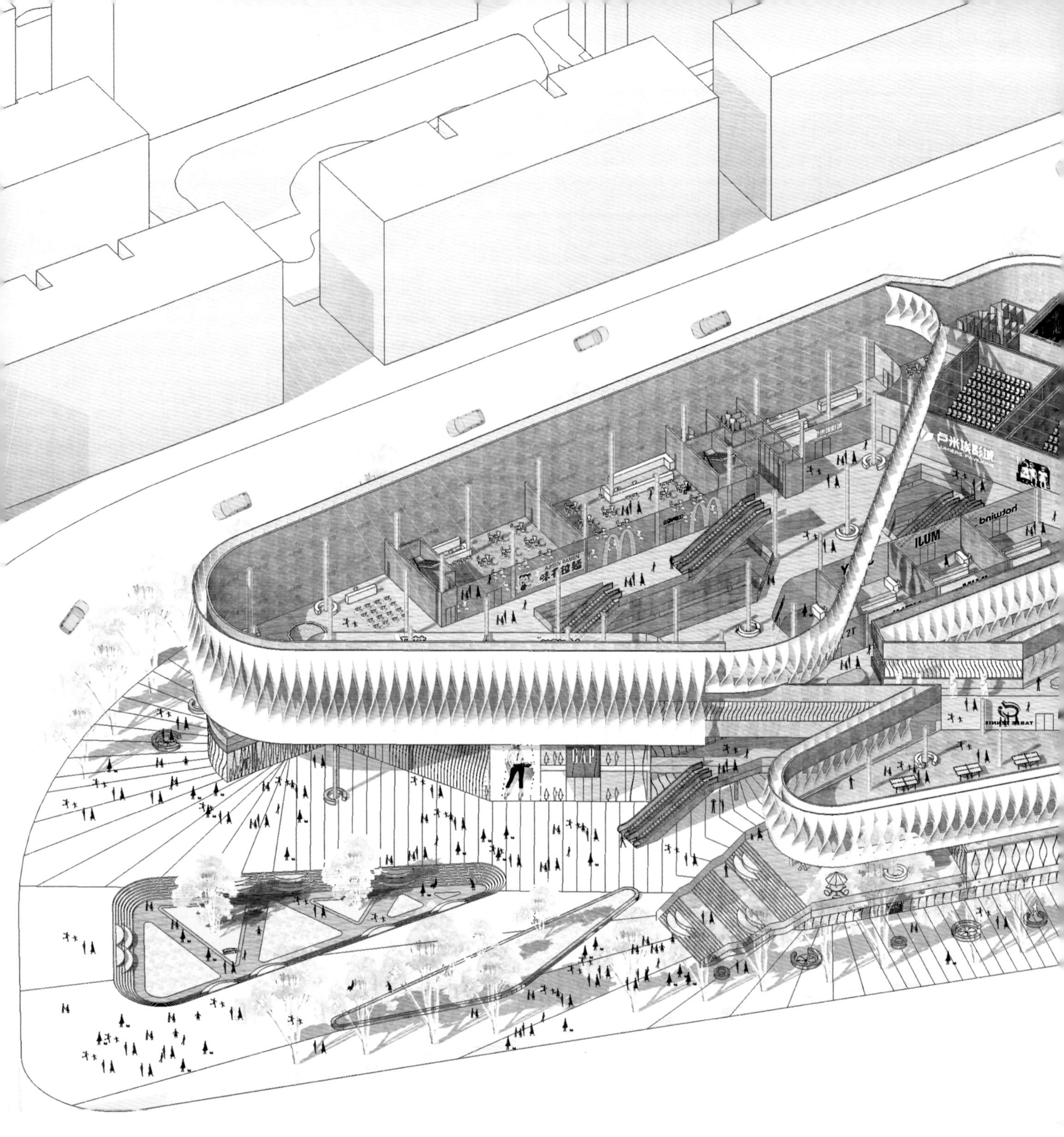

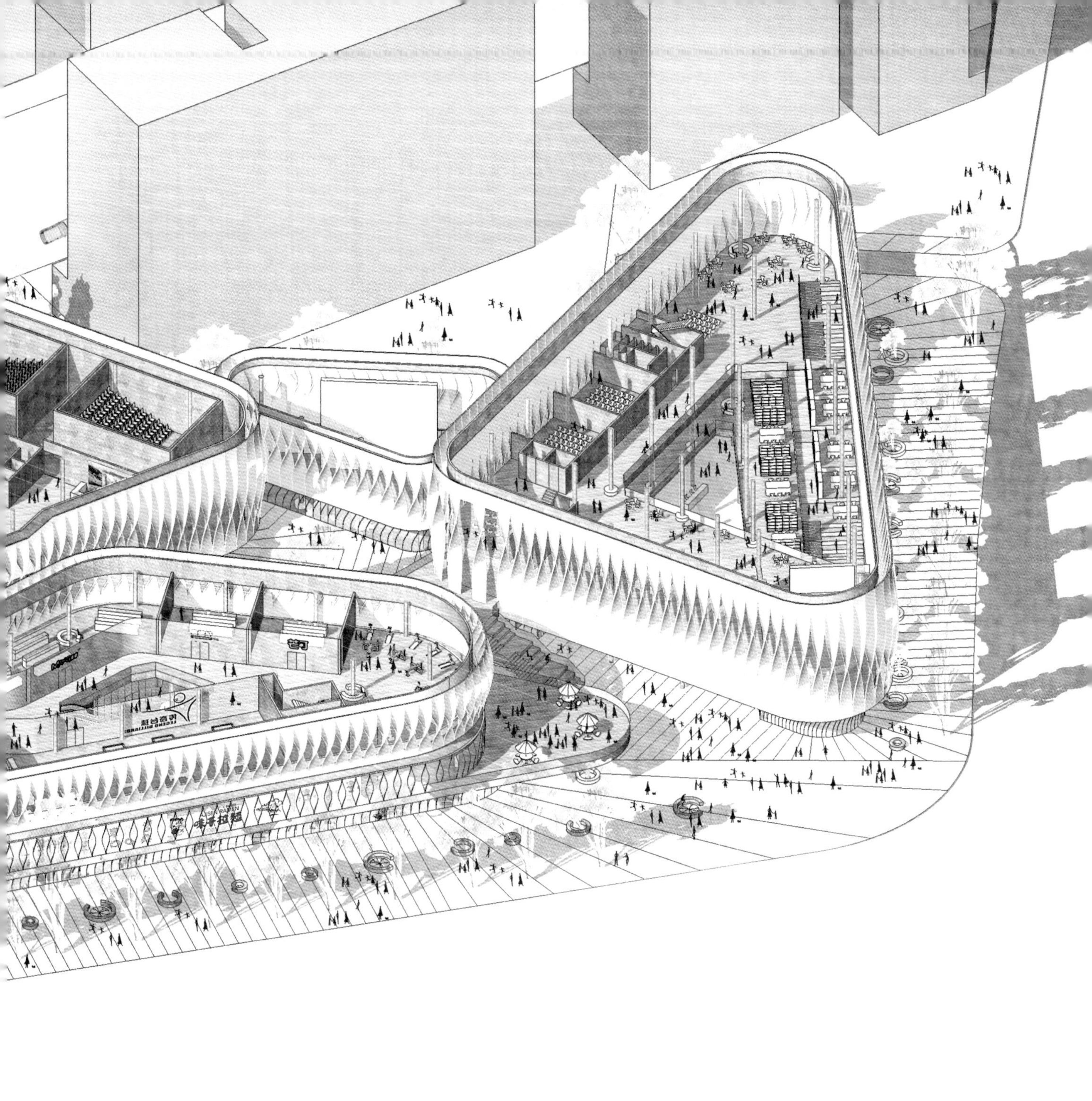

旧城改造城市设计

URBAN DESIGN FOR OLD TOWN RENOVATION

吉国华 胡友培 尹航

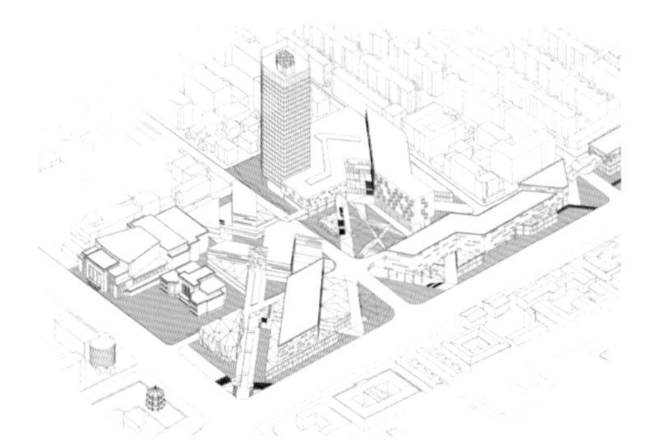

教学目标

1. 着重训练城市空间场所的创造能力，通过体验认知城市公共开放空间与城市日常生活场所的关联，运用景观环境的策略创造城市空间的特征。

2. 熟练掌握城市设计的方法，熟悉从宏观整体层面处理不同尺度空间的能力，并有效地进行图纸表达。

3. 理解城市更新的概念和价值；通过分析理解城市交通、城市设施在城市体系中的作用。

4. 多人小组合伙，培养团队合作意识和分工协作的工作方式。

教学内容

1. 设计地块位于南京市玄武区，总用地约为 6.20 hm^2。地块内国民大会堂旧址、国立美术陈列馆旧址和北侧邮政所大楼可保留，其余地块均需进行更新。地块周边有丰富的博物馆、民国建筑等文化资源，设计应对周边文化环境起到进一步提升作用。地块周边用地情况复杂，设计中需考虑与周边现状的相互影响。

2. 本次设计的总容积率指标为 2.0~2.5，建筑退让、日照等均按相关法规执行。

3. 碑亭巷、石婆婆庵需保留，碑亭巷和太平北路之间现状道路可根据设计进行位置或线形调整。

4. 地下空间除满足单一地块建筑配建的停车需求外，应综合考虑地上、地下城市一体化设计与综合开发。

Training Objective

1. Emphasize the training on ability of creating urban spatial places, and create features of urban space with the strategy of landscape environment through experiencing and perceiving the links between urban public spaces and urban daily living places.

2. Master methodology of urban design, grasp the ability of handling spaces of different dimension at macro and integral level, and achieve effective representation with drawings.

3. Understand the concept and value of urban renovation; understand the role of urban traffic, urban facilities in the urban system through analysis.

4. Form partnership with several group members to cultivate awareness of teamwork and the working mode of collaboration.

Teaching Content

1.Land parcel of the design is located in Xuanwu District, Nanjing, covering an area of 6.20 hm^2 approximately. Site of the former National Assembly Hall, site of the former National Art Gallery and the Post Office Building at north side may be retained, other parts of the land parcel need to be renovated. There are abundant cultural resources such as museums and buildings constructed in the period of the Republic of China around the land parcel, so the design should further improve the cultural environment around it. Land use conditions around the land parcel are very complicated, so mutual influence with surrounding existing conditions must be taken into account in the design.

2.Gross plot ratio of the design is 2.0~2.5, and building setback and sunlight value shall comply with relevant laws and regulations.

3.Beiting Lane and Shipopo Nunnery are to be retained, and location and route of the existing road between Beiting Land and North Taiping Road may be adjusted according to the design.

4.For underground space, besides meeting the associated parking demand of buildings on the single land parcel, overall consideration shall be taken for integrated design and comprehensive development of urban spaces above and under the ground.

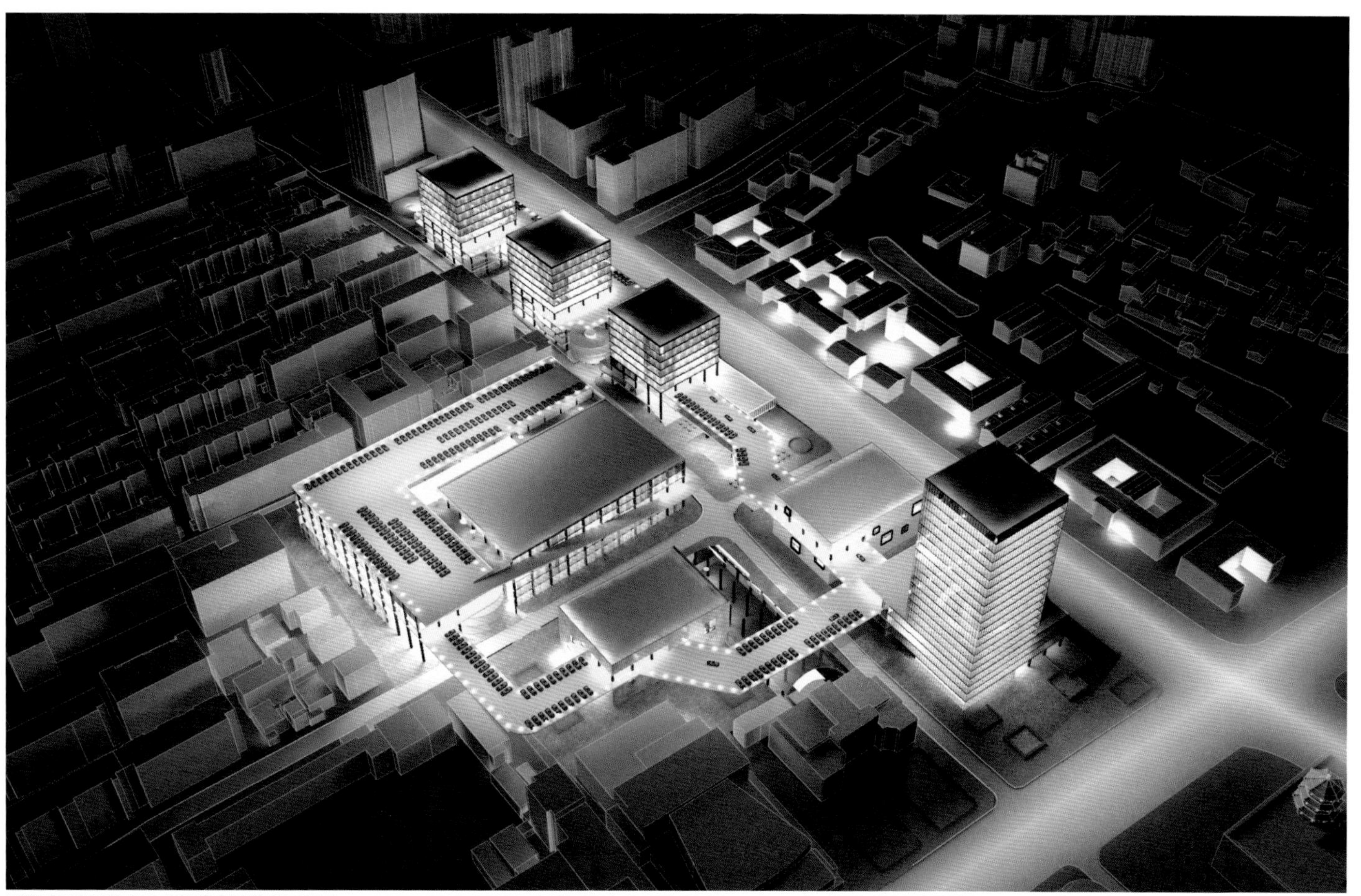

本设计基于对老城区的深入调研，针对周边居民区、娱乐街区突出的停车与交通问题进行专项研究，并最终创造出一个将新建街区停车交通与旧街区相互分离的高架交通体系，以此为中心，解决街区功能、交通问题，甚至希望围绕这个高架交通体系来形成独特的街区建筑氛围。

The design is based on the in-depth investigation on old town, aims to conduct specialized research on the outstanding parking and traffic problems of surrounding residential quarters and recreational blocks, and in the end to create an elevated traffic system that separates parking and traffic systems of the new block with that of the old block with each other, and solve functional and traffic problems of the block by centering on the system, and even expect to shape unique architectural atmosphere of the block around the elevated traffic system.

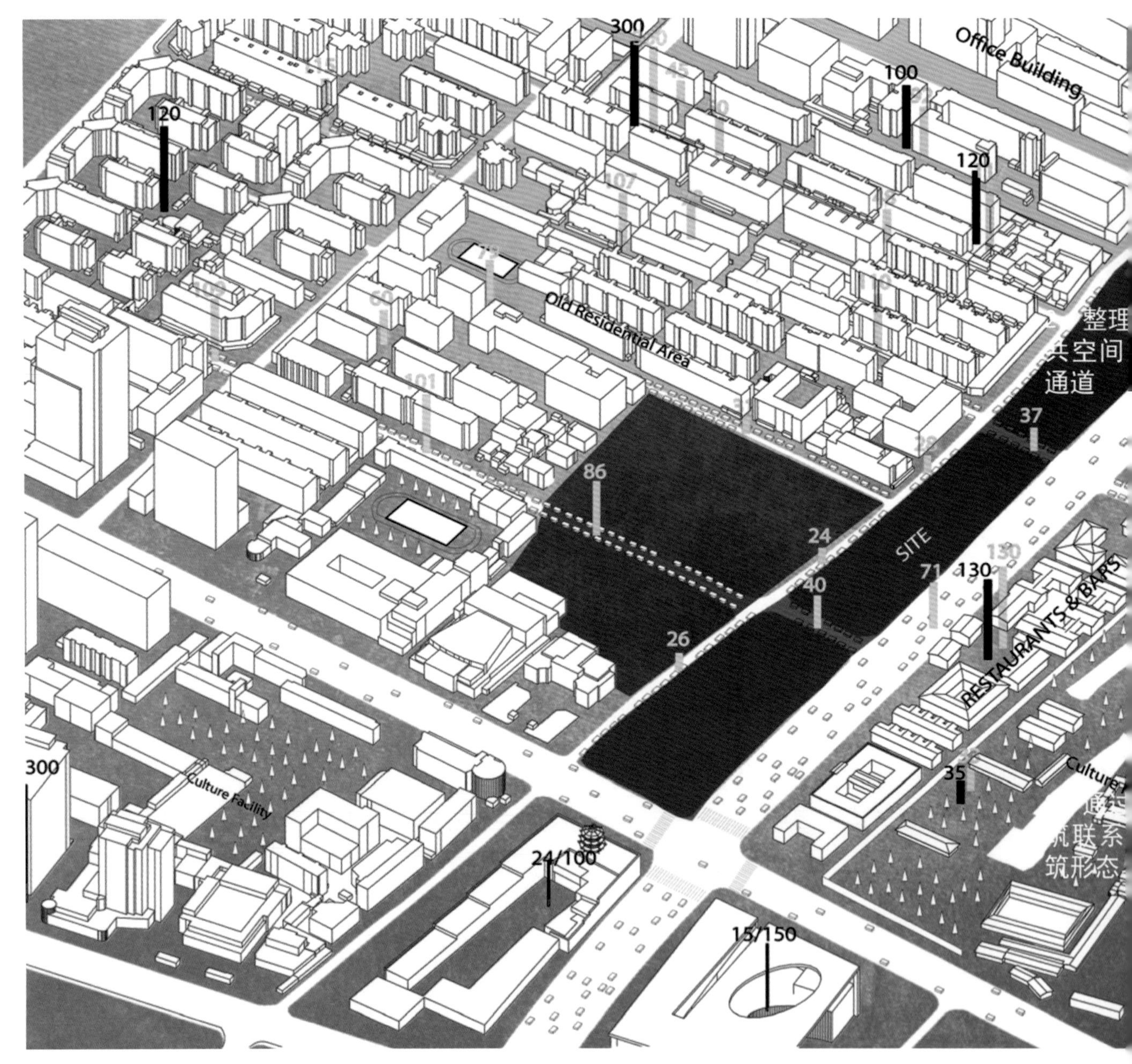

对周边街区的交通停车问题进行深入调研，得出停车需求和现状的缺口。

Conduct in-depth research on parking and traffic problems of adjacent blocks, and get to know the gap between parking demand and existing parking spaces.

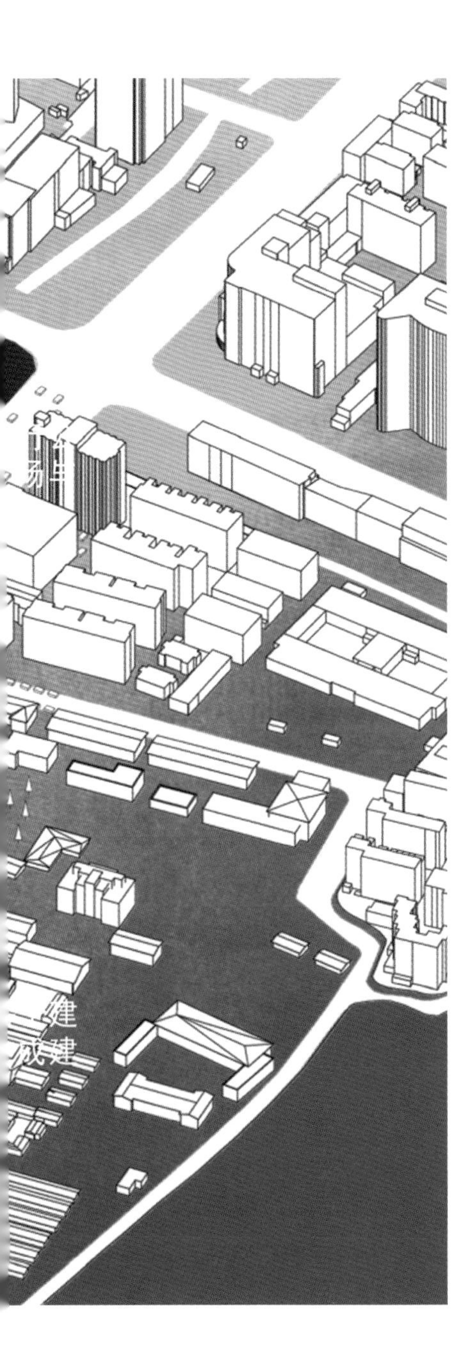

流线分析

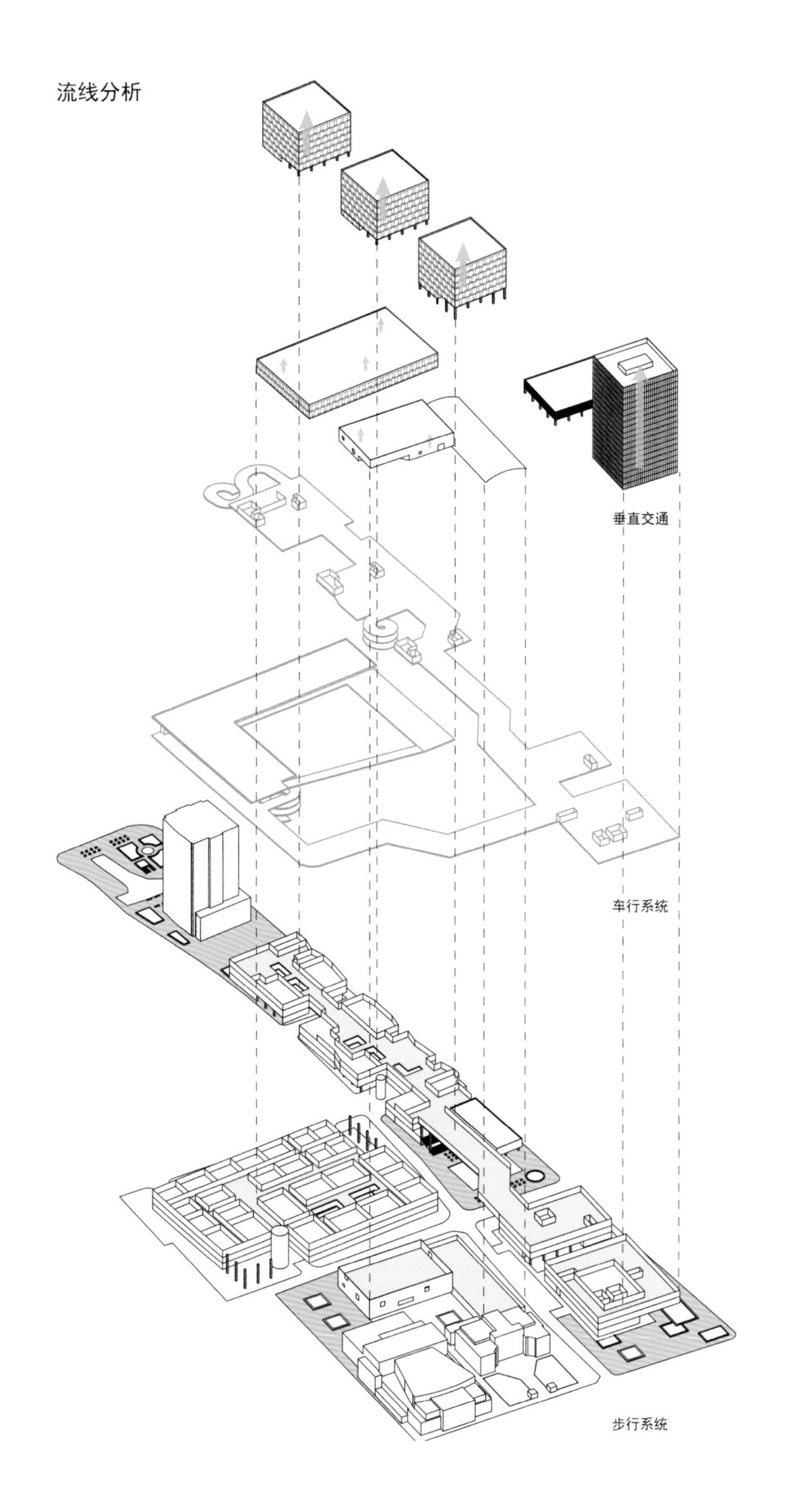

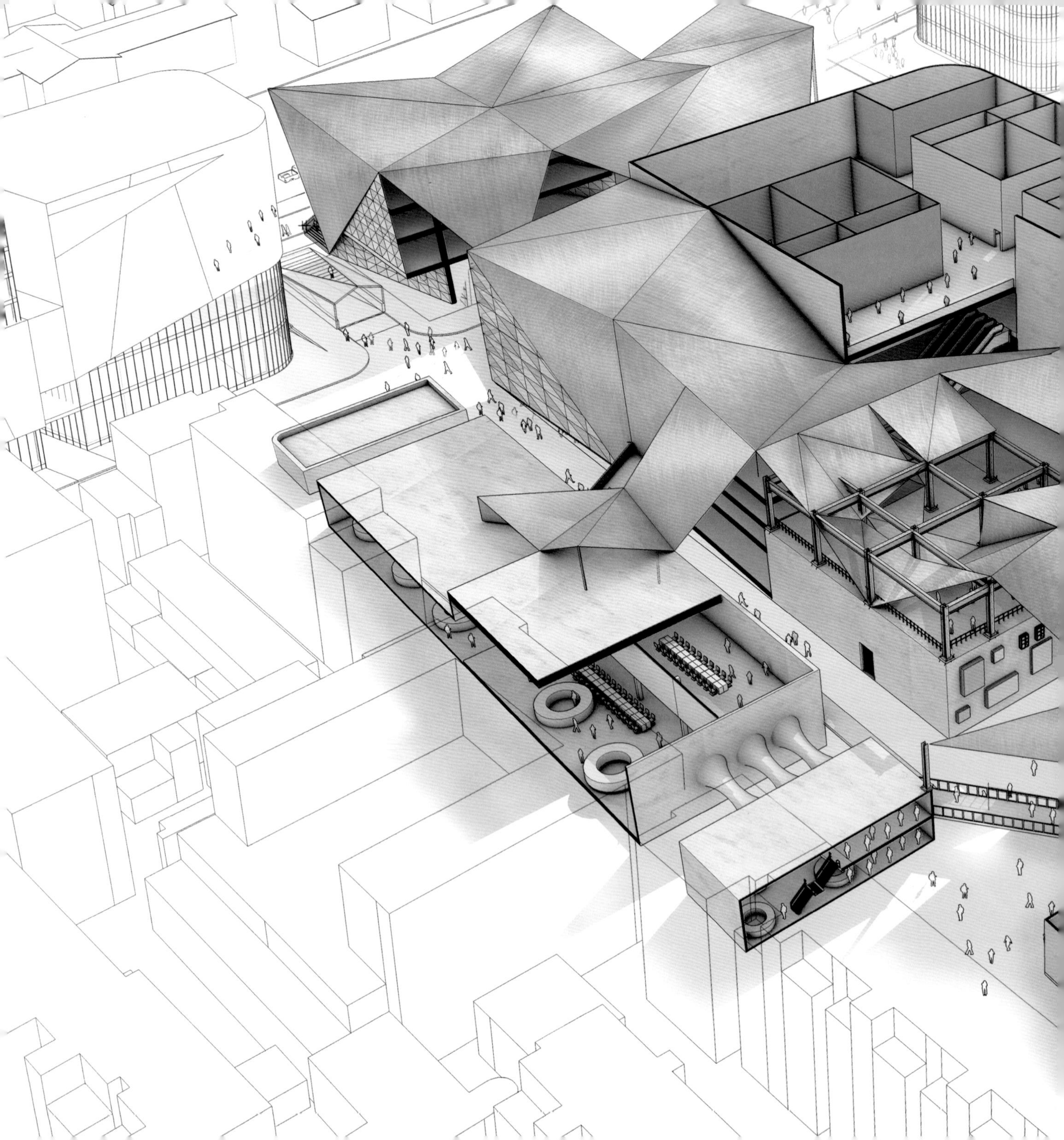

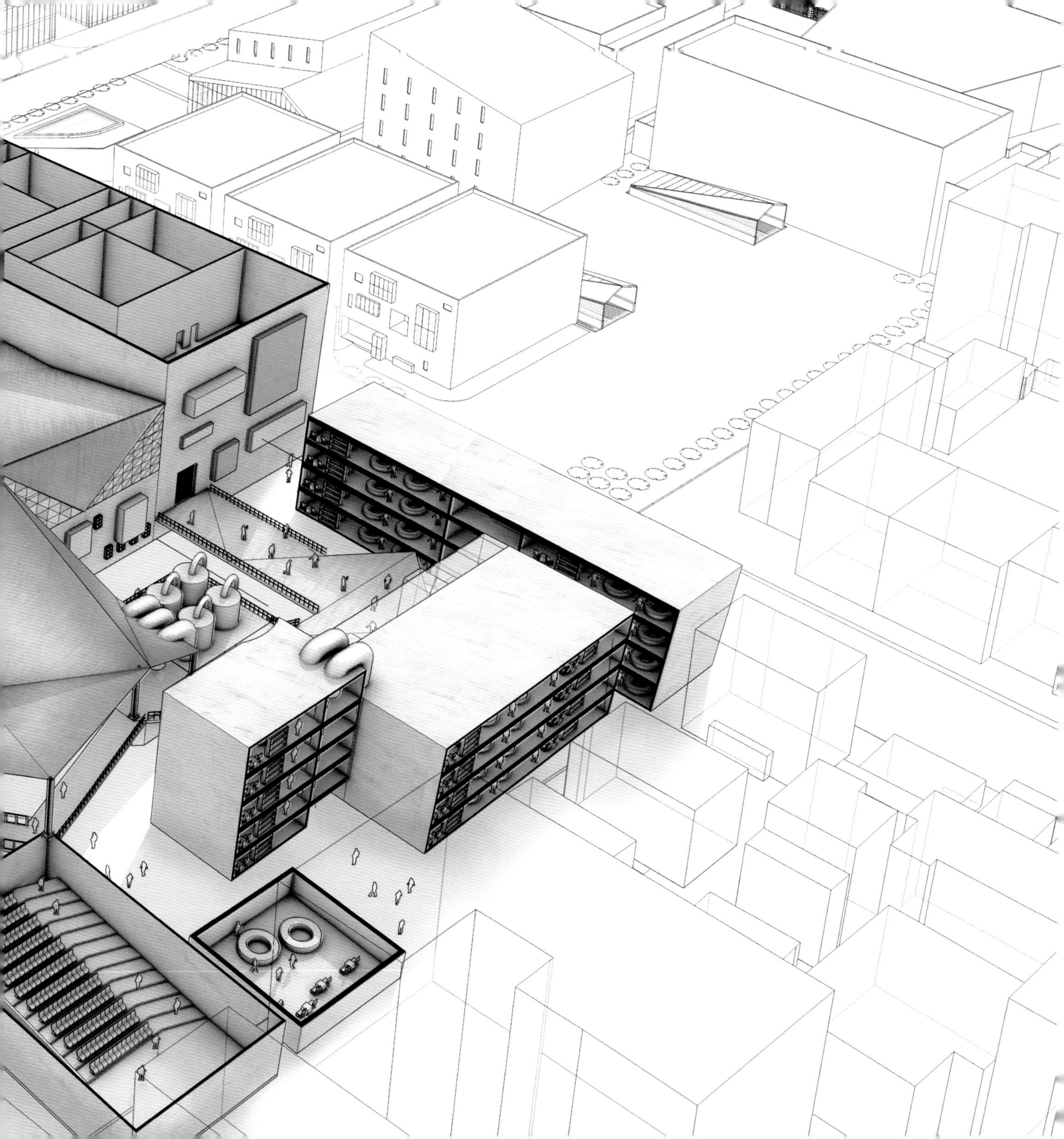

中心的张拉膜屋顶系统起着对文化、娱乐功能的引导作用。

The tensioned membrane roofing system at center assumes the guiding role for cultural, recreational functions.

0 AM
大型聚会
屋顶观光台
6 AM
创意工作坊
创意咖啡厅
12 PM
会议室
办公室
汇演中心
18 PM
酒吧
音乐现场
24 PM

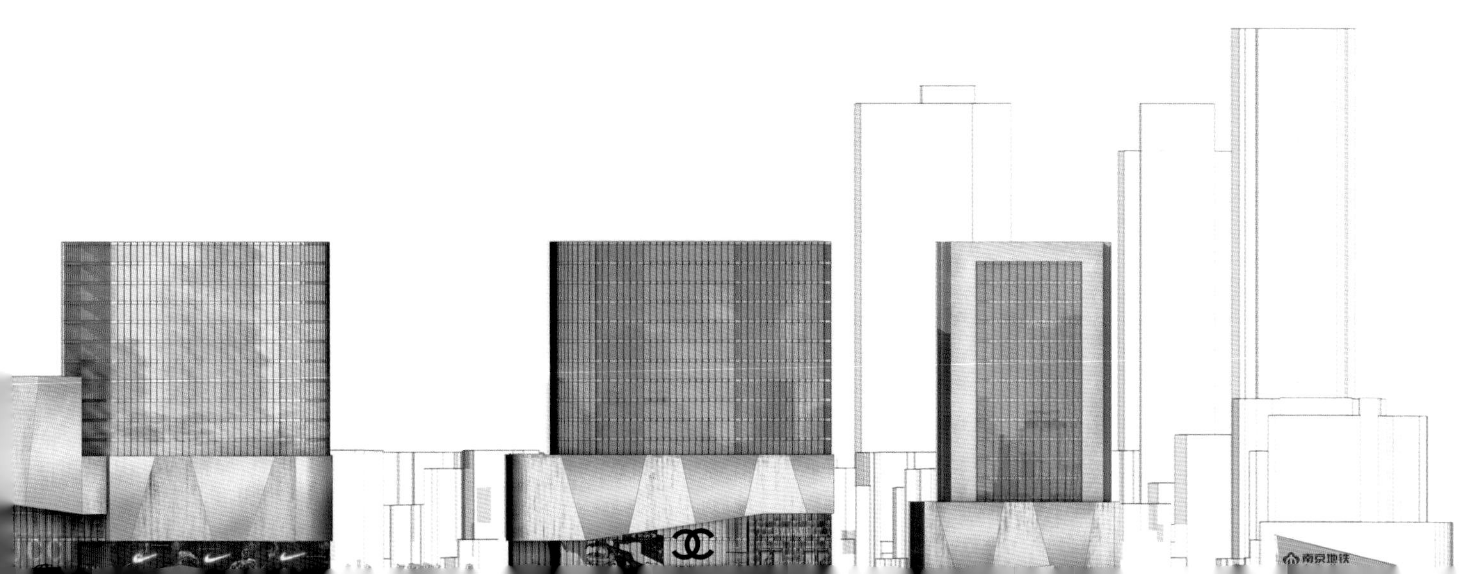

南京地铁

微筑.预制混凝土构建设与研究

MINI-BUILDING. CONSTRUCTION AND RESEARCH OF PRECAST CONCRETE STRUCTURES

冯金龙 周凌 胡友培

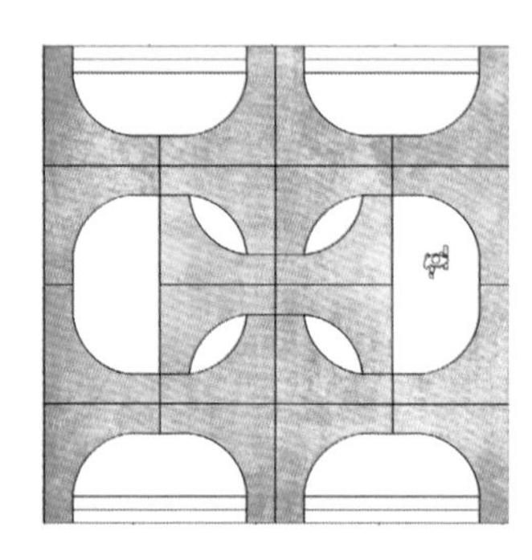

教学目标

结合设计竞赛的要求，通过学习预制混凝土装配建造体系的相关知识，探索预制混凝土装配体系的设计可能与诗意表达。

研究主题

预制混凝土建造体系的多重建筑学意义——建构的美学意义、设计方法论的意义、建构的社会文化意义、人居环境意义。

教学内容

以一个微型建筑为设计对象，通过现场调研和预制工厂考察了解预制混凝土的材料特性，认识预制混凝土建造体系在施工上的精确性、快速性和可控性；展开有关结构体系与形式、预制技术与施工、材料性能与建构表现方面的设计研究。在形成初步设计方案基础上，设计详细的构造大样，进行贴近真实建造层面的深化设计。

Training Objective

Explore design possibility and poetic presentation of precast concrete assembly system through study of knowledge related to precast concrete assembly and construction system, and in combination with the requirements of design contest.

Research Subject

Multiple architectural meanings of precast concrete construction system – aesthetic meaning of construction, meaning of design methodology, social and cultural meanings of construction, and meaning of human settlements.

Teaching Content

By making a mini building as design object, get to know material characteristics of precast concrete through field research and precast plant investigation, understand accuracy, rapidity and controllability of precast concrete construction system in term of construction; conduct design research on structural system and forms, precast technology and construction, material performance and constructional expression. Design detailed constructional details, and conduct in-depth design close to real construction level based on the formed preliminary design scheme.

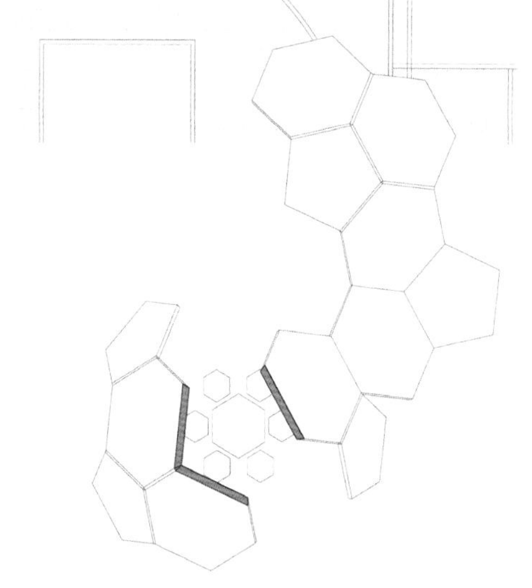

本设计以足球的结构为原型，以正五边形和正六边形为构件，在原结构的基础上进行变体，创造供人休息和供人进行运动的的拱形结构，契合预制混凝土和"运动公园"的主题。

In this design, we use structure of football as prototype, regular pentagon and regular hexagon as members, transform based on original structure to create an arch structure as a resting and sporting place for humans, and to agree with the topic of precast concrete and "sports park".

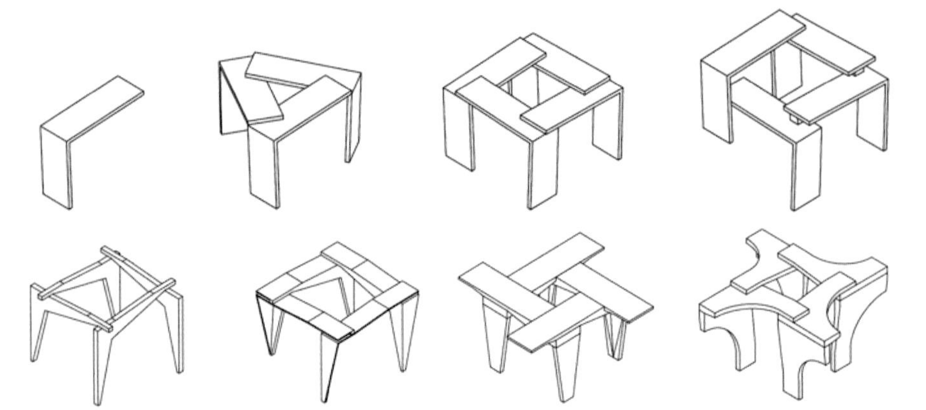

装配式建筑要求构件简洁，建造简单快捷经济。本设计通过优化传统框架结构接近这一目标。减少预制构件种类，简化建造方式，L形构件的互承结构使受力稳定。同时构件不同组合拼接实现多样的功能和空间。

Pre-fabricated building requires simple members, easy, rapid and economical construction. This design aims to approach this objective by optimizing traditional frame structure. Reduce types of precast members, simplify construction method, and stabilize forces with mutually-supported structure of L-shaped members. In the meantime, different ways of assembly and splicing of members can achieve various functions and spaces.

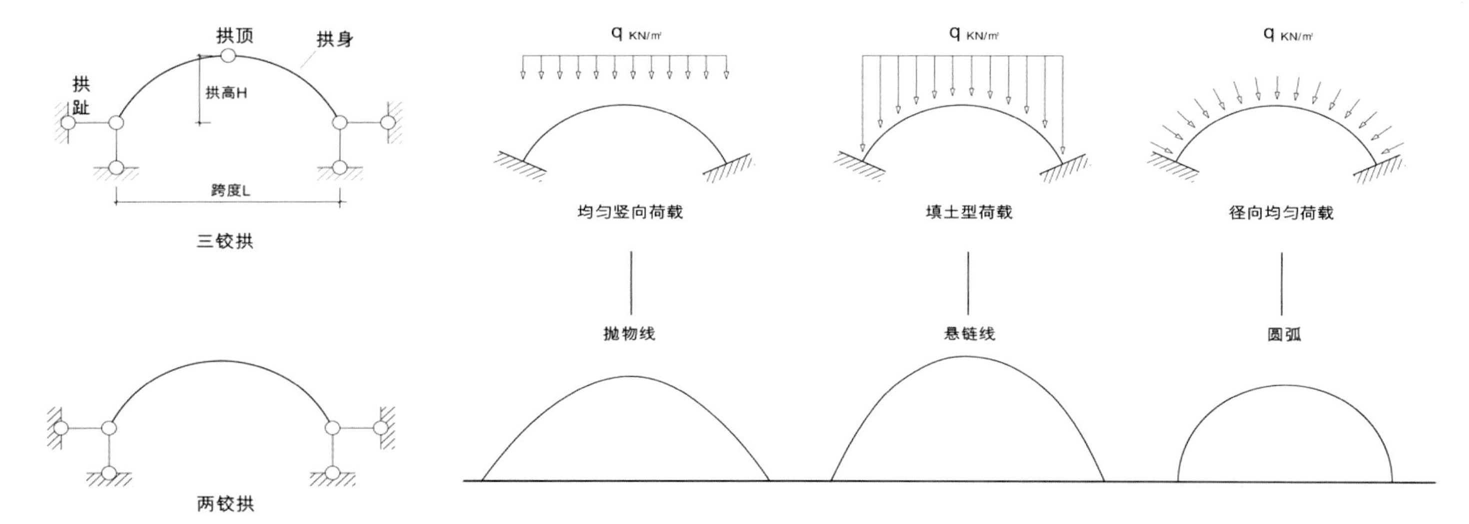

设计中综合考虑合理拱轴线和预制构件节点设计，最终选择接近最合理拱轴线的圆弧作为拱轴线力求降低拱结构内部弯矩。构件单元设计时考虑以直代曲形成的剪力以及构件交接的强度需求在构件两端进行加强，最终形成类似于板梁一体的构件单元。

Through overall consideration of rational arch axis and precast member joint design, the arch close to the most rational arch axis was selected as arc axis finally in the design, so as to reduce internal bending moment of the arc structure. When the member unit was designed, by considering forming shear force by replacing curved elements with straight ones and strengthening strength requirement of member joints at both ends of members, a member unit similar to integrated slab-beam structure was formed in the end.

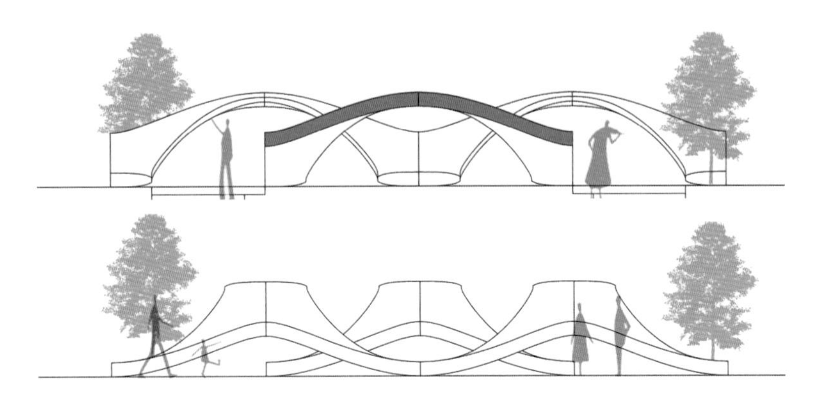

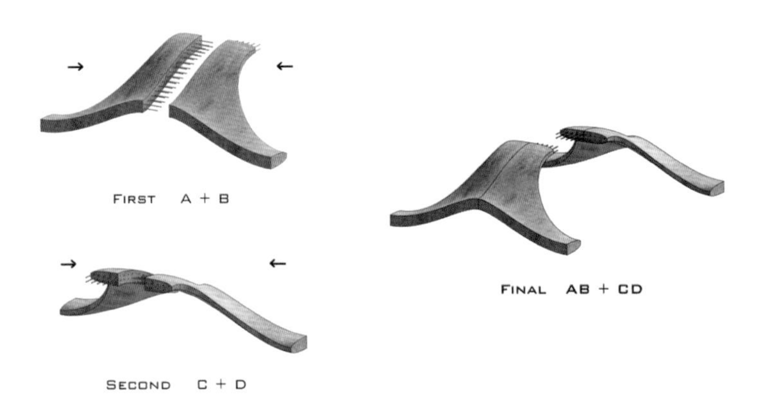

设计从本身功能的运动性考虑，希望创造一些曲面提供极限运动场所。参考了类似编织的原型，上下空间得以最大利用。

Considering the sports nature of its function, the design aims to create some curved surfaces to provide space for extreme sports. Utilization of upper and lower spaces was maximized by referring to prototype similar to weaving structure.

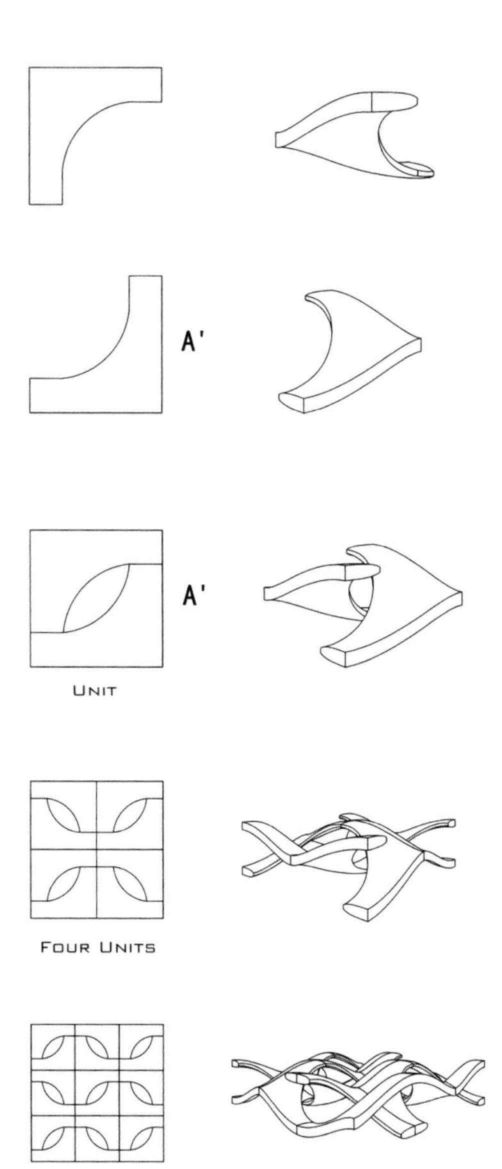

利用参数化得到一个二次曲面单元，拼接得到稳定单元。最后呈现出下有覆盖空间、上有运动场所的微型构筑物。

A quadric surface unit was obtained through parameterization, and a stabilized unit was realized through splicing. Finally a mini structure with lower covered space and upper sports field was presented.

设计基础（一）
Basic Design 1

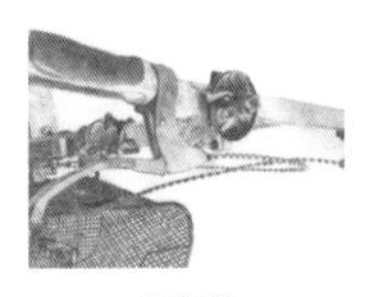
刘宛莹

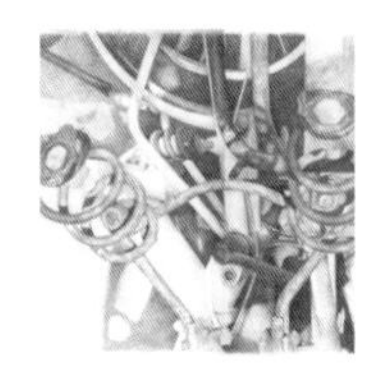
施孝萱

宋宇宁

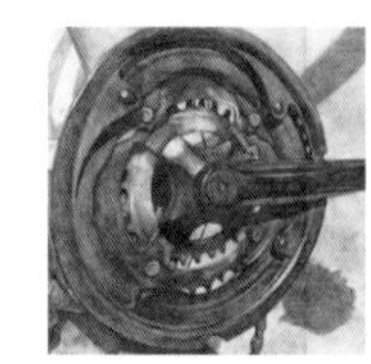
杨云睿

设计基础（二）
Basic Design 2

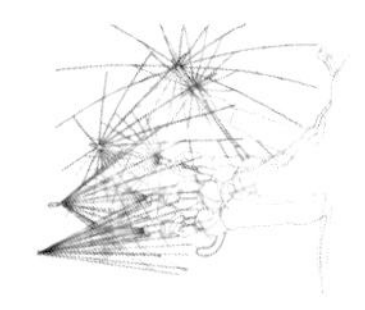
施孝萱

杜孟泽杉

宋宇宁

施孝萱

夏心雨

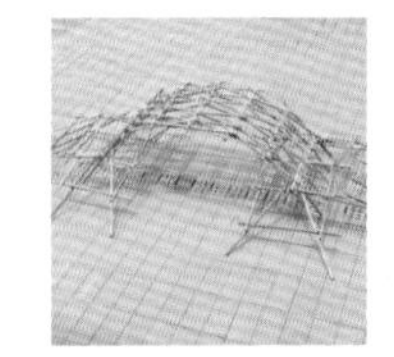
施孝萱、杨云睿、张俊、刘坤龙、马西伯、施少鋆

张珊珊、林宇、胡皓捷、宋宇宁、陈妍霓、曹焱、唐萌、刘畅

谢峰、刘宛莹、完颜尚文、严紫微、刘为尚、宋云龙、夏心雨、蔡英杰

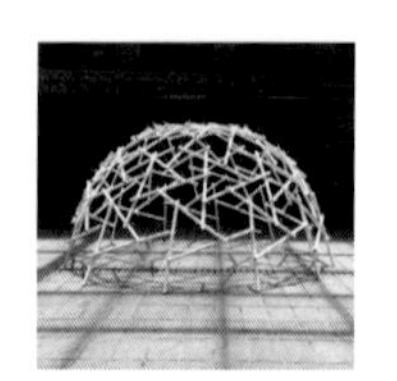
杨钊、杜孟泽杉、兰阳、宋怡、李雪琦、梁晓蕊、卢鼎、尹子晗

风景区茶室设计
Tea House Design

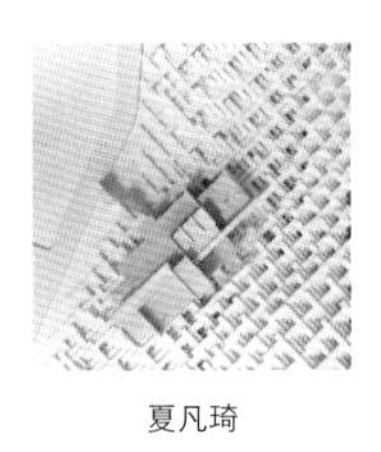
夏凡琦

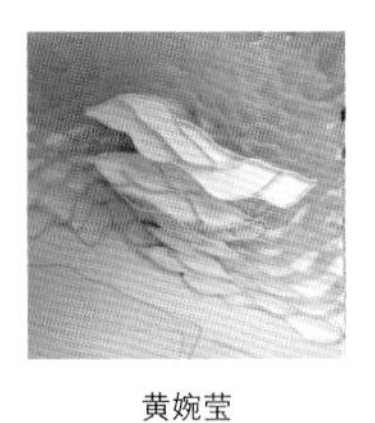
黄婉莹

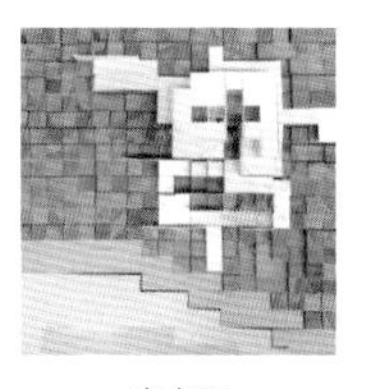
章太雷

曹舒琪

社区商业中心+活动中心设计
Community Bussiness Centre & Activity Centre Design

罗坤

桂喻

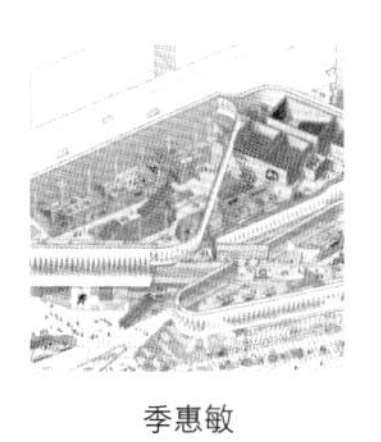
季惠敏

旧城改造城市设计
Urban Design for Old Town Renovation

倪若宁、王潇聆、王思绮、左思、王新宇

崔傲寒、黄雯倩、黎乐源、宋富敏、张黎萌

微筑 · 预制混凝土构建设与研究
Mini-building·Construction and Research of Precast Concrete Structures

《足球之拱》
王却奆

《新城故土》
崔傲寒

《重力 · 拱》
谢忠雄

《拱之间》
吴结松

基本建筑建构研究
CONSTRUCTIONAL DESIGN

傅筱

教学目标

设计概念与构造设计

教学内容与计划（2 人 / 组）

1. 设计概念与构造设计

（1）处理好节点的基本工程技术问题。

（2）根据设计概念研究建造材料的选用和节点设计，在满足基本功能性构造技术的前提下，重点研究超越功能性技术问题的构造设计表达。

时间：6 周　成果：数字模型、概念分析图、构造详图

2. 设计成果整理与表现

（1）1：100 平面、立面图纸，表达深度达到施工图深度。

（2）剖面要求：先以节点大样的深度绘制 1：20 剖面图，然后再绘制表达。空间的单线剖面图 1：100。抛弃施工图局部断面的表达方式。

（3）必须有带节点的空间透视表达。

（4）必须有带节点的三维轴测表达。

时间：2 周，成果：PPT 演示和 A1 展板（不少于 2 张）

Training Objective

Design concepts and structural design

Teaching Content and Program (2 students/team)

1. Design concept and structural design (6 weeks)

(1) Settle the basic engineering technical issues regarding the nodes.

(2) Based upon the design concepts, study the selection of construction materials and the design of nodes, and on the premise of satisfying basic functional construction techniques, place emphasis on studying the expression of structural design beyond technical issues.

Achievements: digital models, conceptual analysis diagrams, detailed constructional drawings

2. Reorganization and expression of design results (2 weeks)

Requirements: The depth of expressions on large-scale plan drawings, elevation drawings and sectional drawings shall reach that of the construction drawings; the detailed drawing of nodes must have sectional nodes, plan nodes and 3-dimensional nodes; the expression must have spatial perspective expressions with nodes and 3-dimensional isometric expressions with nodes.

Achievements: PPT presentations and A1-format exhibition boards (at least 2 pieces)

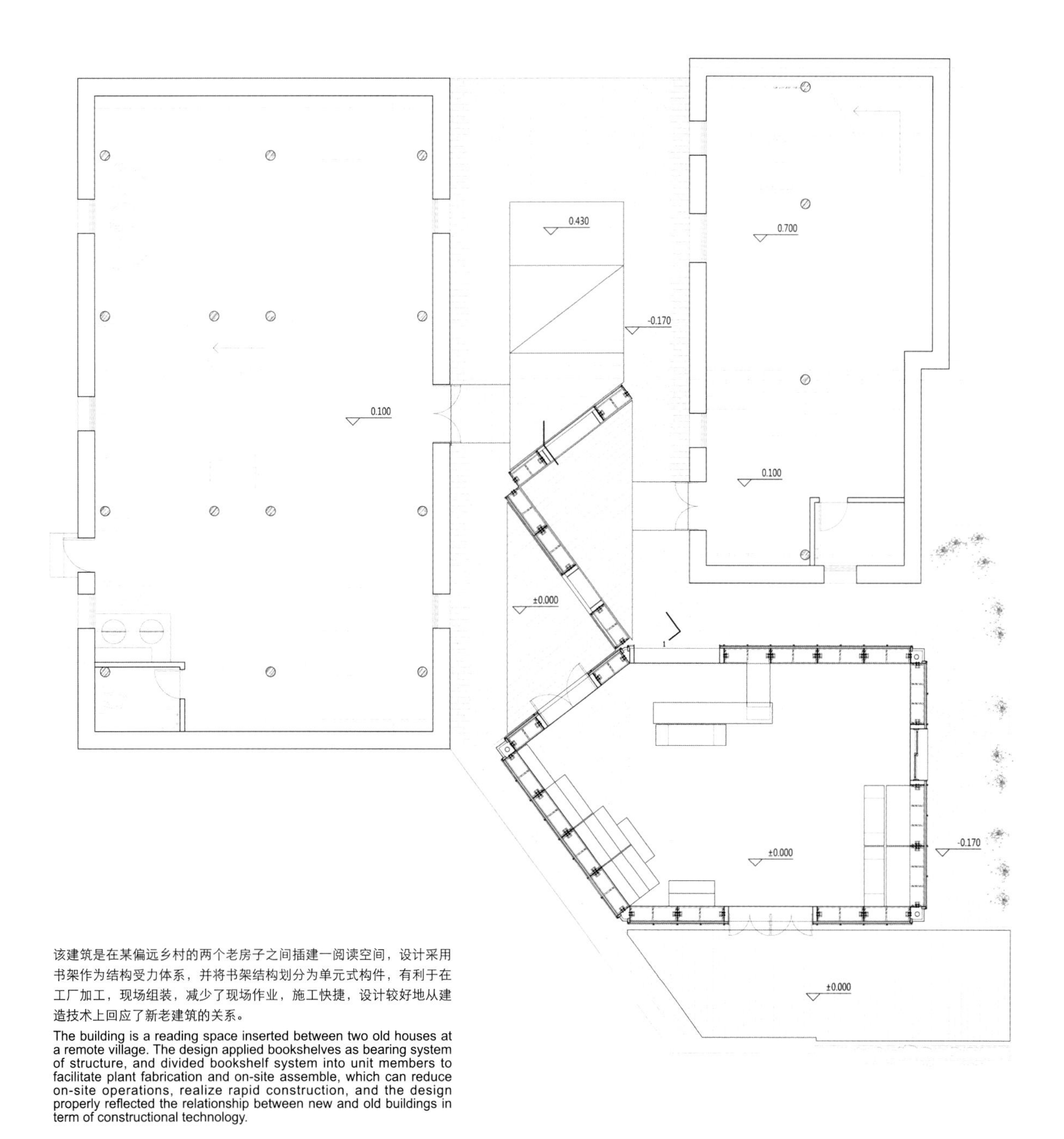

该建筑是在某偏远乡村的两个老房子之间插建一阅读空间，设计采用书架作为结构受力体系，并将书架结构划分为单元式构件，有利于在工厂加工，现场组装，减少了现场作业，施工快捷，设计较好地从建造技术上回应了新老建筑的关系。

The building is a reading space inserted between two old houses at a remote village. The design applied bookshelves as bearing system of structure, and divided bookshelf system into unit members to facilitate plant fabrication and on-site assemble, which can reduce on-site operations, realize rapid construction, and the design properly reflected the relationship between new and old buildings in term of constructional technology.

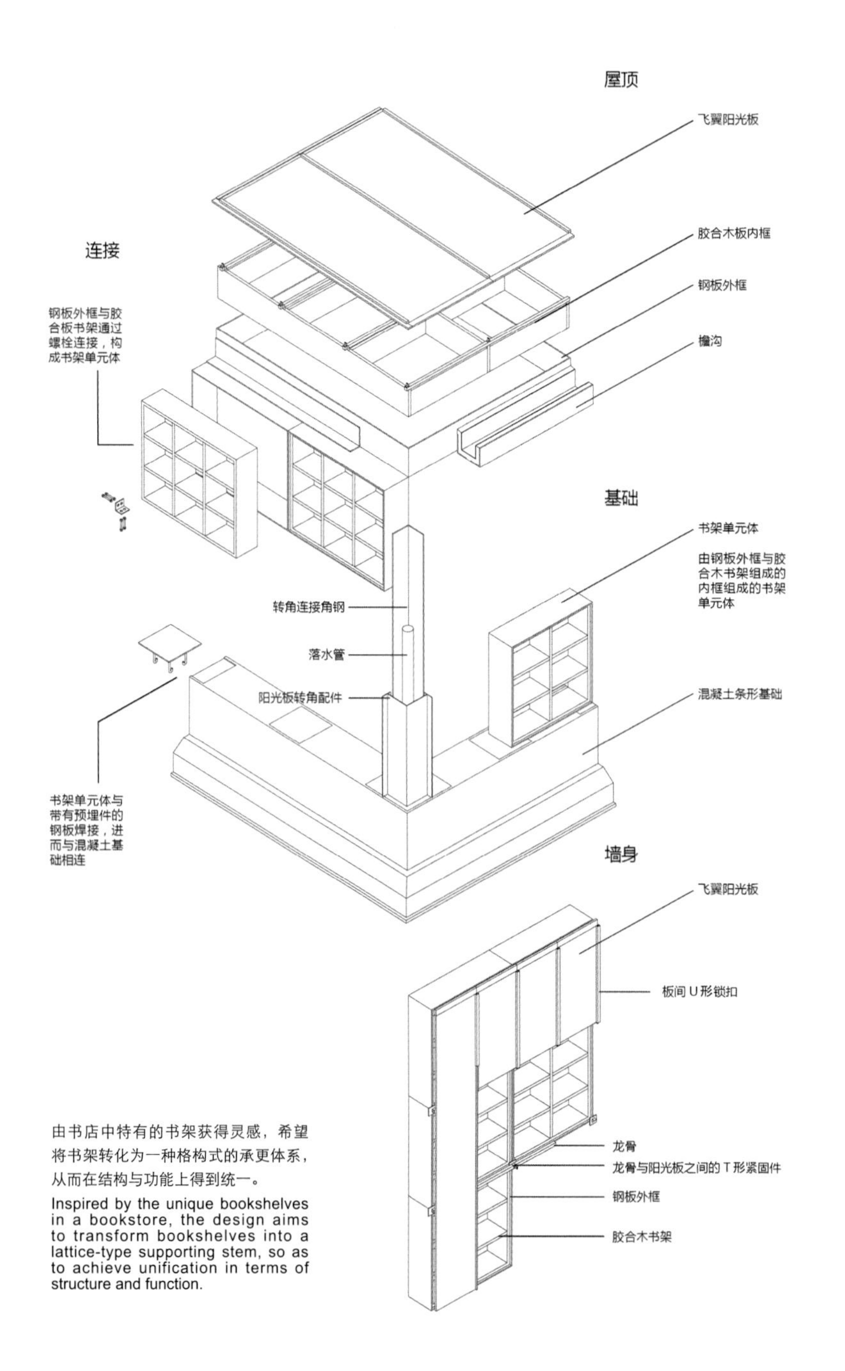

由书店中特有的书架获得灵感，希望将书架转化为一种格构式的承更体系，从而在结构与功能上得到统一。

Inspired by the unique bookshelves in a bookstore, the design aims to transform bookshelves into a lattice-type supporting stem, so as to achieve unification in terms of structure and function.

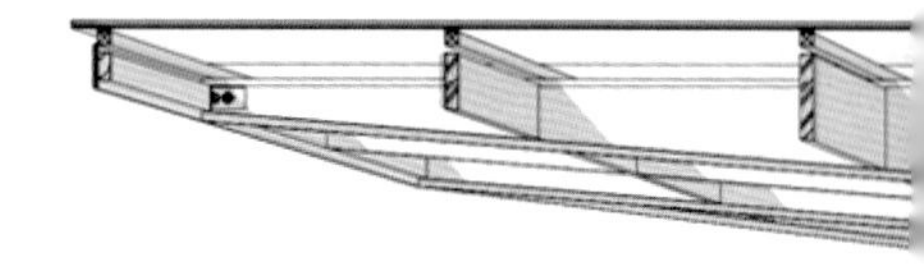

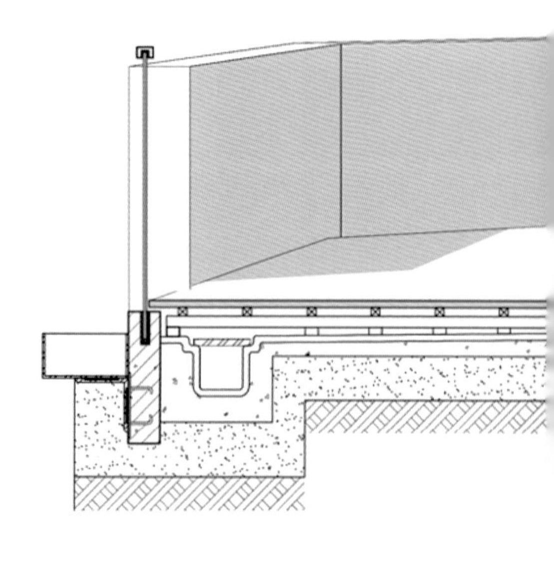

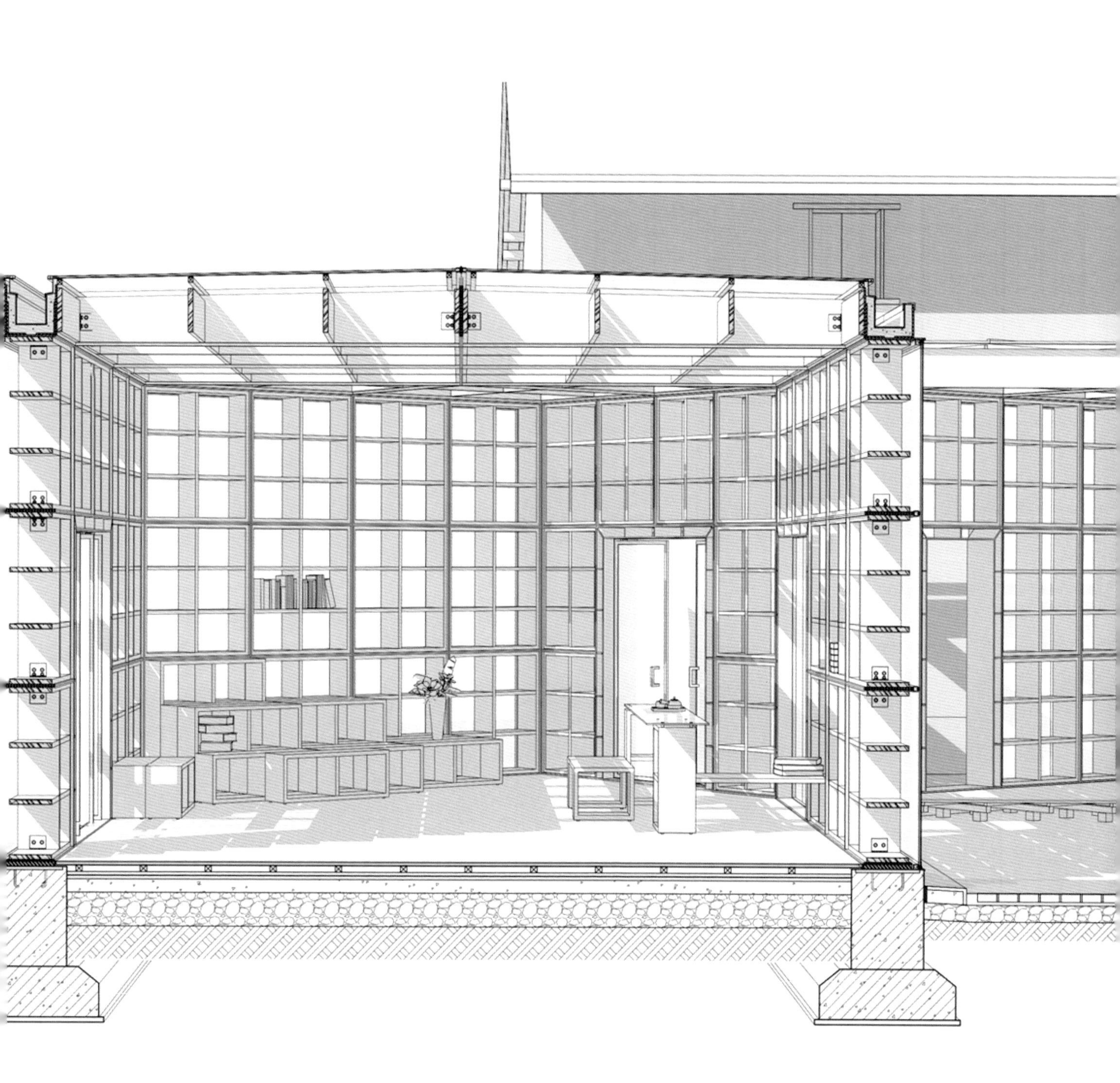

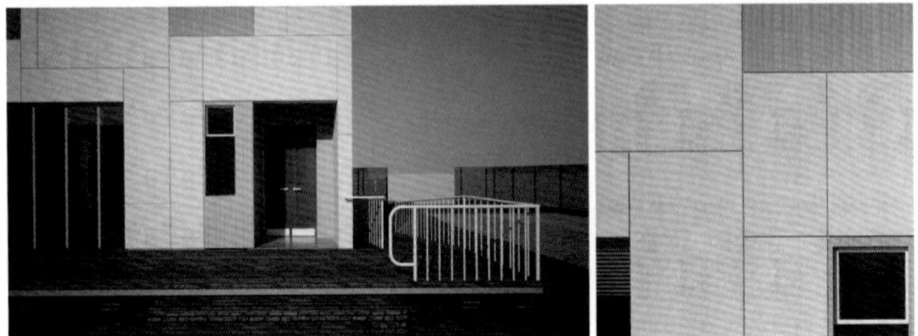

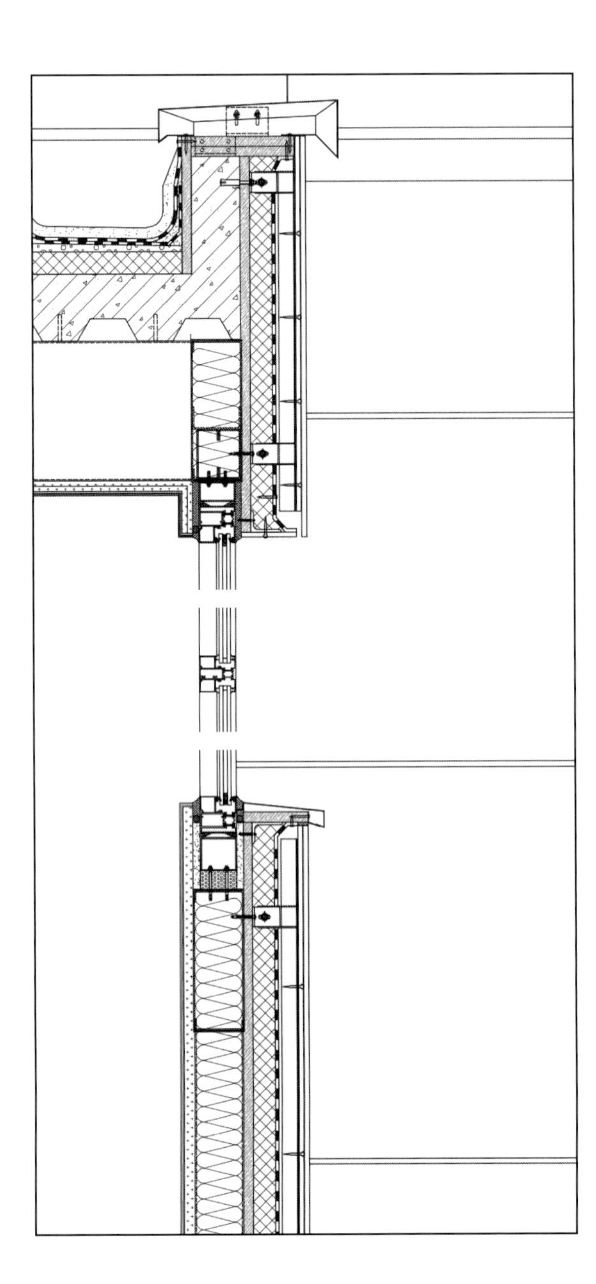

该设计研究了干挂水泥基板材在建筑表皮的应用，设计较为深入细致地研究了材料划分与立面洞口的对位关系，从而建立起材料与建筑尺度的协调关系。

The Application of dry-fastened cement-based panels on building cladding was studied in the design, and in-depth and detailed study on material classification and corresponding relations of openings in elevation was conducted, so as to establish coordinated relations between materials and building dimensions.

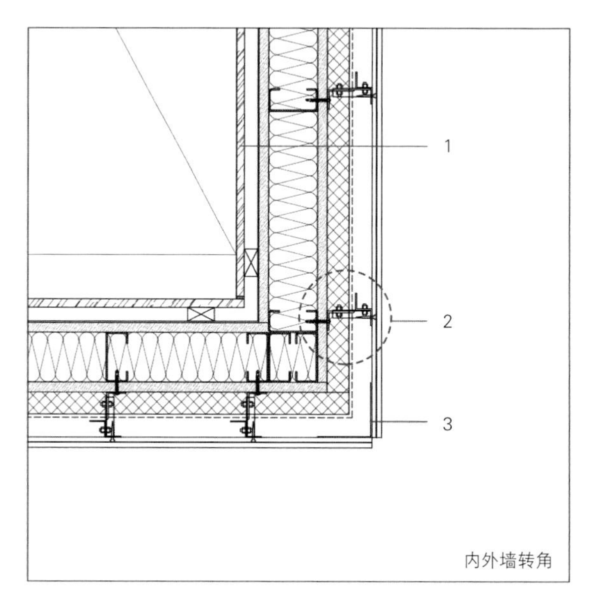
1
2
3
内外墙转角

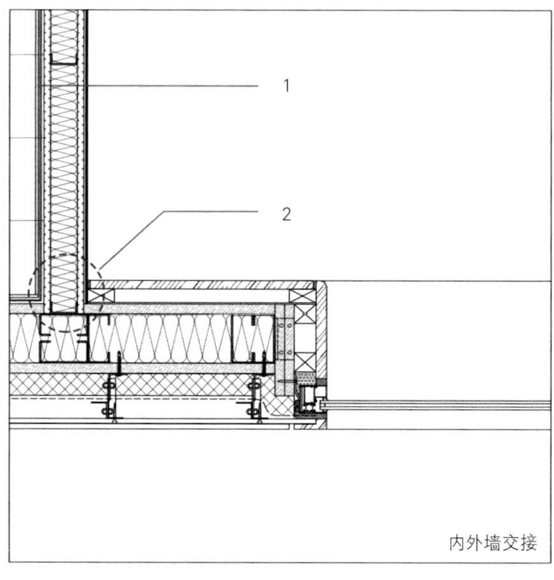
1
2
内外墙交接

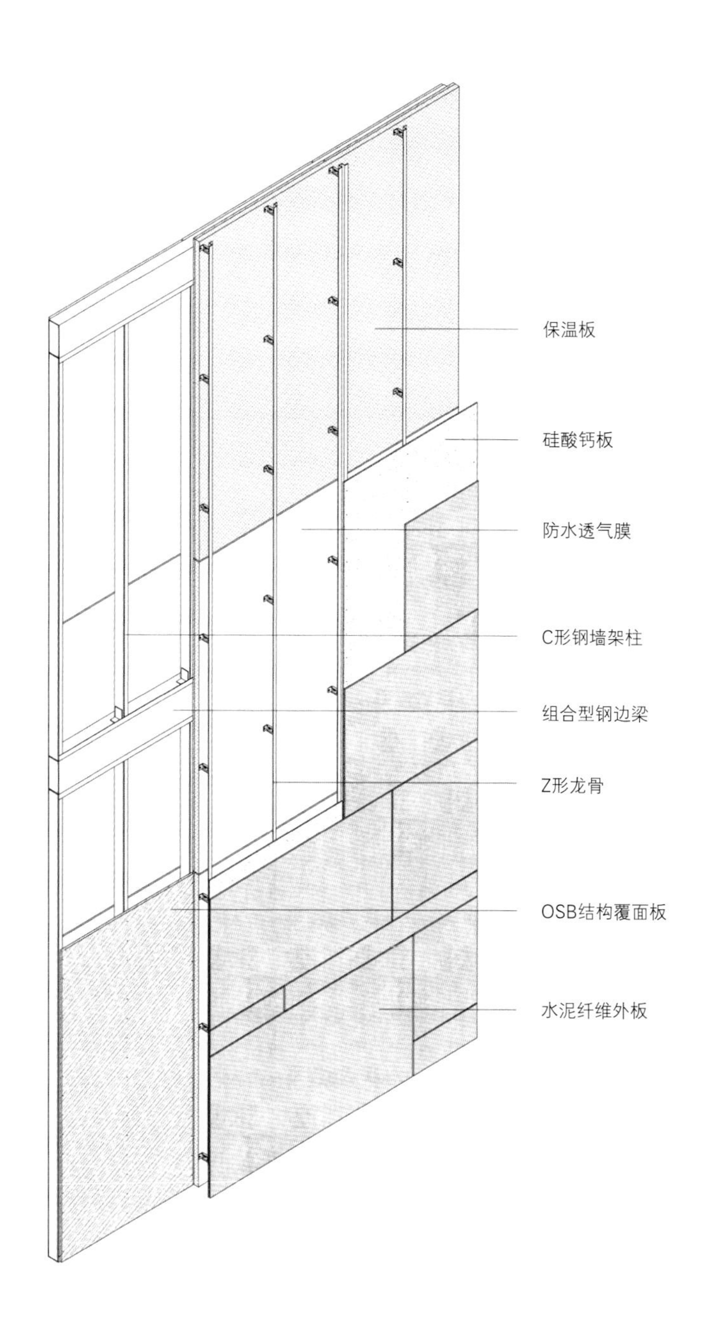
保温板
硅酸钙板
防水透气膜
C形钢墙架柱
组合型钢边梁
Z形龙骨
OSB结构覆面板
水泥纤维外板

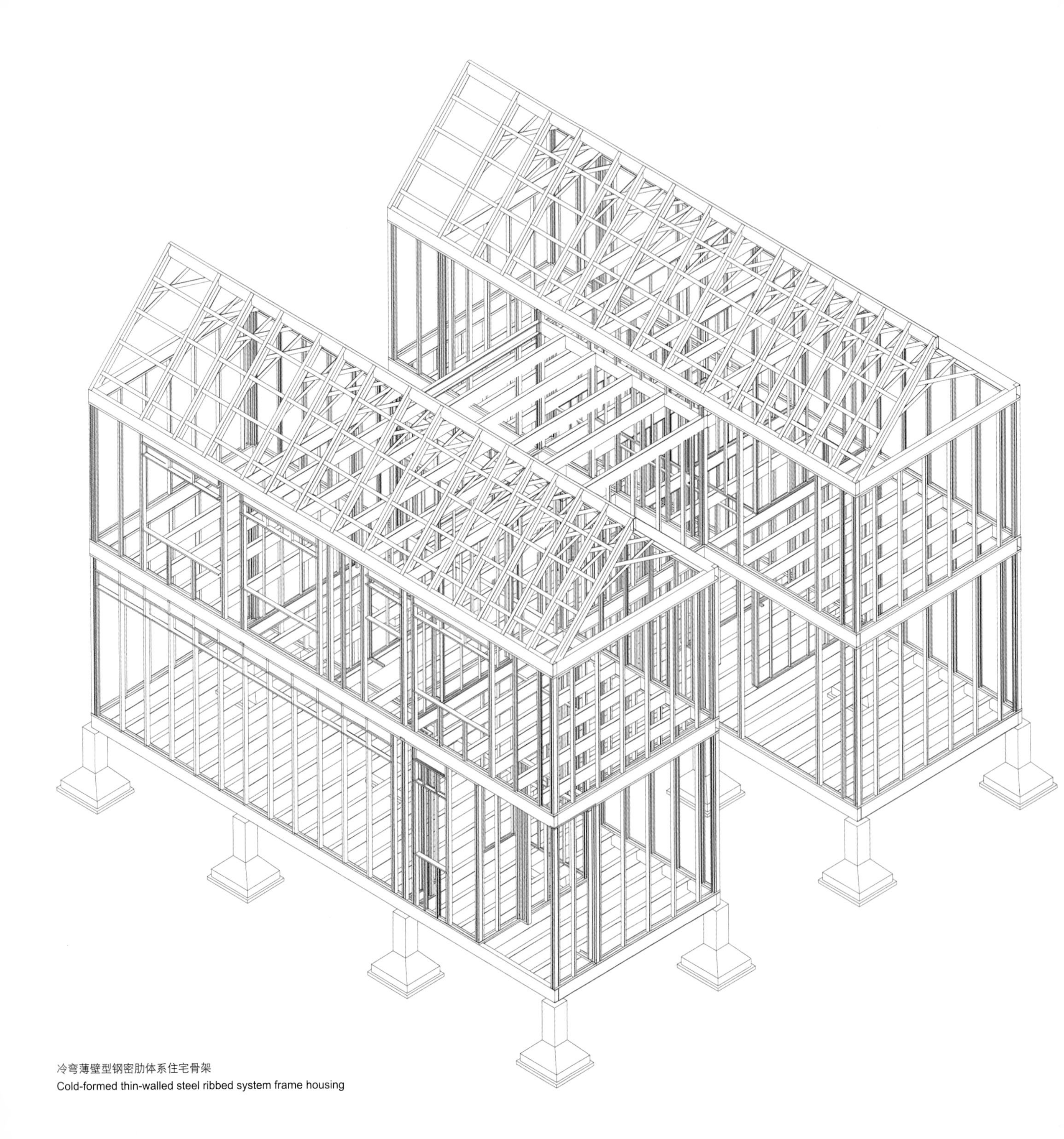

冷弯薄壁型钢密肋体系住宅骨架

Cold-formed thin-walled steel ribbed system frame housing

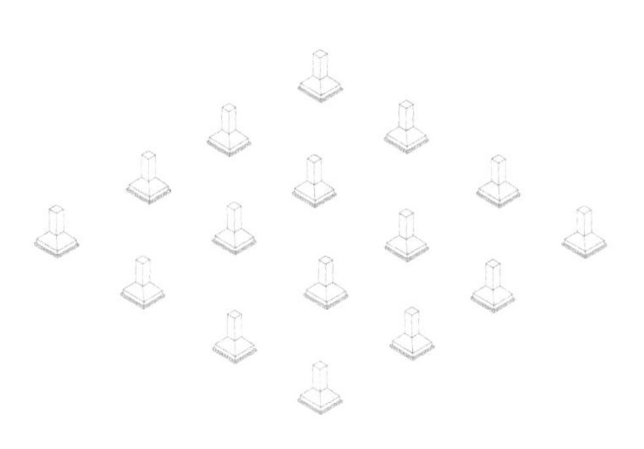

第一步：钢筋混凝土独立基础
Step 1: Independent reinforced concrete foundation

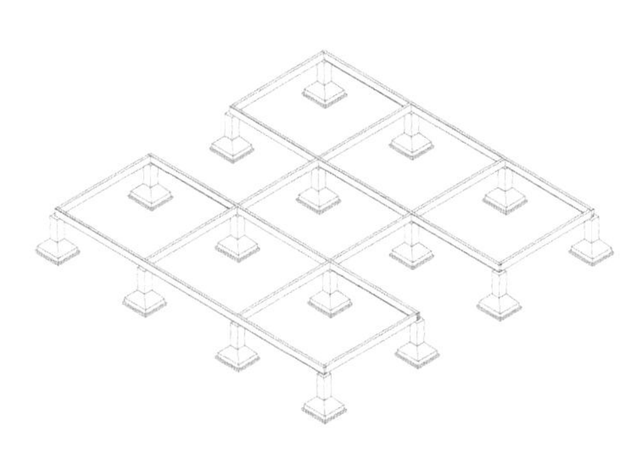

第二步：2C形钢组合地梁
Step 2: Grade beams assembled with 2 C-shaped steels

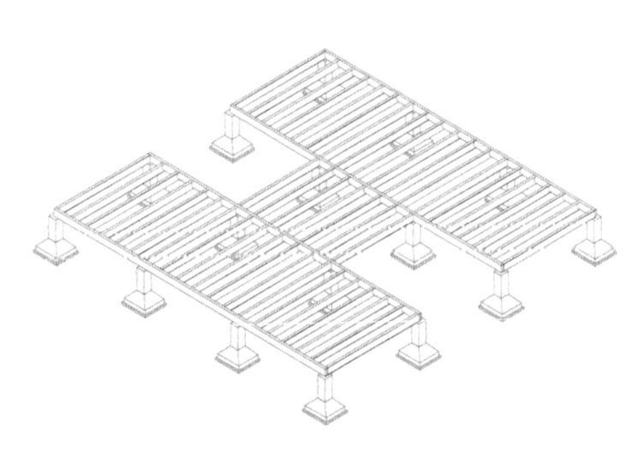

第三步：C形钢次梁
Step 3: Secondary beams of C-shaped steels

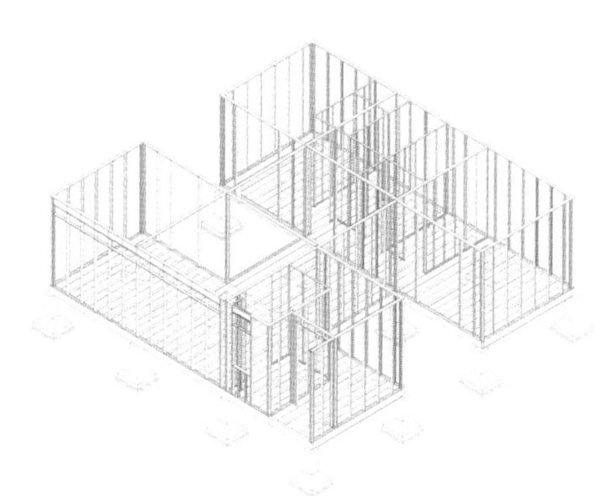

第四步：一层架柱C90×40×20×2.5，带底梁、顶梁
Step 4: First-floor trestle-type columns C90×40×20×2.5, with floor beams, top beams

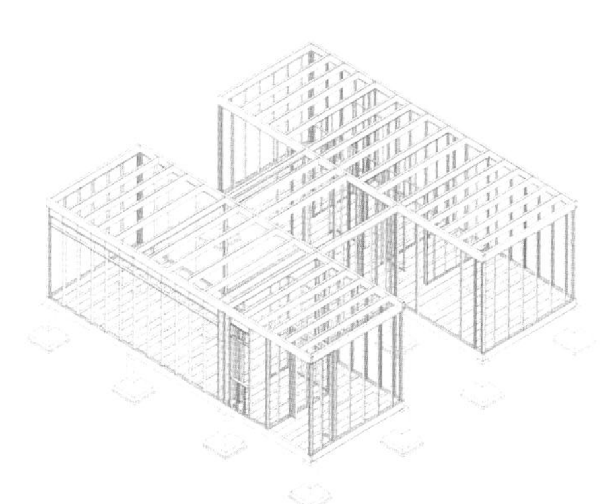

第五步：楼盖梁、边梁C250×90×20×2.5+U255×90×2.5
Step 5: Superstructure beams, boundary beams C250×90×20×2.5+U255×90×2.5

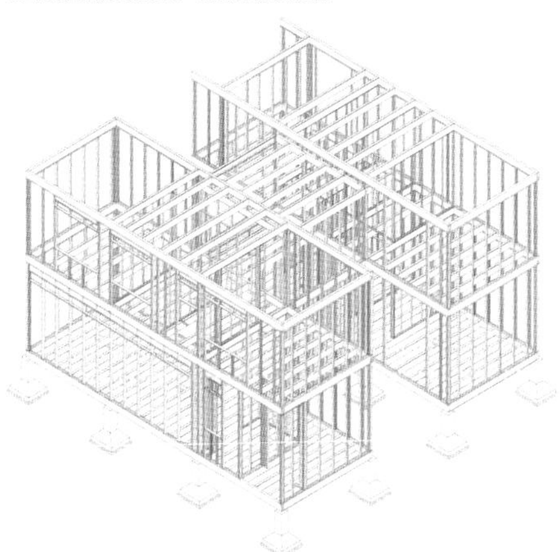

第六步：二层结构
Step 6: Second-floor structure

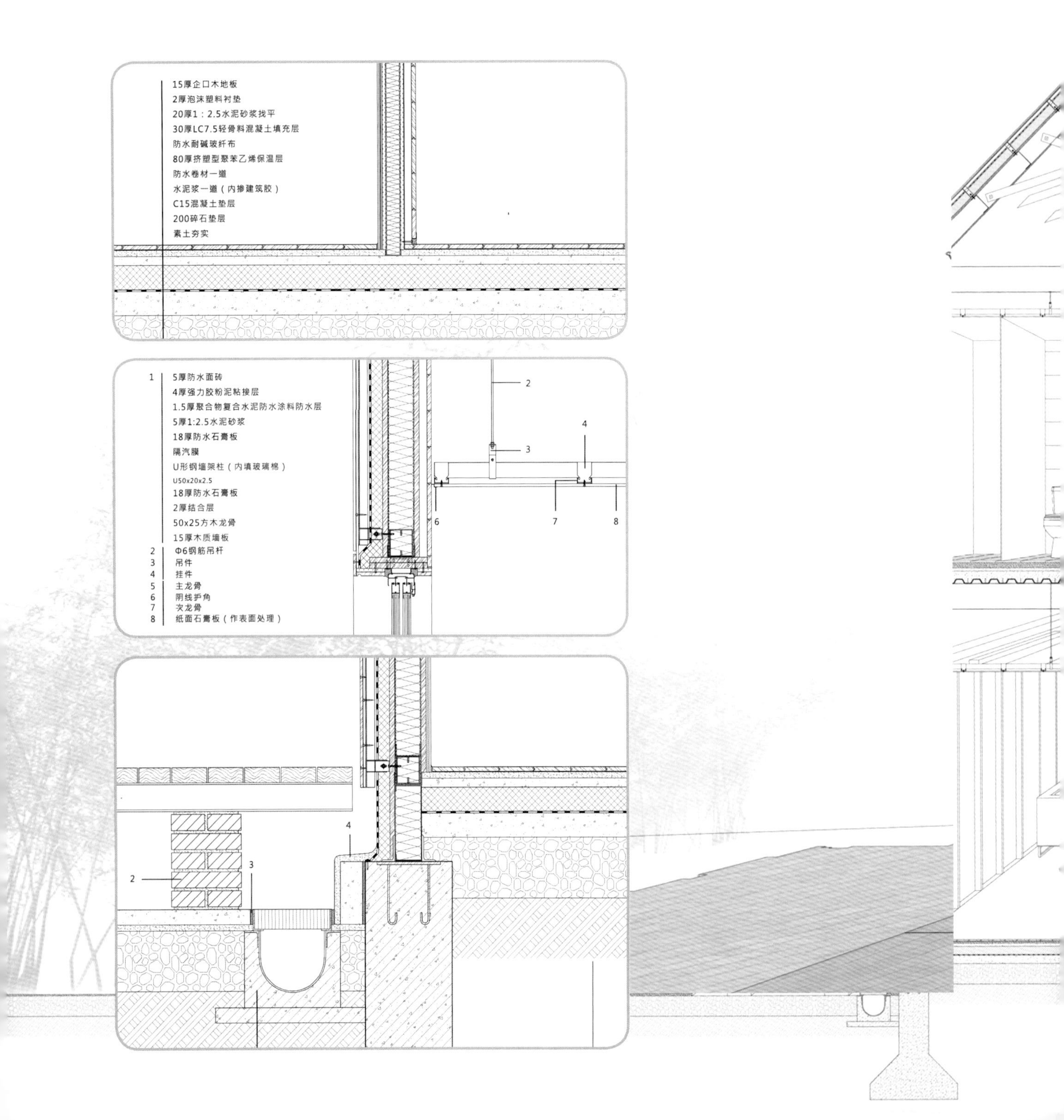
15厚企口木地板
2厚泡沫塑料衬垫
20厚1：2.5水泥砂浆找平
30厚LC7.5轻骨料混凝土填充层
防水耐碱玻纤布
80厚挤塑型聚苯乙烯保温层
防水卷材一道
水泥浆一道（内掺建筑胶）
C15混凝土垫层
200碎石垫层
素土夯实
1 5厚防水面砖
4厚强力胶粉泥粘接层
1.5厚聚合物复合水泥防水涂料防水层
5厚1:2.5水泥砂浆
18厚防水石膏板
隔汽膜
U形钢墙架柱（内填玻璃棉）
U50x20x2.5
18厚防水石膏板
2厚结合层
50x25方木龙骨
15厚木质墙板
2 Φ6钢筋吊杆
3 吊件
4 挂件
5 主龙骨
6 阴线护角
7 次龙骨
8 纸面石膏板（作表面处理）
2
3
4
6
7
8
4
2
3

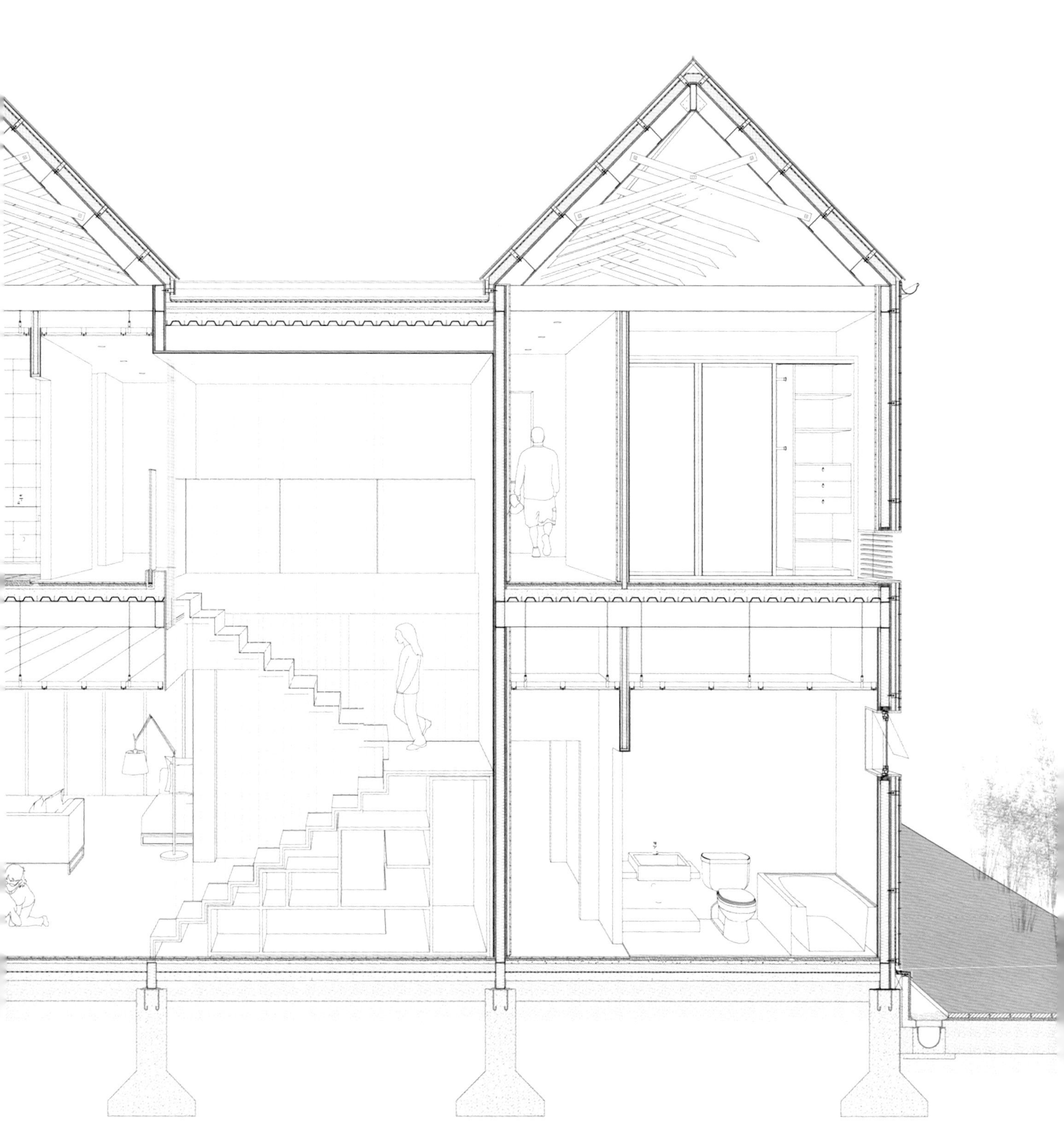

城市空间设计 · 层与界

FROM VOLUMN TO SPACE·MULTILAYER & INTERFACE

丁沃沃

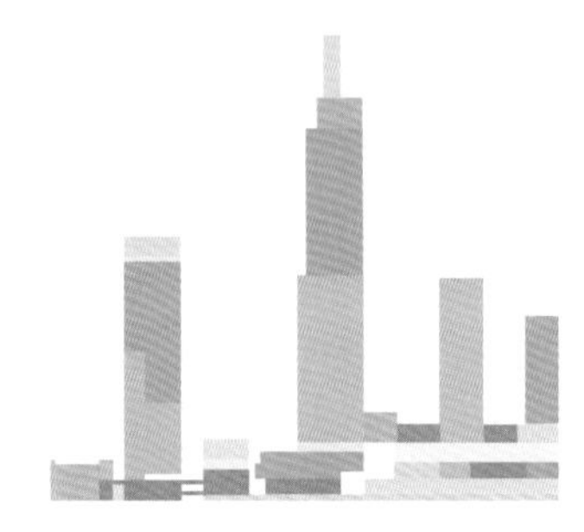

教学目标

本课题拟通过高密度城市建筑与城市空间设计实验，了解城市建筑角色和城市物质空间的本质，初步掌握城市建筑与城市空间塑造之间的关系。此外，通过城市空间设计练习进一步深化空间设计的技能和方法。

研究主题

以“层”“茎”“界”与“空”作为操作载体，通过设计操作，探索高密度城市的构成潜力，认知在高密度城市环境中城市建筑设计元素的转化。

教学内容

1. 以南京下关沿江地块作为设计实验的场所。通过场地地理条件认知、案例研习、设计分析和设计实验，探讨高效城市空间的建构方法和内涵。

2. 通过设计拓展图示技能与表现方法，从空间意向出发建构城市物质空间的层与界面，由“空间意向”“扫描”出城市空间的“层”与“界”。

Training Objective

This course aims to understand the nature of roles of urban buildings and urban physical space, and preliminarily grasp relations between urban building and urban space creation through experimental design of high-density urban buildings and urban space. In addition, it aims to further deepen skills and methodology of space design through practices of urban space design.

Research Subject

Explore composition potential of high-density city, and understand transformation of urban building design elements in high-density city environment through design operations by using Multilayer, rhizome, interface and opening as operation carriers.

Teaching Content

1. Apply a riverside plot at Xiaguan, Nanjing as the site for the design experiment. Explore tectonic methodology and implication of efficient urban space through cognition to geological field conditions, case study, design analysis, and design experiment.

2. Develop drawing presentation skills and expression methods through design, construct multilayer and interface of urban physical space from space image, and “scan” multilayer and interface of urban space from “space image”.

设计地块位于南京市下关，西临长江，东枕狮子山，是历史上重要的物资集散中心。地块交通便利，内存有码头、火车站及其他重要的历史建筑。快速城市发展下，城市规划加大了该地块的城市密度（容积率8～10）。设计要求综合多项场地条件，实验更高密度（在规划基础上容积率翻倍）的条件下营造高品质城市空间的可能性。

The plot of the design is located at Xiaguan, Nanjing, which is close to Yangtze River in the west, adjacent to Lion Rock in the east, and an important goods distribution center in history. The plot features convenient transportation, has dock, railway station and other important historical buildings. With rapid development of the city, urban planning increased the urban density of the plot (with plot ratio 8~10). The design requires to integrate multiple field conditions, and to experiment the possibility of creating quality urban space under high-density conditions (double the plot ratio on basis of the planning).

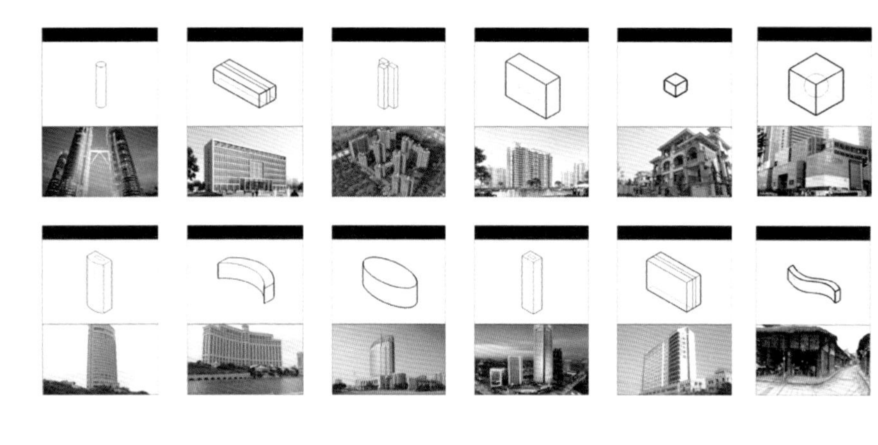

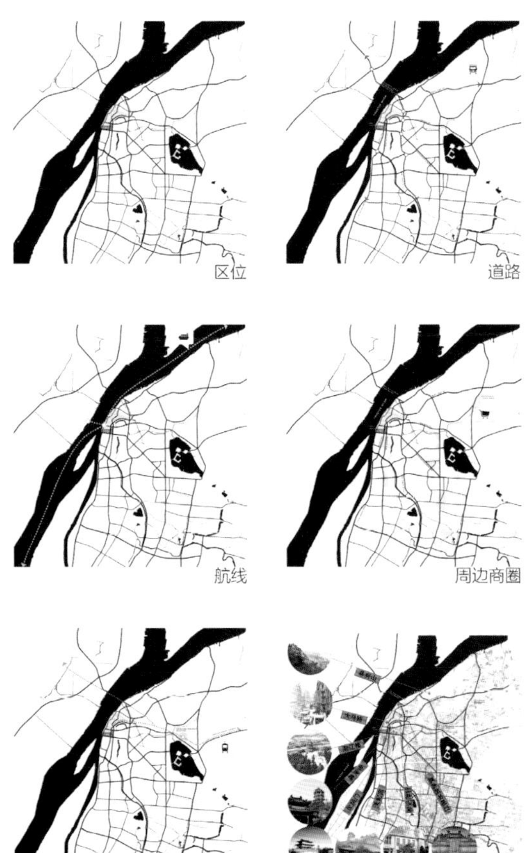

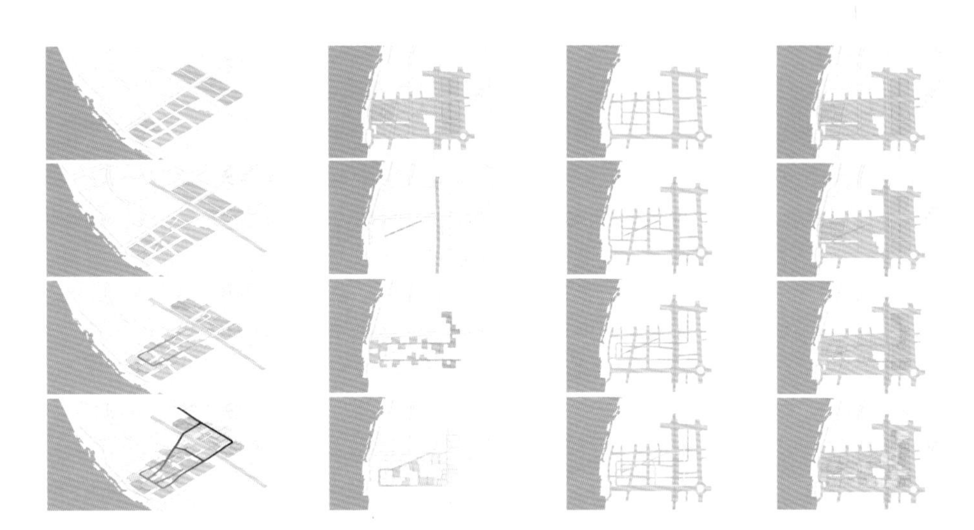

案例研习要求从多个立体城市案例中总结功能在建筑高度上的分布规律，考量每个场地中地块的划分以及置入建筑的可能性，链接街区、地块、指标与形态。

The cast study requires students to summarize distribution rules of functions at building height from several dimensional city cases, examine division of plot at each site as well as possibility of placing buildings, and link blocks, plots, indicators and morphology.

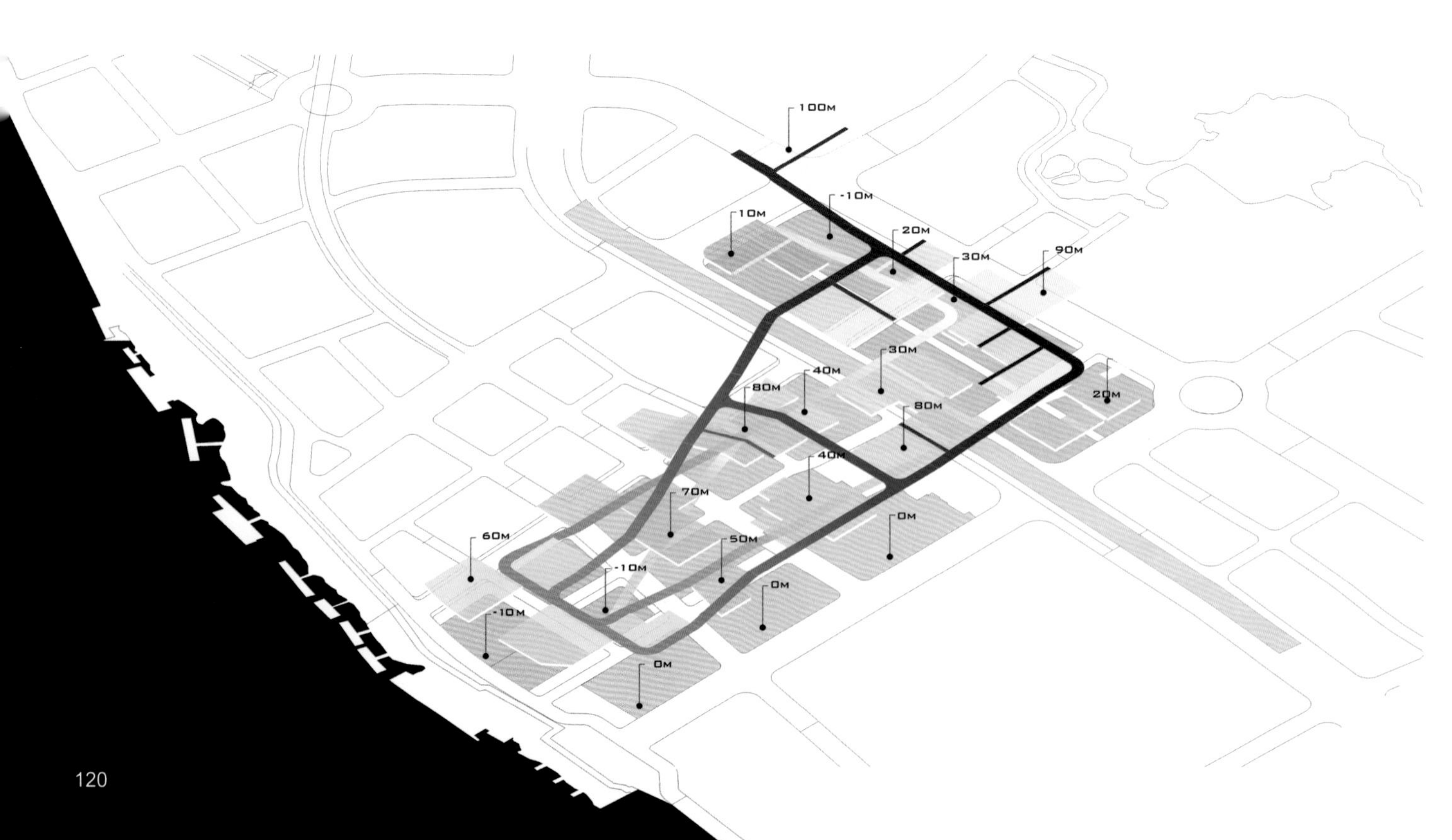

CHILID

SLE-EP LUNCH DINNER SLE-EP GAME COMPUTER INTERNET DRAMA TV QUIZ LECTURE DRAWING READING SPORT HOMEWORK COMPUTER STUDY HOMEWORK

0 1 2 3 4 5 6 7 8 9 10 11 12 13 14 15 16 17 18 19 20 21 22 23 24

ADULT

SLEEP WORK WORK ENTERTAINMENT MEAL

SLE-EP LUNCH DINNER SLEEP SHOPPING INTERNET SHOPPING GYM DRAMA STUDY TEA REPORT INTERNET KTV BATH MEETING KTV PUB PUB MEETING STUDY HOUSEHOLD MOVIE BATH WORK CONSULT WORK OVERTIME WORK OVERTIME

0 1 2 3 4 5 6 7 8 9 10 11 12 13 14 15 16 17 18 19 20 21 22 23 24

ELDERLY

SLEEP HOUSEHOLD HOUSE MEAL

SLE-EP CHATTING READING NEWSPAPER WATCHING TV SHOPPING PLAYING CARDS DANCING EXERCISE WALK CHILDHOLD STUDY CHILDHOLD WORK DANCING WALK HOUSEHOLD

0 1 2 3 4 5 6 7 8 9 10 11 12 13 14 15 16 17 18 19 20 21 22 23 24

CULTURE FUNCTION

COMMERCIAL FUNCTION

SOCIAL FUNCTION

0 1 2 3 4 5 6 7 8 9 10 11 12 13 14 15 16 17 18 19 20 21 22 23 24

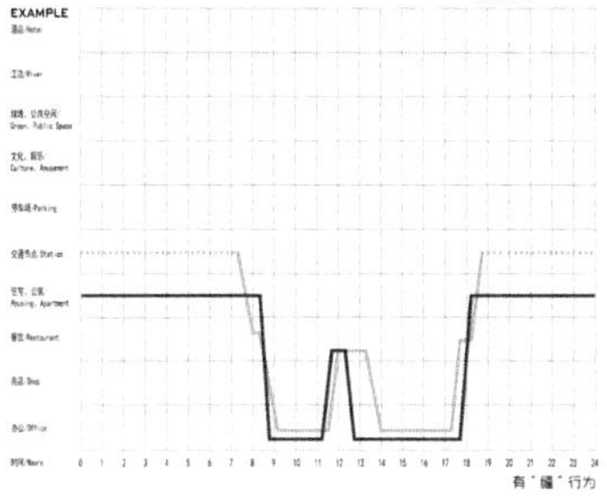

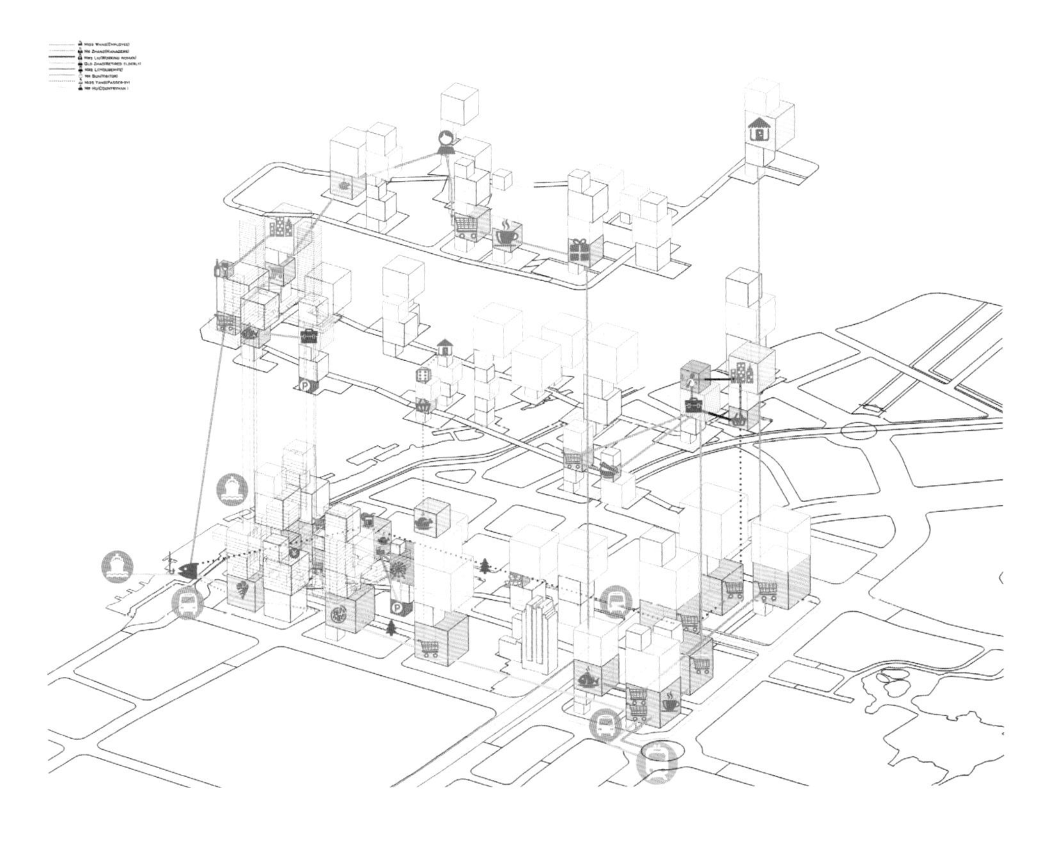

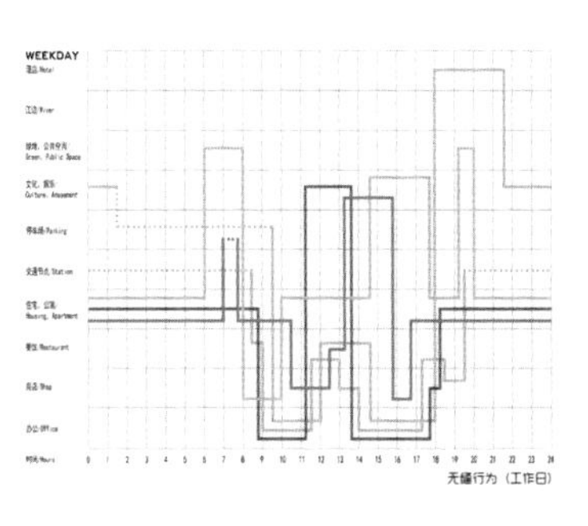

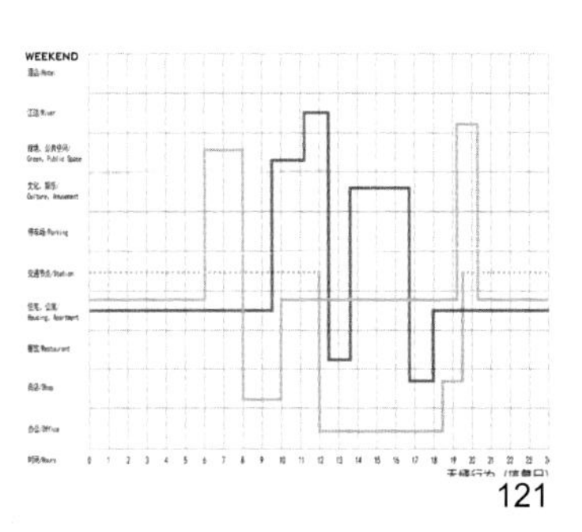

VOLUME

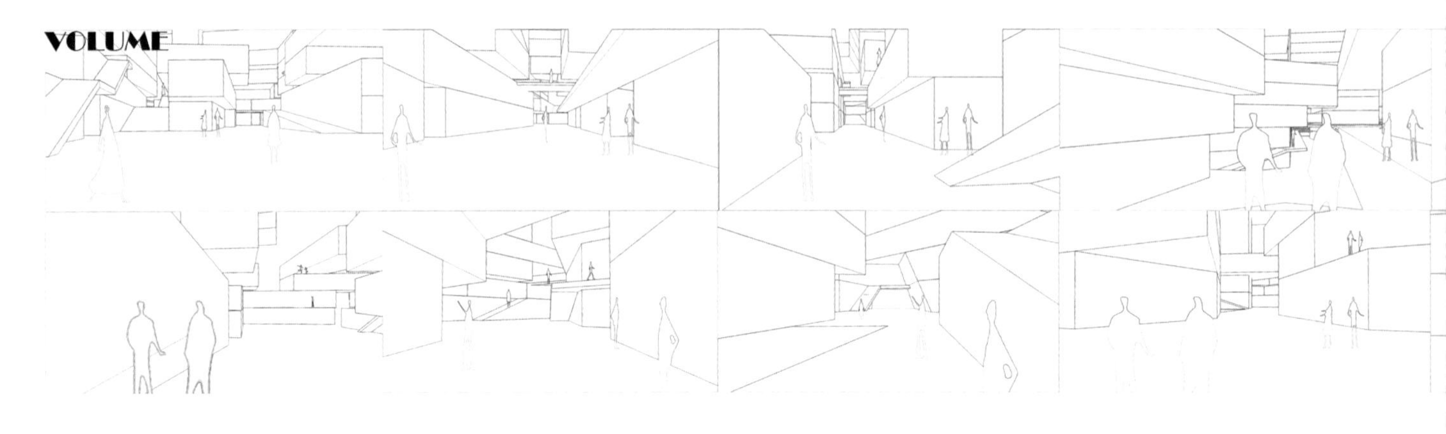

概念1：“无缝”的城市：“无缝”是“功能”的无缝，亦是“时间”的无缝。通过“层”的设计，在场地不同高度上划分地块，控制划分后地块的高度、形状、尺度与功能；通过合理的车行与人行道路的连接，使得功能体块之间实现无缝连接；通过界面的设计，实现对流线的控制：狭长的“引导”式界面、镜框式的“展现”式界面等使功能空间按照人的行为序列有机的串联。

Concept 1: Seamless City: "Seamless" refers to the seamless state of "functions", and also the seamless state of "time". Through the design of "multilayer", divide the plot at different height of the site, and control height, shape, dimensions and functions of the plot; achieve seamless connection between functional blocks through rational connection of vehicle lanes and pedestrian paths; realize control on circulations through interface design: long, narrow "guiding" interfaces, and mirror-frame-type "presentation" interfaces enable organic link of functional spaces as per human behaviors.

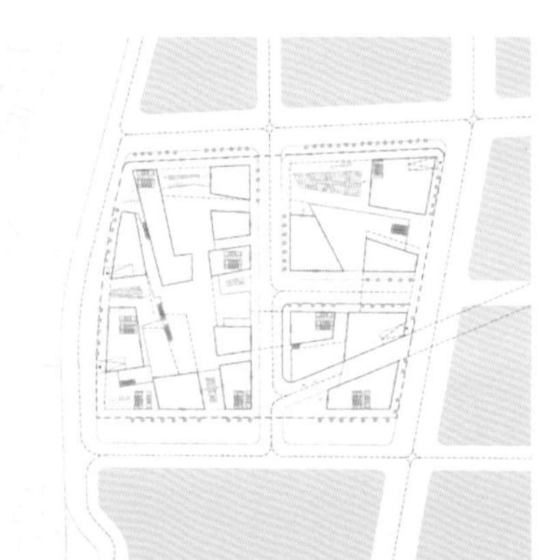

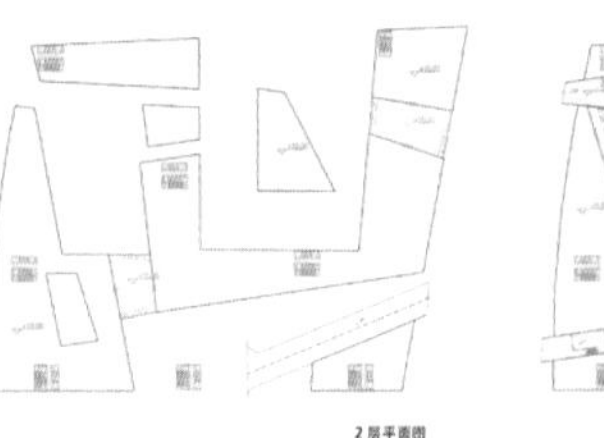

2层平面图

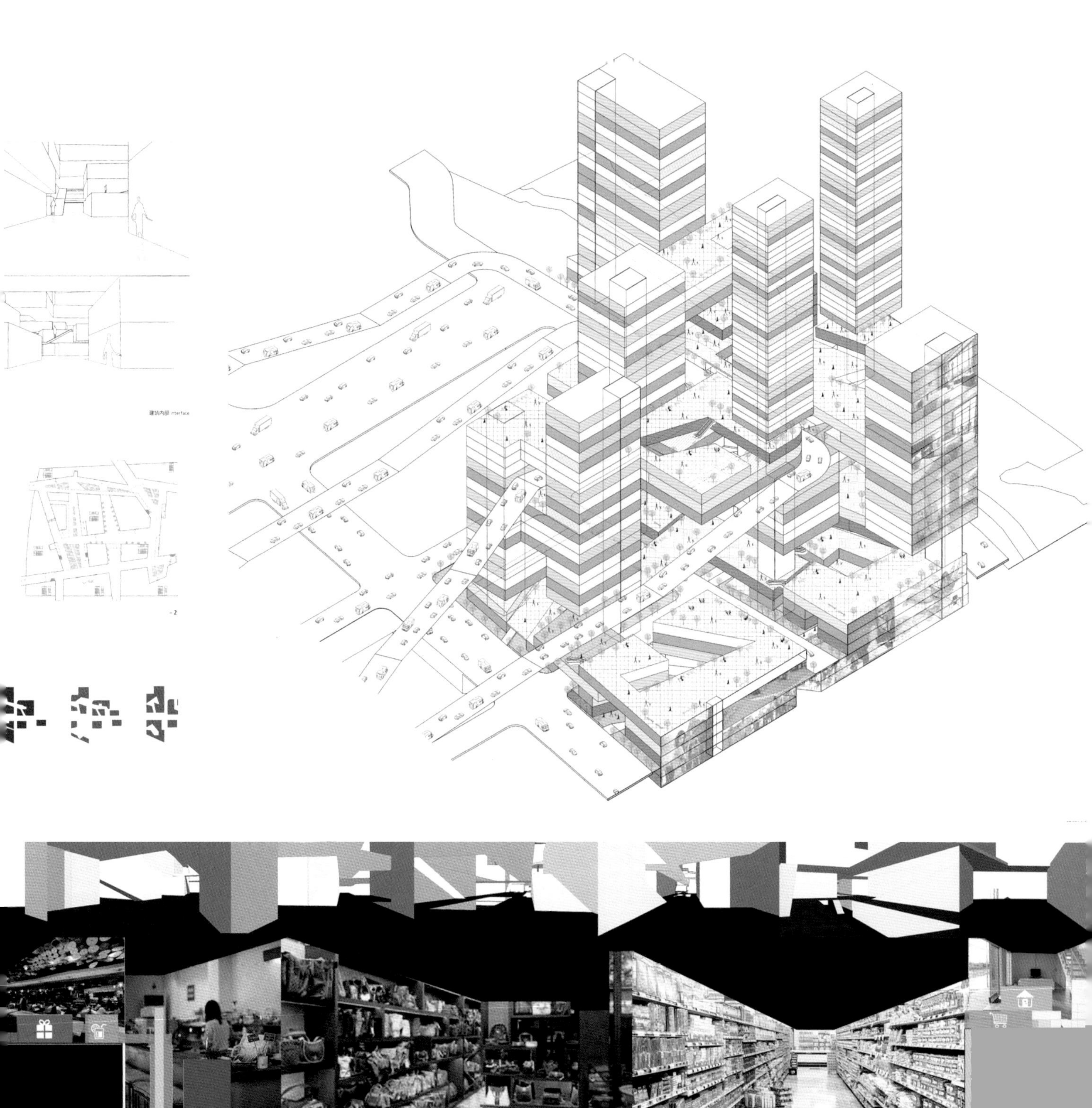
建筑内部 interface
-2

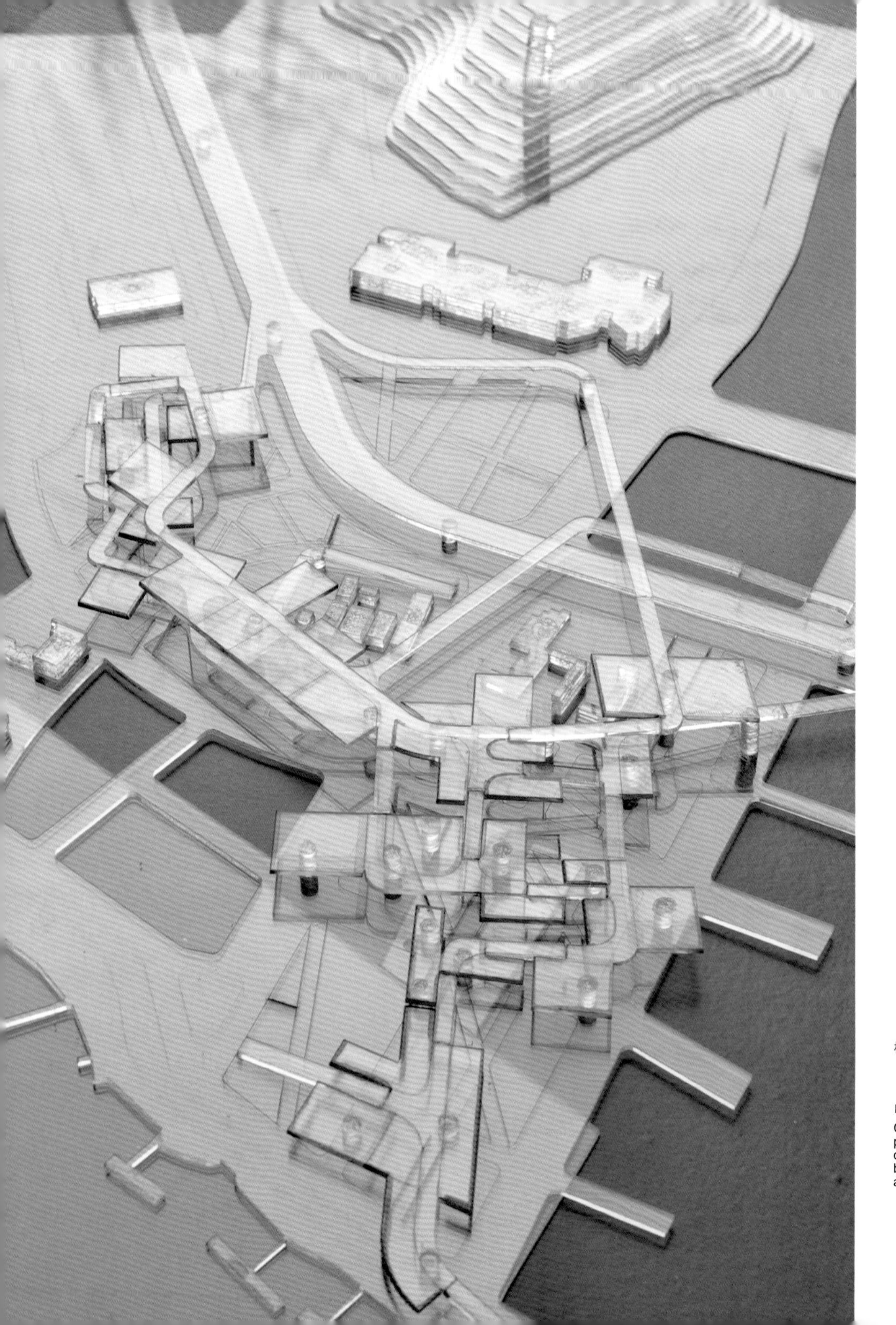

概念2：景观最大化：通过场地周边景点连线，切分地块“层”级；通过建筑形体变化，设计每一层级看到景观的“界”。实现场地内部观景最大化，在高楼林立中看得见山，望得见水。

Concept 2: Landscape Maximum: Split “multilayer” of plot with connecting lines of scenic spots around the plot; design the interface seen at each layer through change of building shape. Maximize landscapes within the plot, and allow hills and waters to be seen amid high-rise buildings.

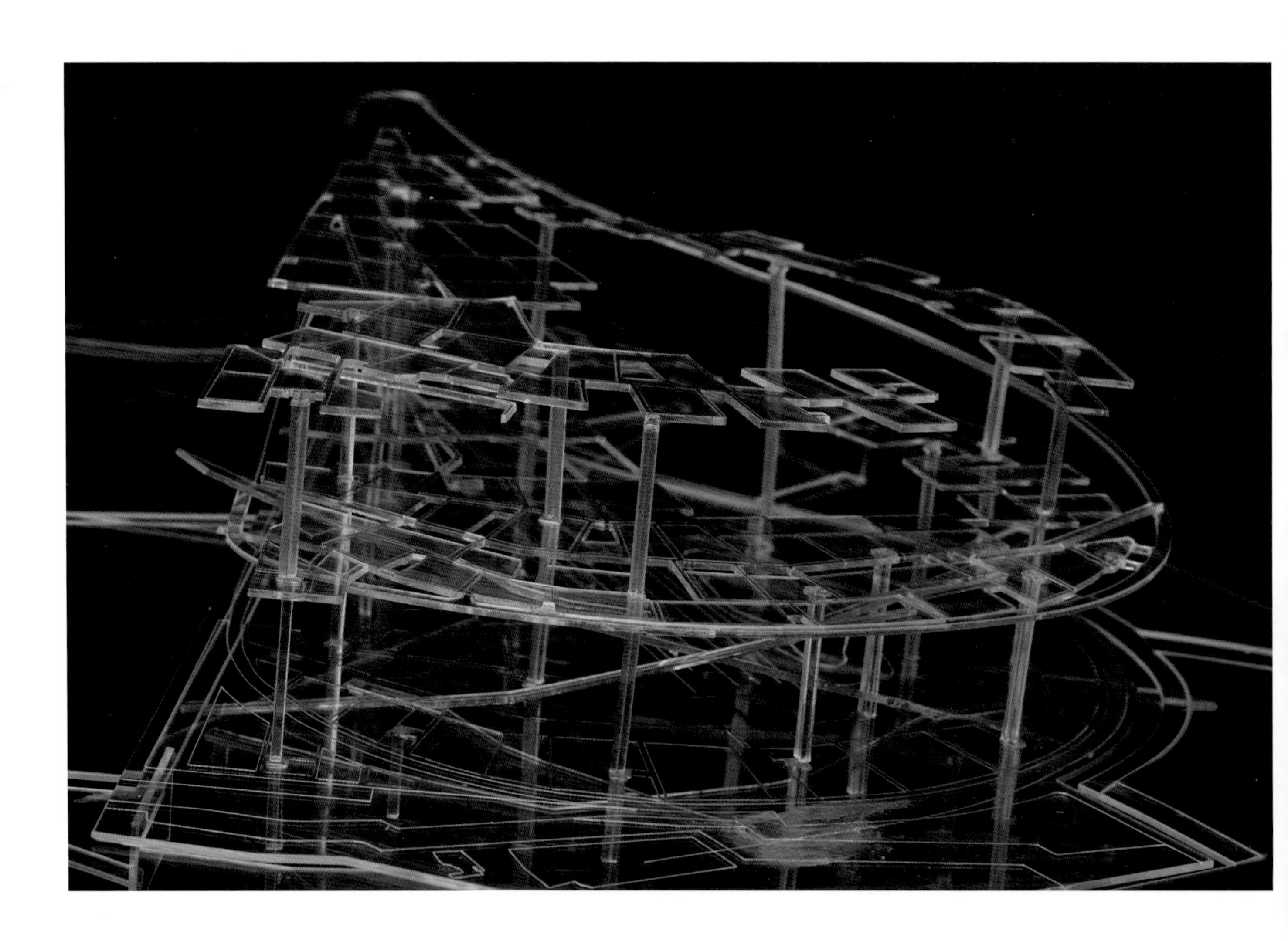

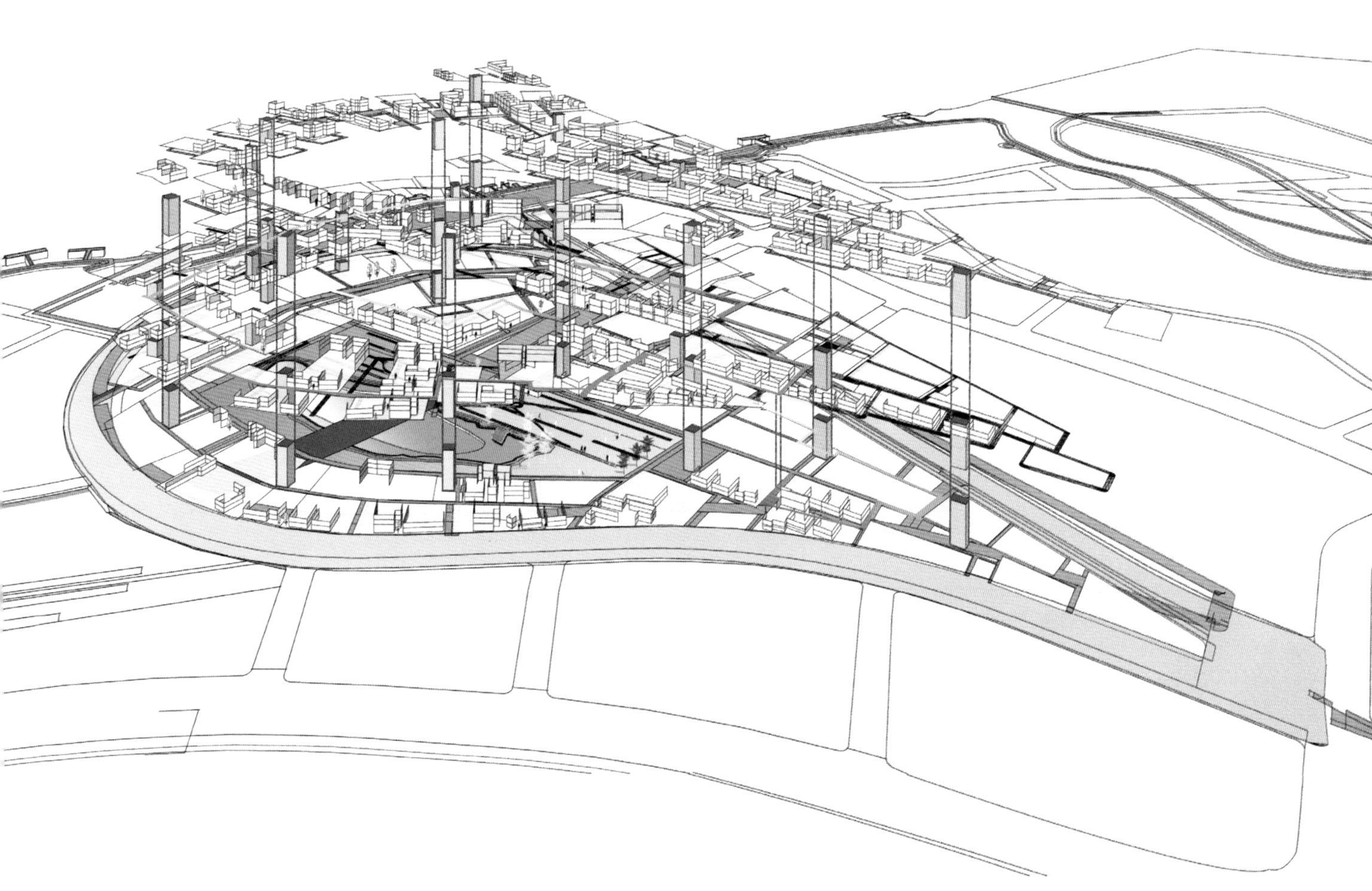

概念3: 再现：设计在容积率翻倍的情况下，通过统计与运用原有街巷结构的连接方式、方向、长度、覆盖率、建筑密度、街巷高宽比等参数，在立体空间上“再现”传统街巷尺度与肌理。

Concept 3: Reappearance: With doubled plot ratio, "reappear" dimensions and texture of traditional streets and lanes with three-dimensional space by counting and applying connection mode, direction, length, coverage ratio, building density, height-width ratio of streets and lanes and other parameters of existing structure of streets and lanes.

电影建筑学

CINEMATIC ARCHITECTURE

鲁安东

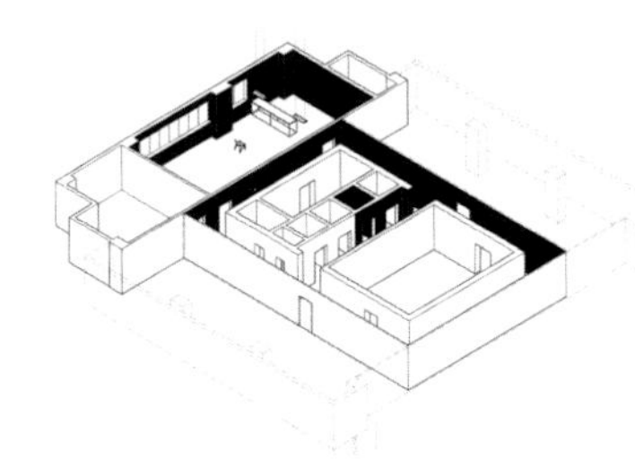

教学目标

在本课程中，电影被视做一种独特的空间感知和思想交流的媒介。我们将学习如何使用电影媒介来对建筑和城市空间进行微观地分析与研究。

教学步骤

第一阶段：空间分析训练：在寻常空间中发现一个带有电影性的片段，拍摄一个运动镜头和一个静止镜头或设计一个相机运动路径对它进行表达；设计一个利用空间结构性、场景性或辅助性特征的空间事件，并使用连续性剪辑加以表达；设计一次游历性的建筑漫步，并使用连续性剪辑从不同镜头角度对建筑漫步进行观察和表达。

第二阶段：场所分析训练：挑选一个带有强烈场所感的环境，对该场所进行深度观察，把握它的关键特征，并运用蒙太奇手法加以表达。或挑选一个带有强烈空间氛围的环境，根据空间氛围设计和编排剧情，主要的行为或事件应符合场所特征。

第三阶段：叙事分析训练：体验和拍摄一个陌生现场。运用影像语言将身体感受、空间氛围、情感表达结合起来。在对陌生场所的再现中表达个体的“诗意的经验”。在城市中寻找一个异化的建筑，通过影像语言以该建筑为例表述自己对“建筑异化”这一命题的理解。

Training Objective

In this course, film is considered as a unique medium for spatial perception and thought communication. We will learn how to use the medium of film to conduct micro analysis and research on building and urban spaces.

Teaching Procedure

Stage 1: Spatial analysis training: Find out a fragment with film nature in ordinary space, and represent it by shooting a motion shot and a static shot or design a camera moving path; design a spatial event that utilizes the structural, scenic or supplementary feature of space, and represent it with continuous editing; design an architectural promenade of traveling nature, and observe and represent the promenade from different camera angles with continuous editing.

Stage 2: Place analysis training: Select a setting with strong place sense, observe the place deeply, grasp its key features, and represent it with montage approach. Or select a setting with strong space sense, design and arrange story according to spatial ambience, and main actions or events shall match features of the place.

Stage 3: Narrative analysis training: Experience and film a strange site. Combine body sensation, spatial ambience and emotion with cinematic language. Represent individual “poetic experience” in reappearance of the strange place. Find an alien building in the city, and represent your understanding on the topic of “building alienation” with cinematic language by taking this building as an example.

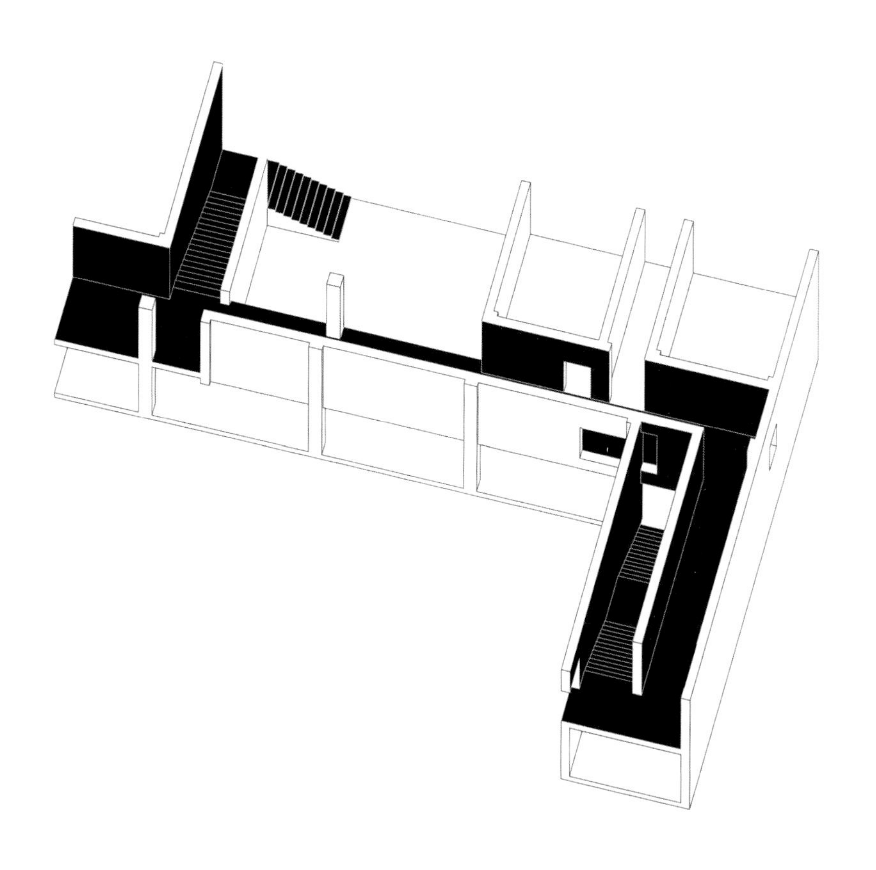

短片《对一个楼梯的空间注记》利用了两部楼梯带来的循环运动的可能性，表现了空间的结构性特征产生的电影性，并通过对人物运动的调度加以呈现。

The short film *Spatial Notation to a Stair*, which expressed the cinematics generated by structural features of space with possibility of circulatory movement brought by two stairs, and presented it by allocating movement of characters.

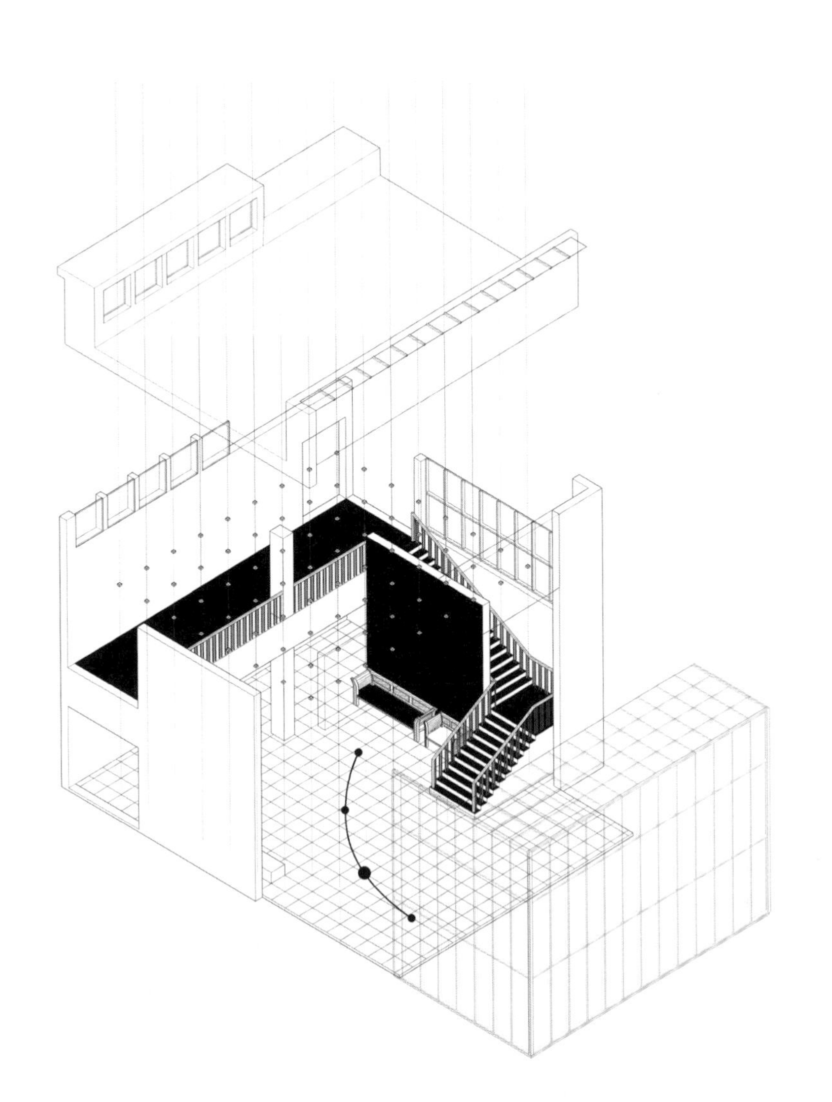

短片《格网》从阿伦·雷乃导演的影片《去年在马里昂巴德》（1961年）中的一个楼梯镜头出发，利用了拍摄场景（门厅）中带有的强烈几何秩序。
The short film *Grid*, which started from a stair scene in the film *Last Year at Marienbad* directed by Alain Resnais, and utilized the strong geometric order in the scene (hallway).

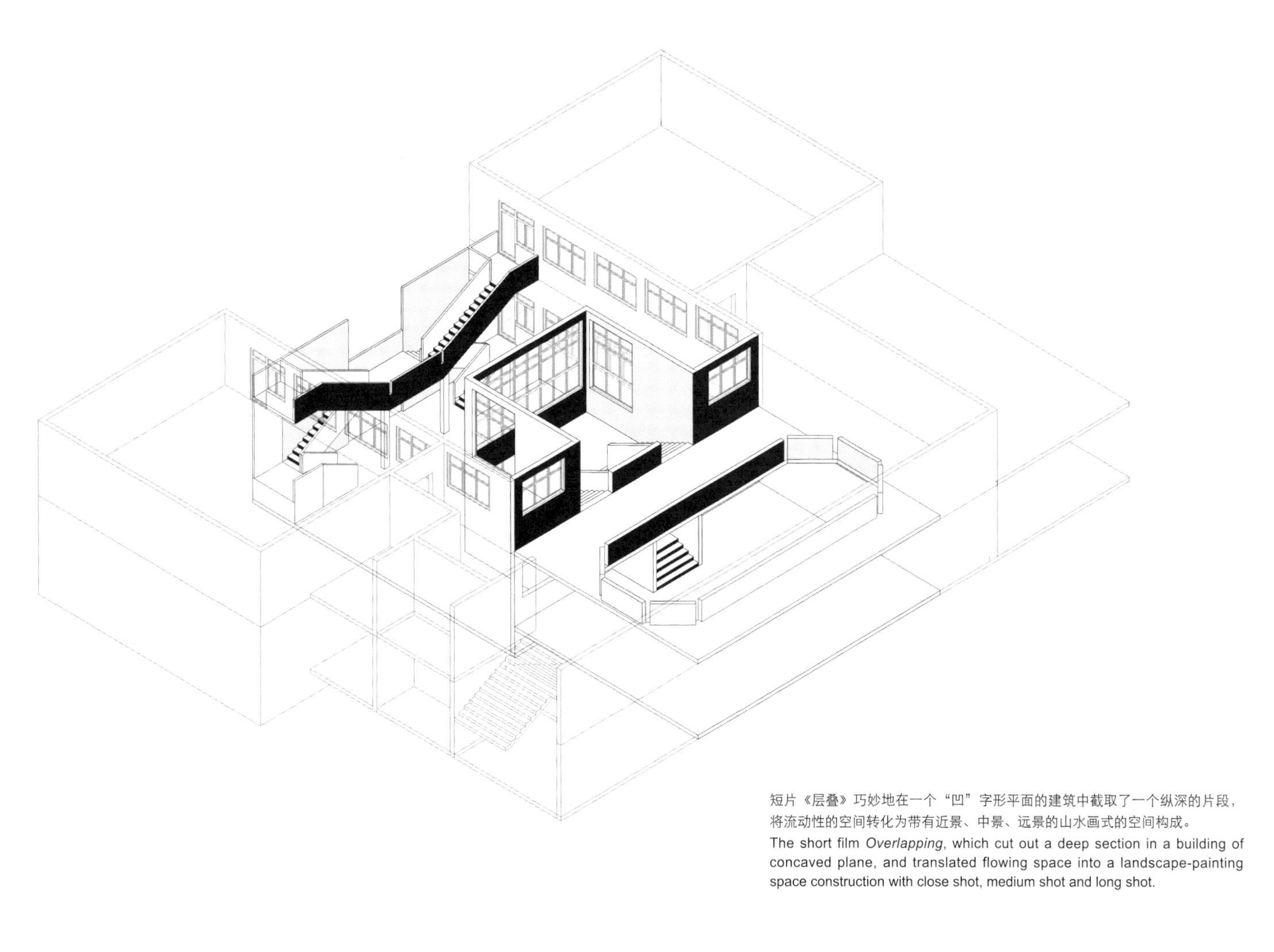

短片《层叠》巧妙地在一个“凹”字形平面的建筑中截取了一个纵深的片段，将流动性的空间转化为带有近景、中景、远景的山水画式的空间构成。

The short film *Overlapping*, which cut out a deep section in a building of concaved plane, and translated flowing space into a landscape-painting space construction with close shot, medium shot and long shot.

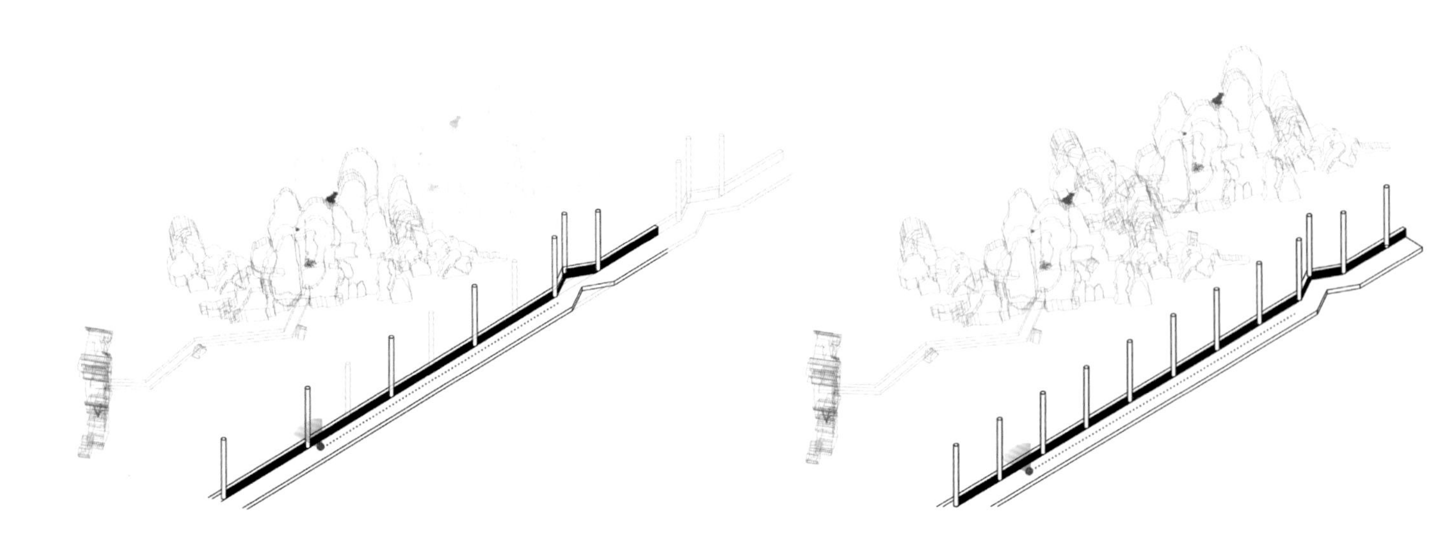

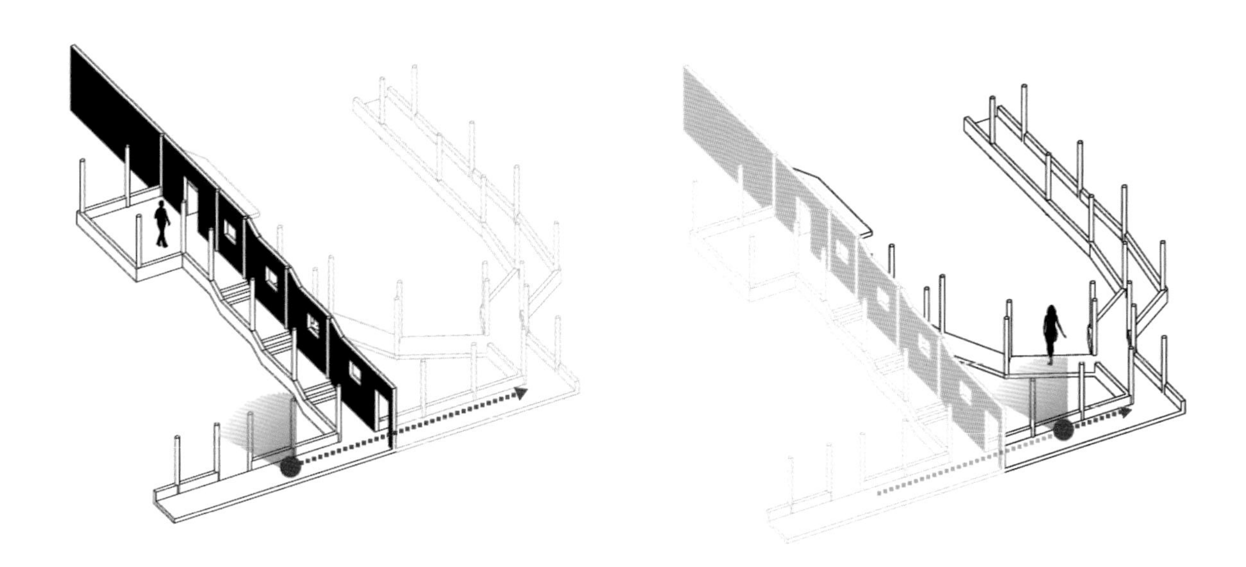

设计工作坊
THE SHORT STUDIO

更轻的结构

阿尔贝托·普尼亚莱、索菲亚·科拉贝拉、童滋雨

本设计工作坊关注于格网壳体的参数化设计、找形、优化和建造。设计强调材料、结构和形式的完整结合。其中，双曲面是设计的几何基础，通过找形步骤确定格网壳体的形体，并利用Karamba对形体的受力进行模拟计算和优化。

社会—空间图示方法

塔特亚娜·施耐德、窦平平、周立涛

本课程旨在探索复杂的社会与空间关系的形象化表达方式。学生分为 5 个小组，每组 3 人，分别关注南京市中心的一个片区并表达日常性的且处于持续变化中的碰撞性现实——这在传统的空间再现方式中往往被我们忽视。我们不仅关注客观物质存在、社交性使用及其背后的理论，也试验形象化表达现实因素的方式，并行表达新兴复杂的体制和不断变化的空间。

VDS 绿色建筑设计

张建舜、迈克尔·佩尔肯、郜志、秦孟昊、尤伟

绿色建筑设计平台 (VDS) 是对建筑能源与环境系统进行整合、协调及优化的数字设计平台。学生在跨学科与跨文化背景的环境下，学习运用 VDS 对气候与场地、建筑外形、体量与朝向、内部构型、外部围护结构与环境控制系统等进行综合分析与性能模拟，探索新的设计理念与策略。

How to Make Things Lighter

Alberto Pugnale, Sofia Colabella, Tong Ziyu

The design studio focused on parametric design, form-finding, optimization, technology and fabrication of gridshells. Material, structure and form are integrated in the design. Double-curvature is the key. 'Form-finding' is the process to get it. Karamba was implemented as a tool to develop the form and optimize the geometry of gridshells.

Mapping Socio—spatial Practices

Tatjana Schneider, Dou Pingping, Zhou Litao

The workshop intends to test how complex socio-spatial relationships can be visualised. We will focus on five distinct urban areas within the central area of the city of Nanjing, and visualise what we find: the everyday and continually changing contexts of these collisional realities – things that are often overlooked in conventional representations of space. Working in groups of 3, we'll engage with the physical realities, the social use and the underlying theories and will experiment with how one can visualise parallel realities, emergent and complex systems as well as space in flux.

Virtual Design Studio for Green Building

Zhang Jianshun, Michael Pelken, Gao Zhi, Qin Menghao, You Wei

This course introduces students to the methodologies and the use of the advanced digital platform "Virtual Design Studio" that has been developed in order to achieve an integrated, coordinated, and optimized design of buildings and their environmental systems. It provides a cross disciplinary environment and cross cultural environment for learning and exploring new design concepts and strategies through integrated analysis and performance simulations of effects of climate and site, form, massing and orientation, internal configuration, external enclosure, and environmental control system on building performance using VDS.

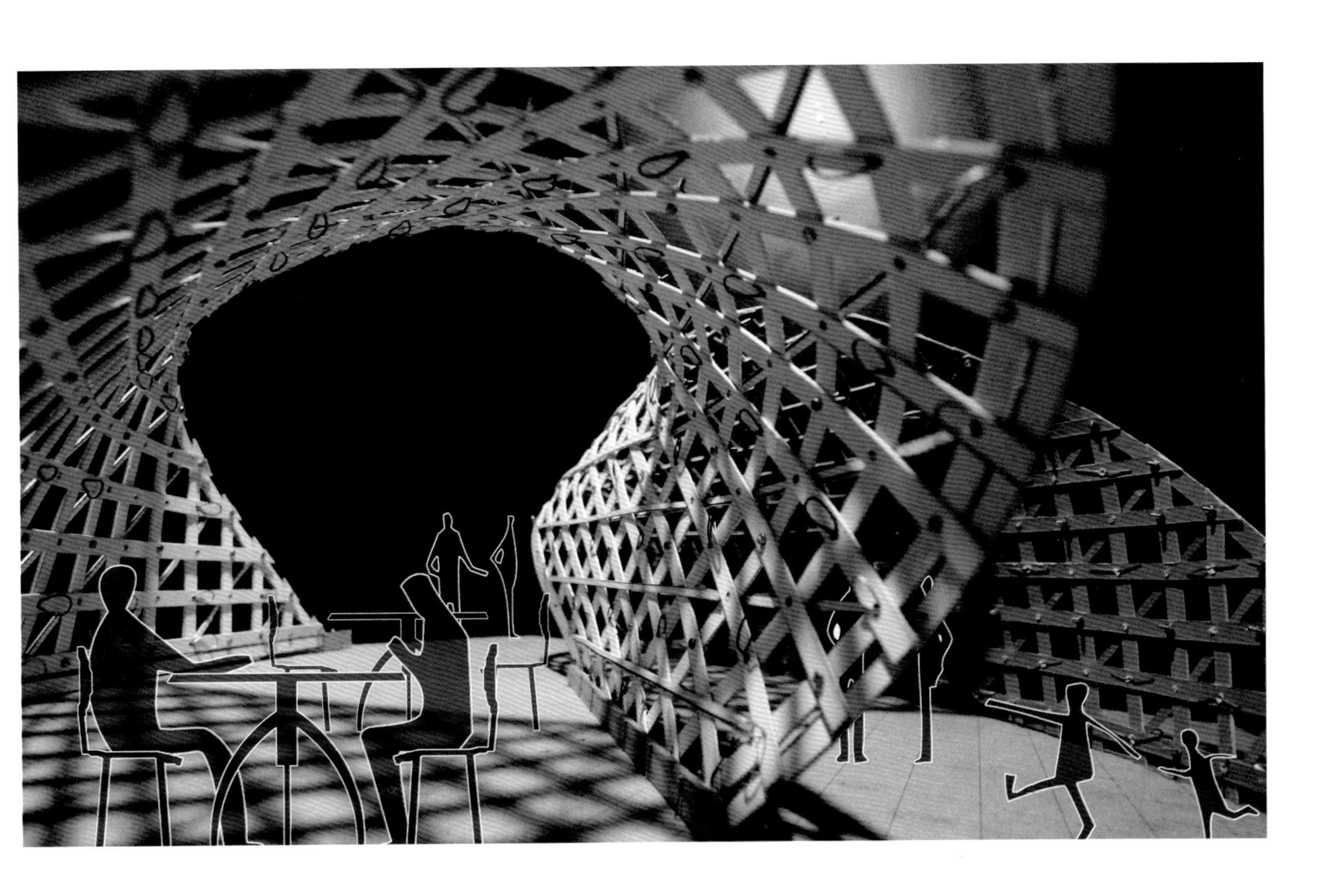

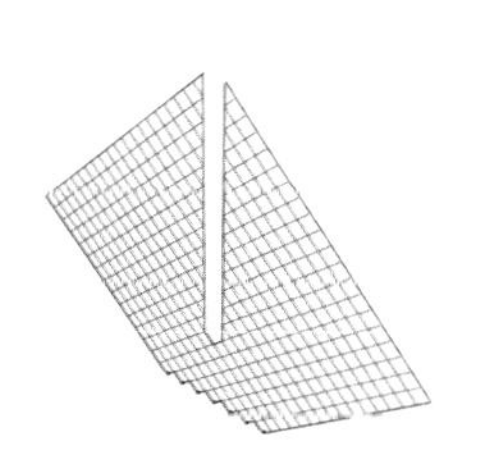

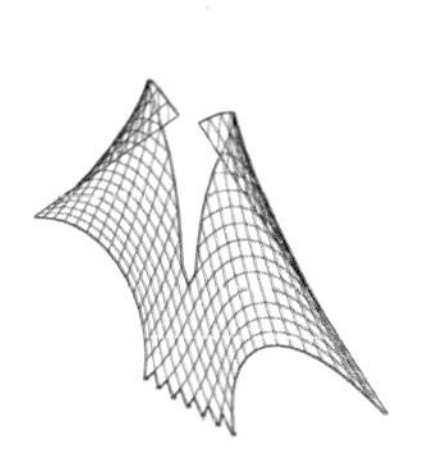

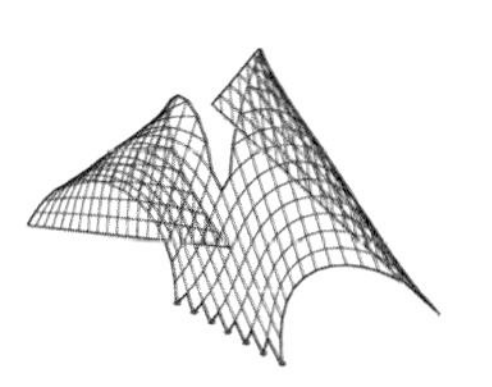

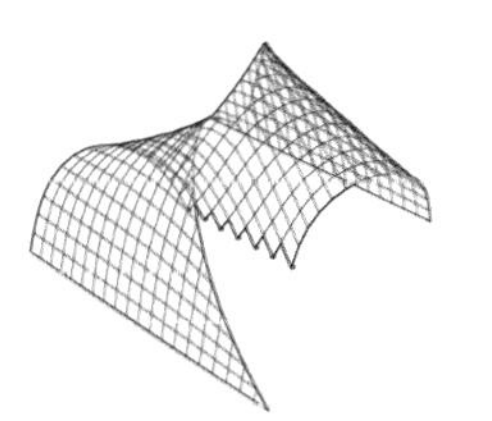

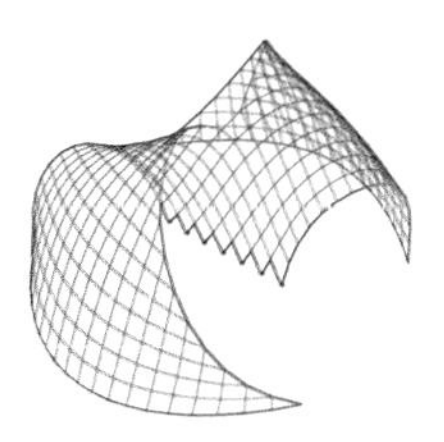

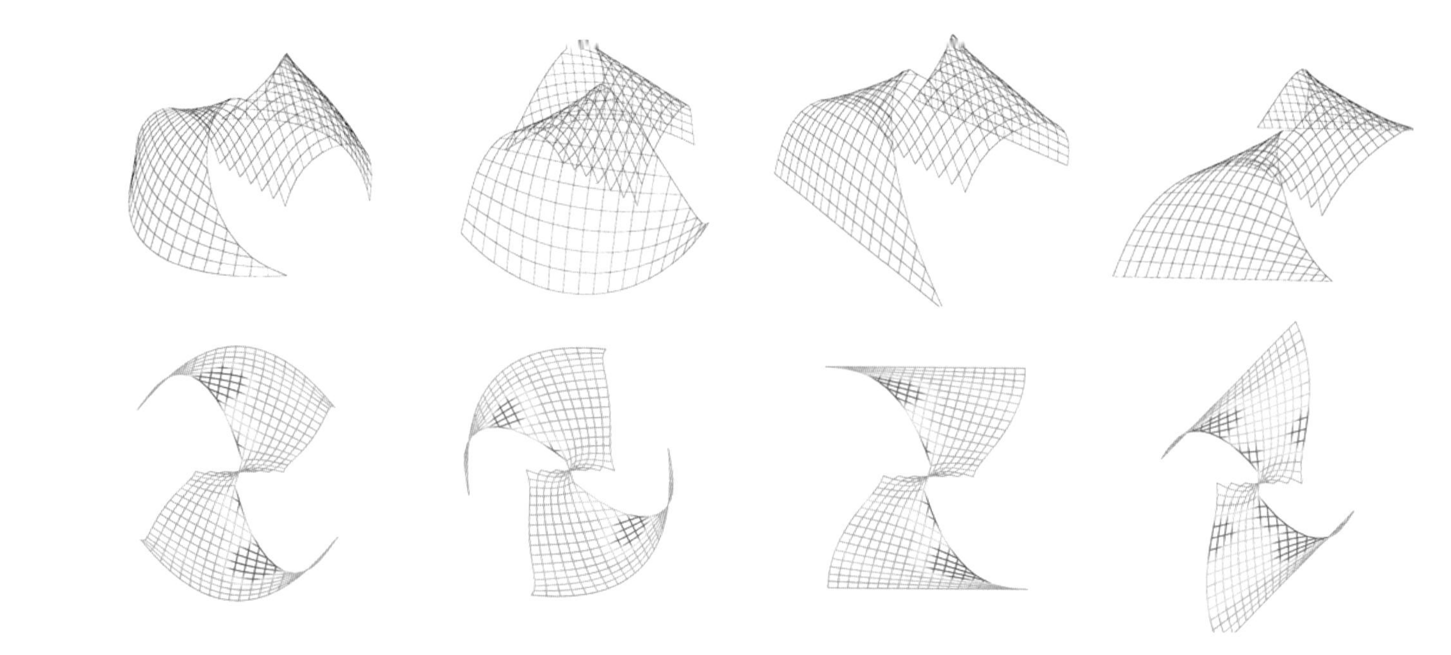

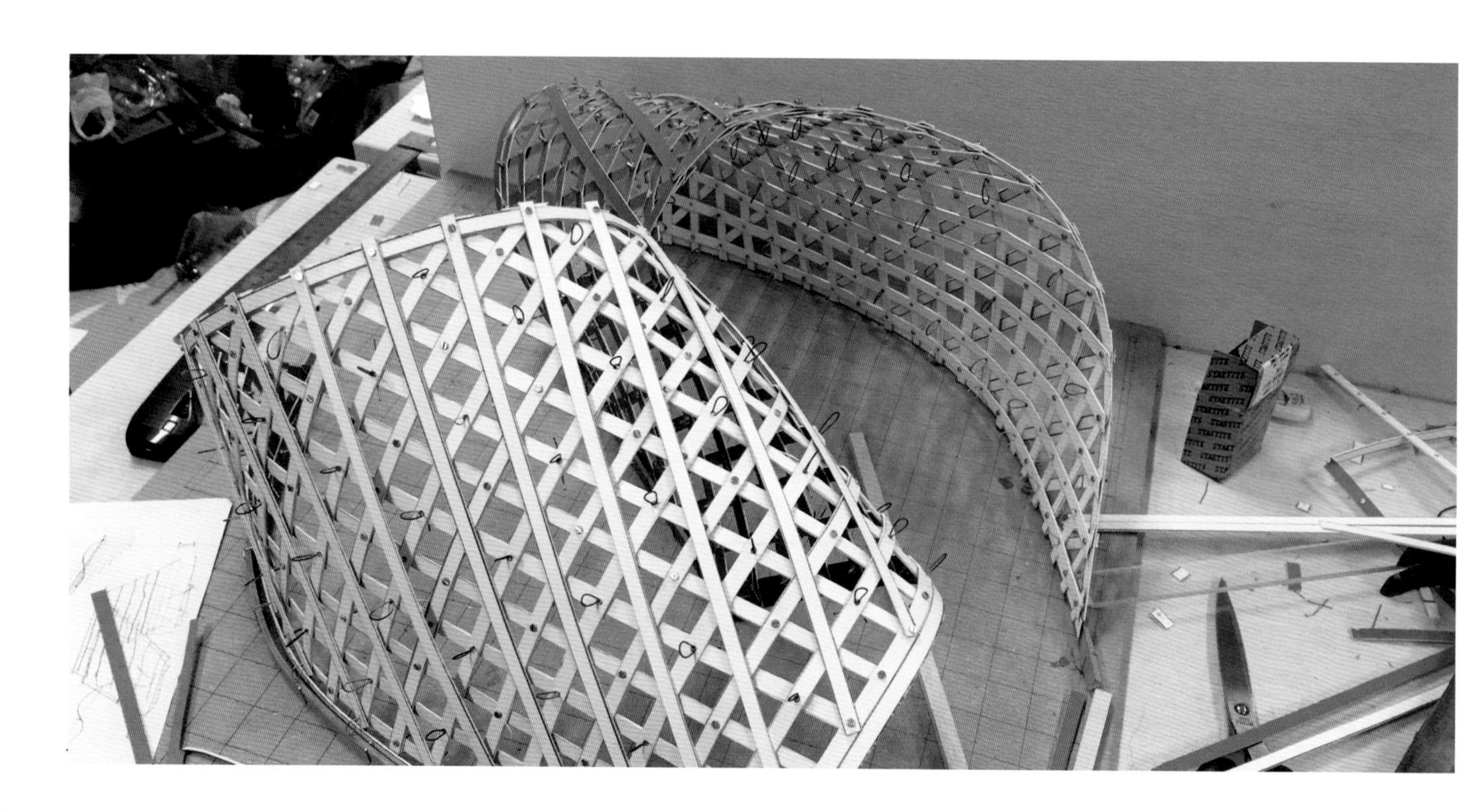

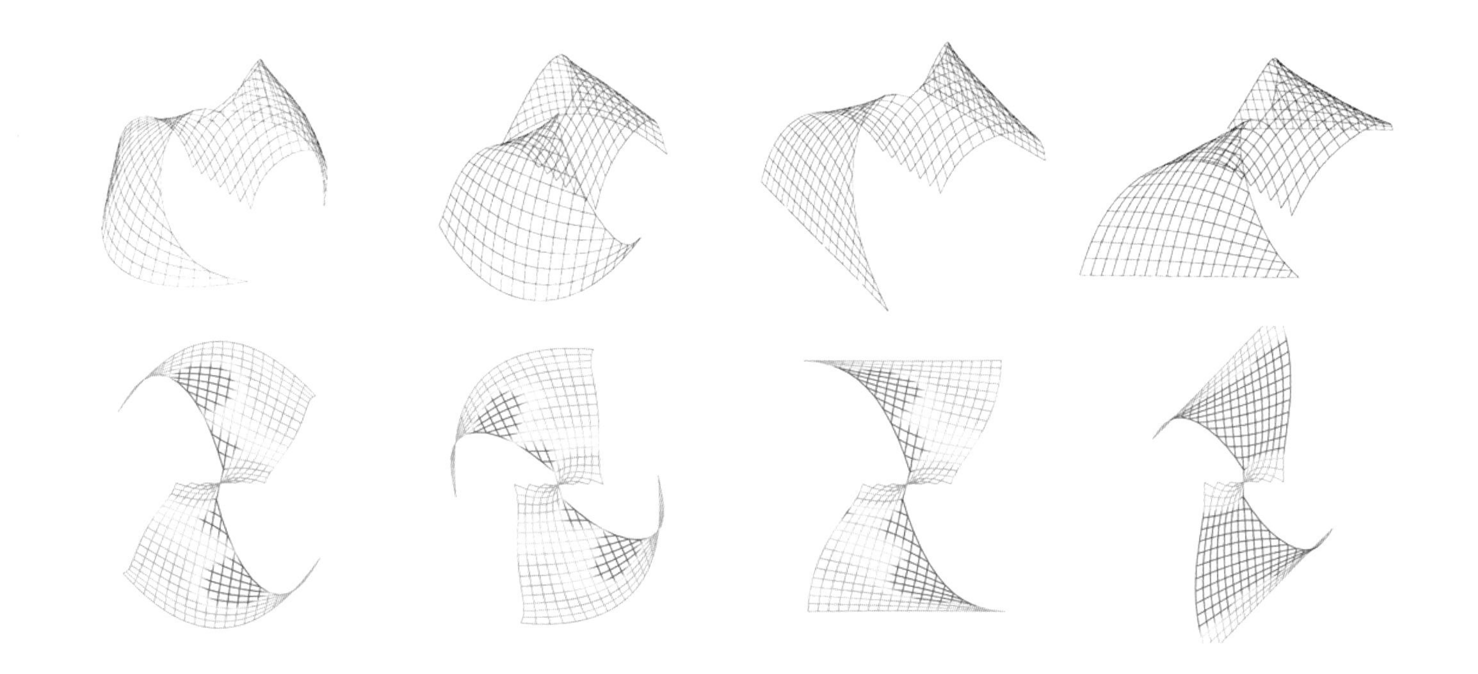

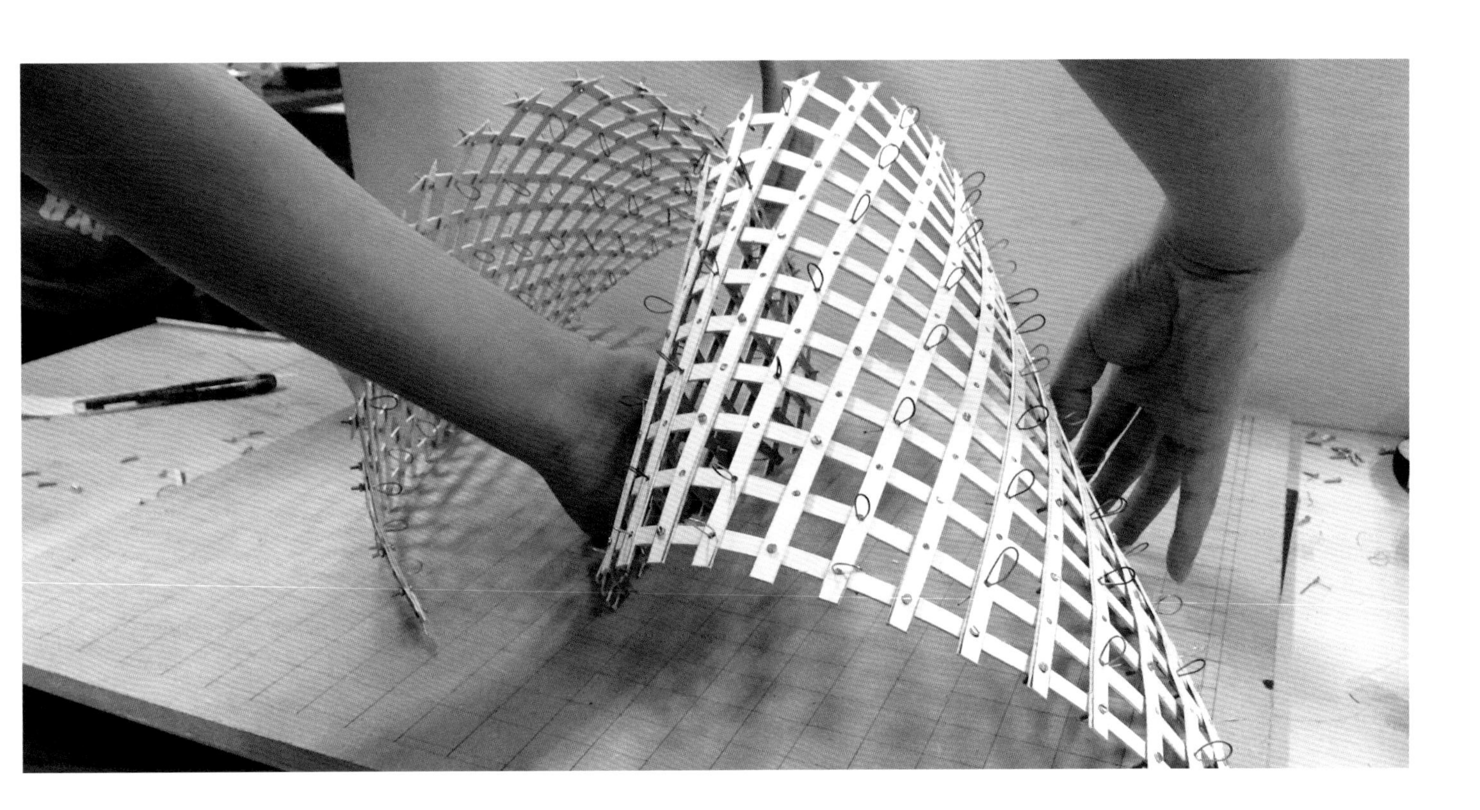

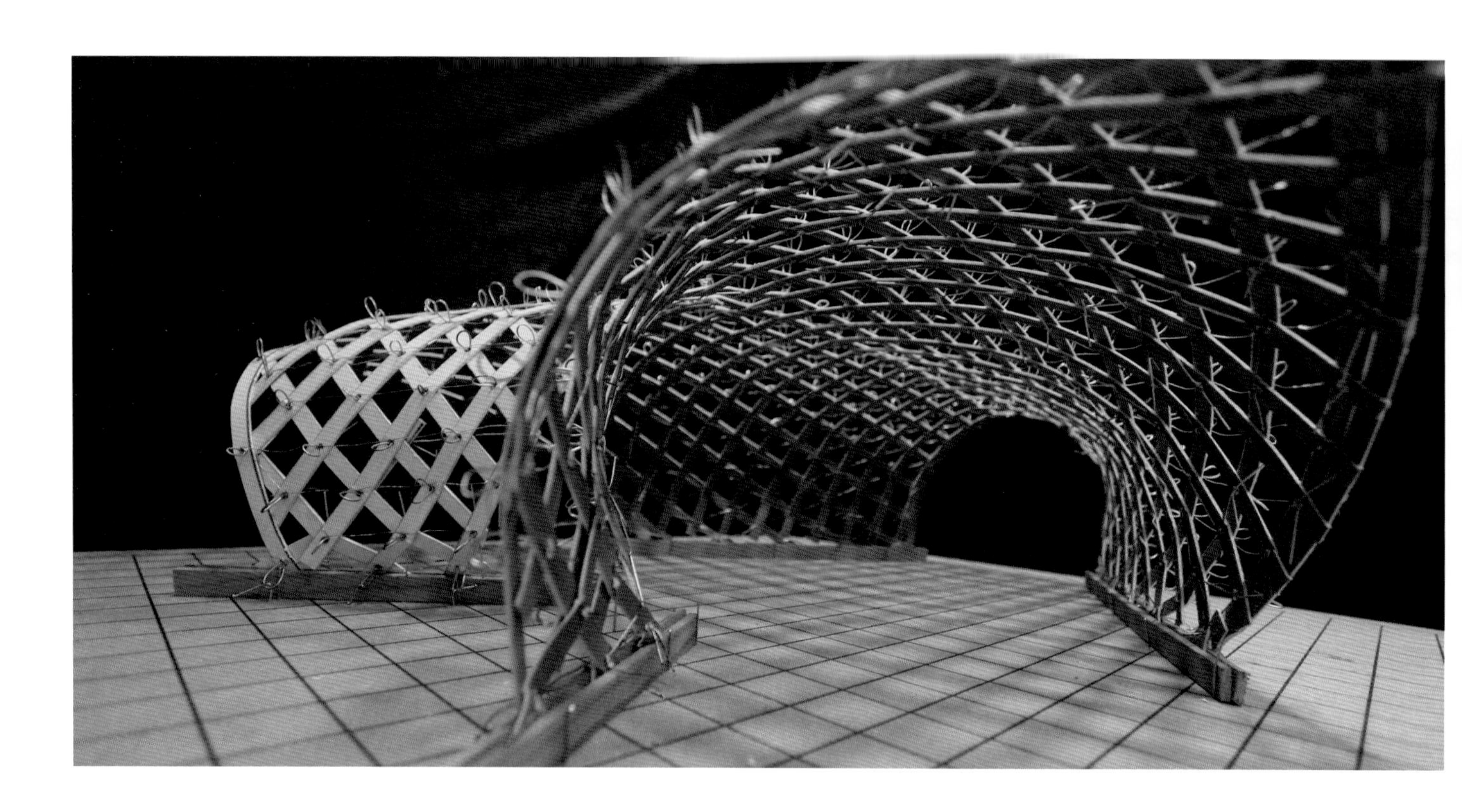

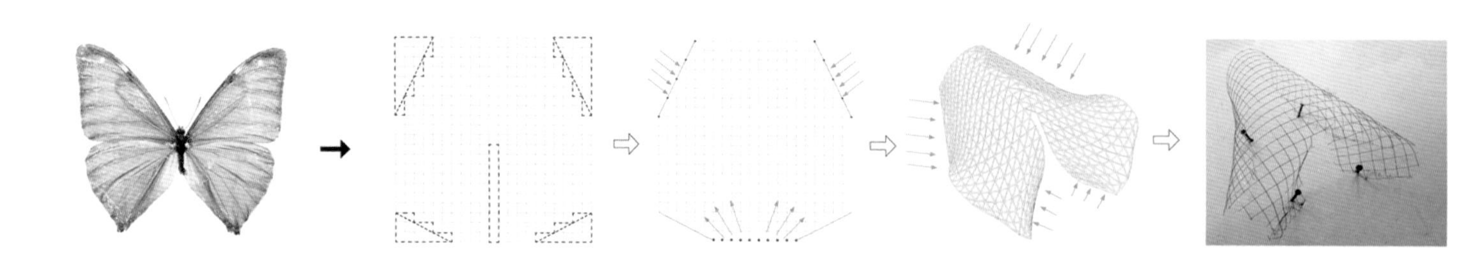

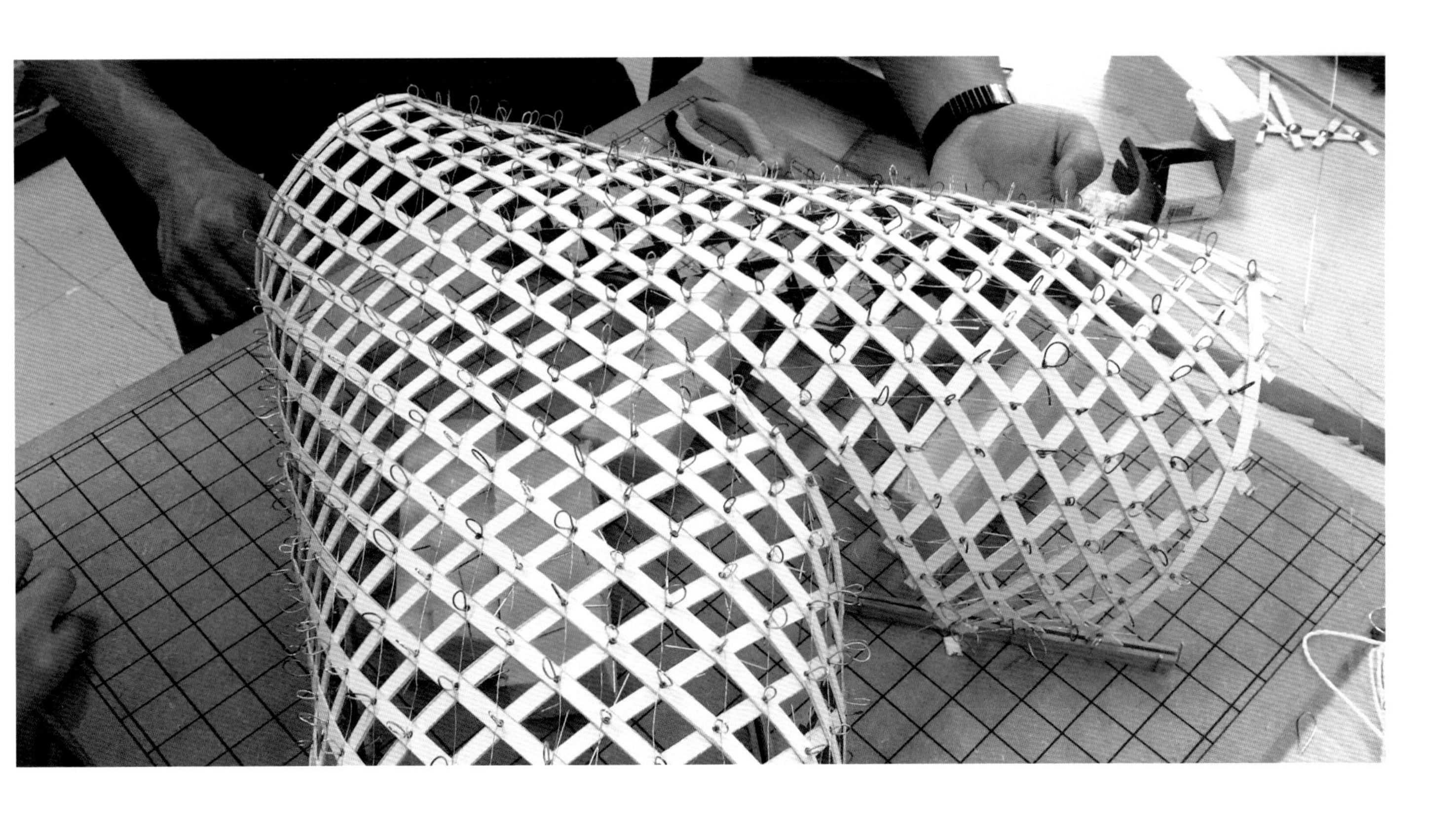

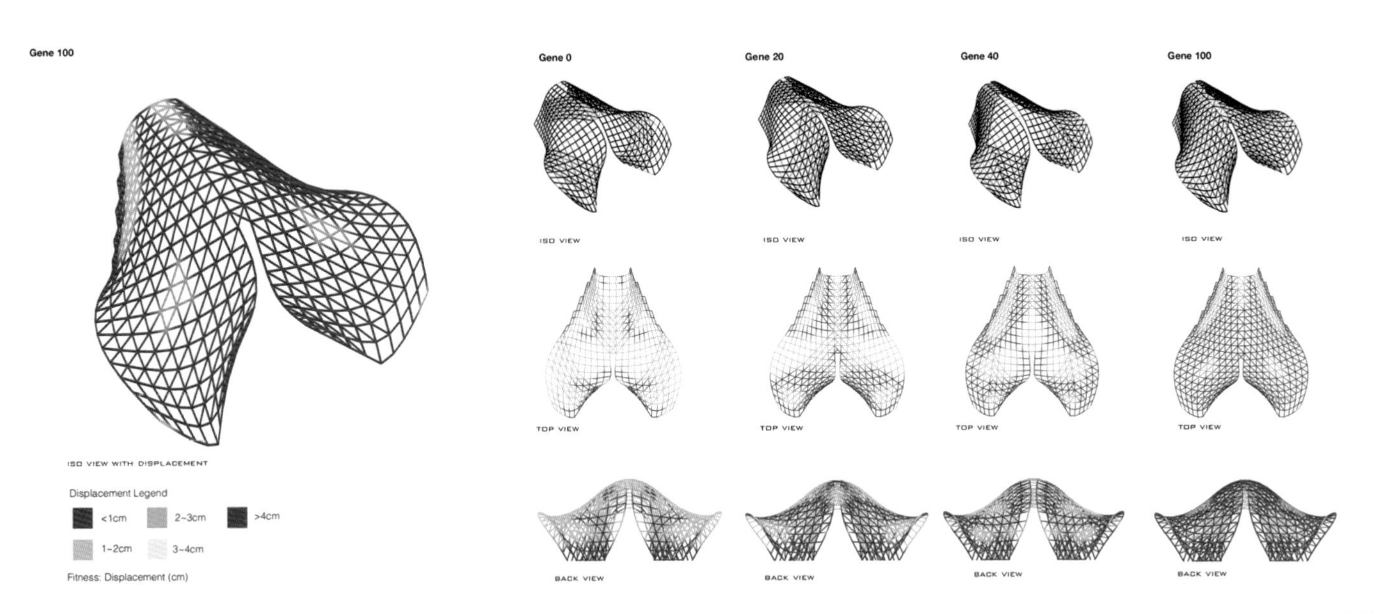
Gene 100
ISO VIEW WITH DISPLACEMENT
Displacement Legend
<1cm
2~3cm
>4cm
1~2cm
3~4cm
Fitness: Displacement (cm)
Gene 0
Gene 20
Gene 40
Gene 100
ISO VIEW
TOP VIEW
BACK VIEW

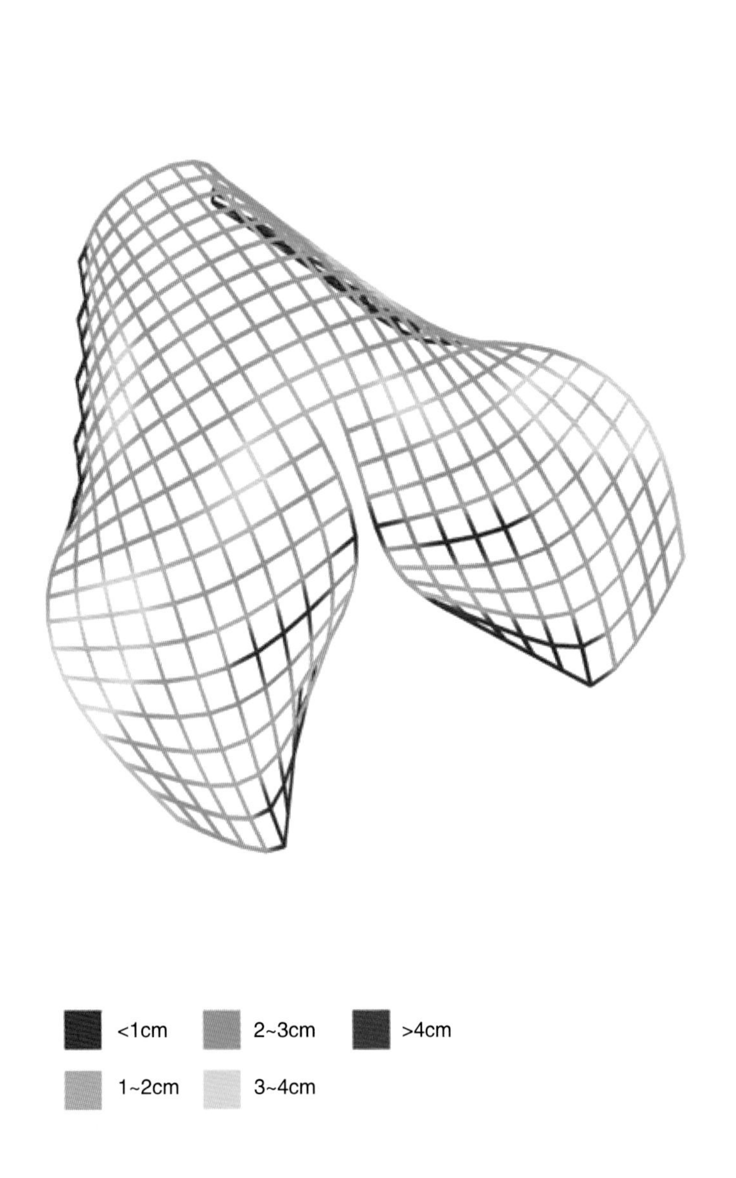

<1cm
2~3cm
>4cm
1~2cm
3~4cm

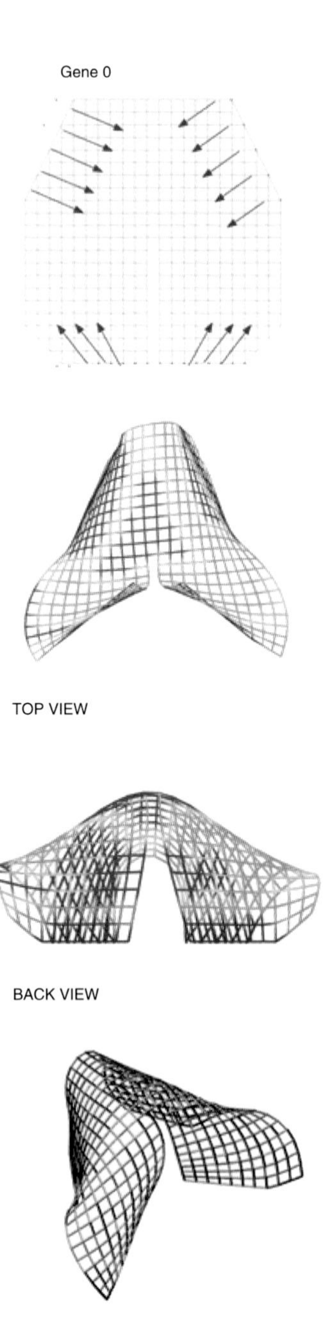

Gene 0
TOP VIEW
BACK VIEW

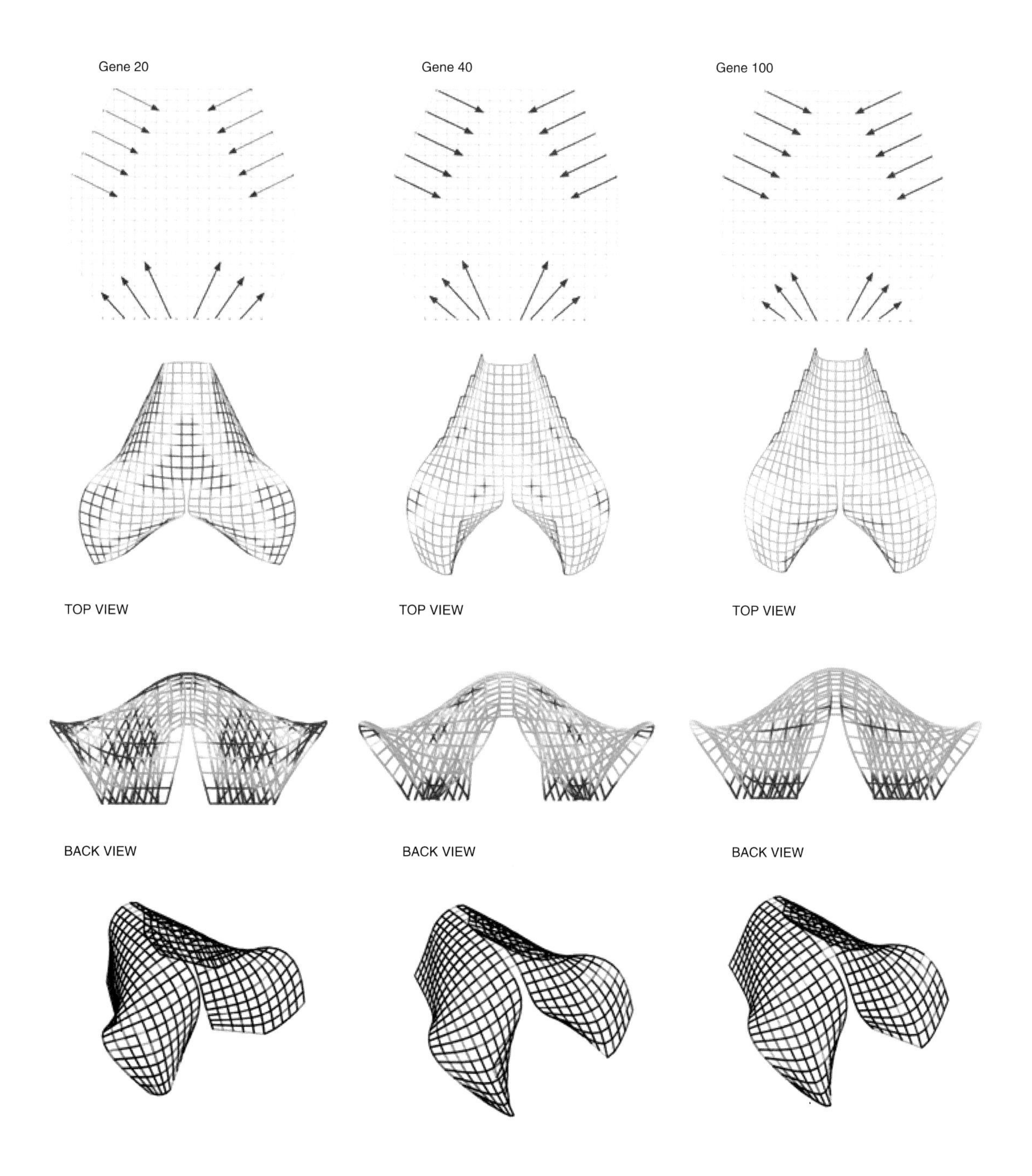
Gene 20
Gene 40
Gene 100
TOP VIEW
TOP VIEW
TOP VIEW
BACK VIEW
BACK VIEW
BACK VIEW

¥2800/month
¥2500/month
GET OUT!!!
¥27000/m²
A7
A7
2F
2F
¥3000/month
GETTTTT OUTTTTT !!!
This is my own house.
We live here for 20 years.
HOTEL
Advertisements for Making fake testimony which is illegal
Students like the fried Chicken here very much
There's a vegetable market near here
Main Supermarket Here

47㎡
price:2.1million
price 44680 元/㎡
living room
kitchen
bedroom
toilet
bedroom
The blue shed was added during the Youth Olympic Games, the original condition was disordered. The old residents said here was used to be a field with a few small houses.
price 41601元/㎡
76.9㎡
price 3.2million
People lived in this area can let their children go to Liqiue primary School and No.29 primary school.
Who built these additional building?
These additional buildings were built by the landlords.
The government brushed the walls yellow during the Youth Olympic games.
The landlords were farmers used to live there, they got the house as the subsidies of the demolition.
乙粉馆
Are these additional buildings illegal?
Yes, but the government doesn't manage this.
哈尔滨水饺

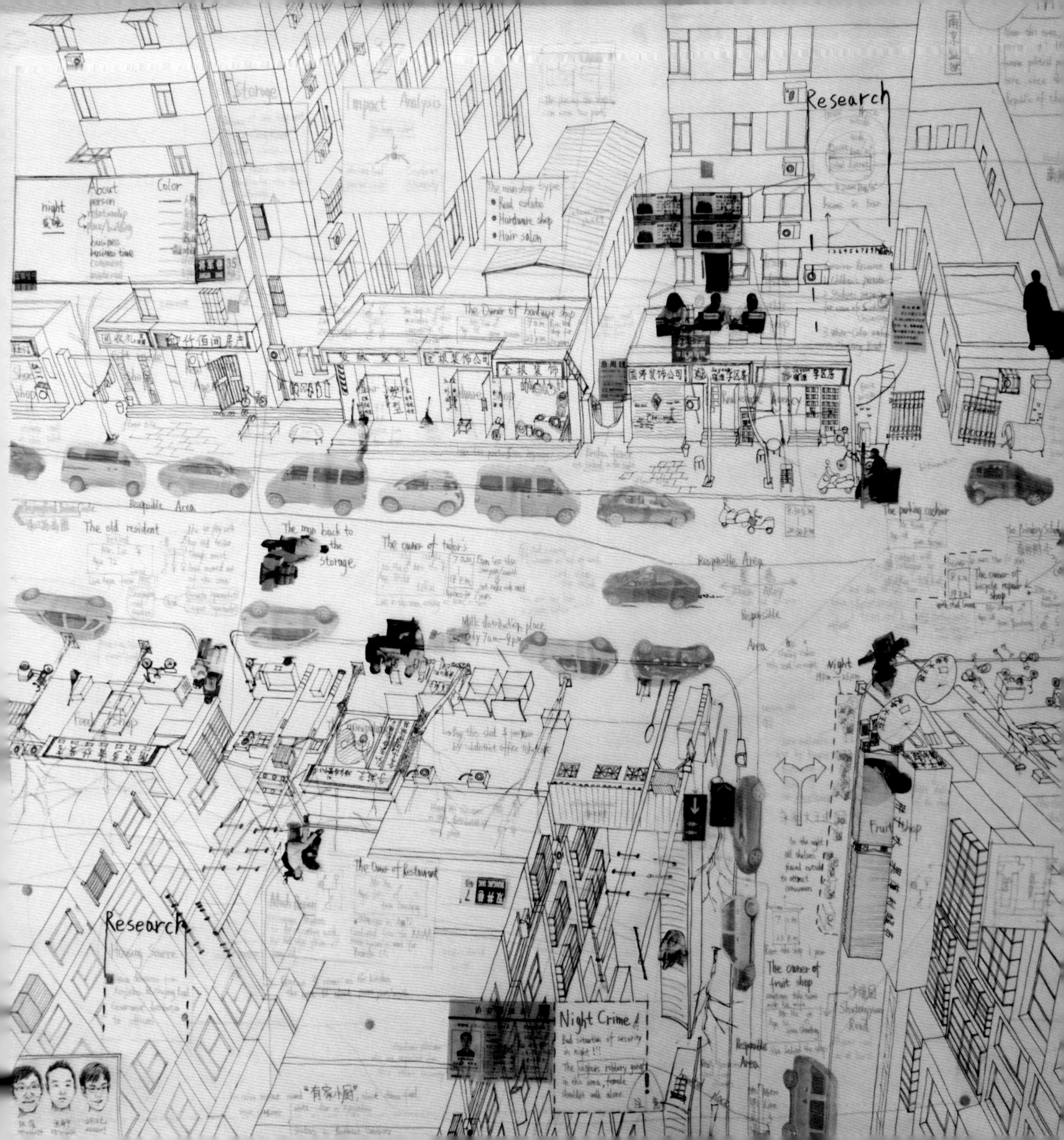

Research
Impact Analysis
Storage
About
Color
night
The main shop type
● Real estate
● Hardware shop
● Hair salon
The Owner of hardware shop
Responsible Area
The old resident
The man back to the storage
The owner of tailor's
Milk distribution place
The parking cashier
The Owner of Restaurant
Research
Night Crime!
Bad situation of security in night!!
The owner of fruit shop
Fruit Shop

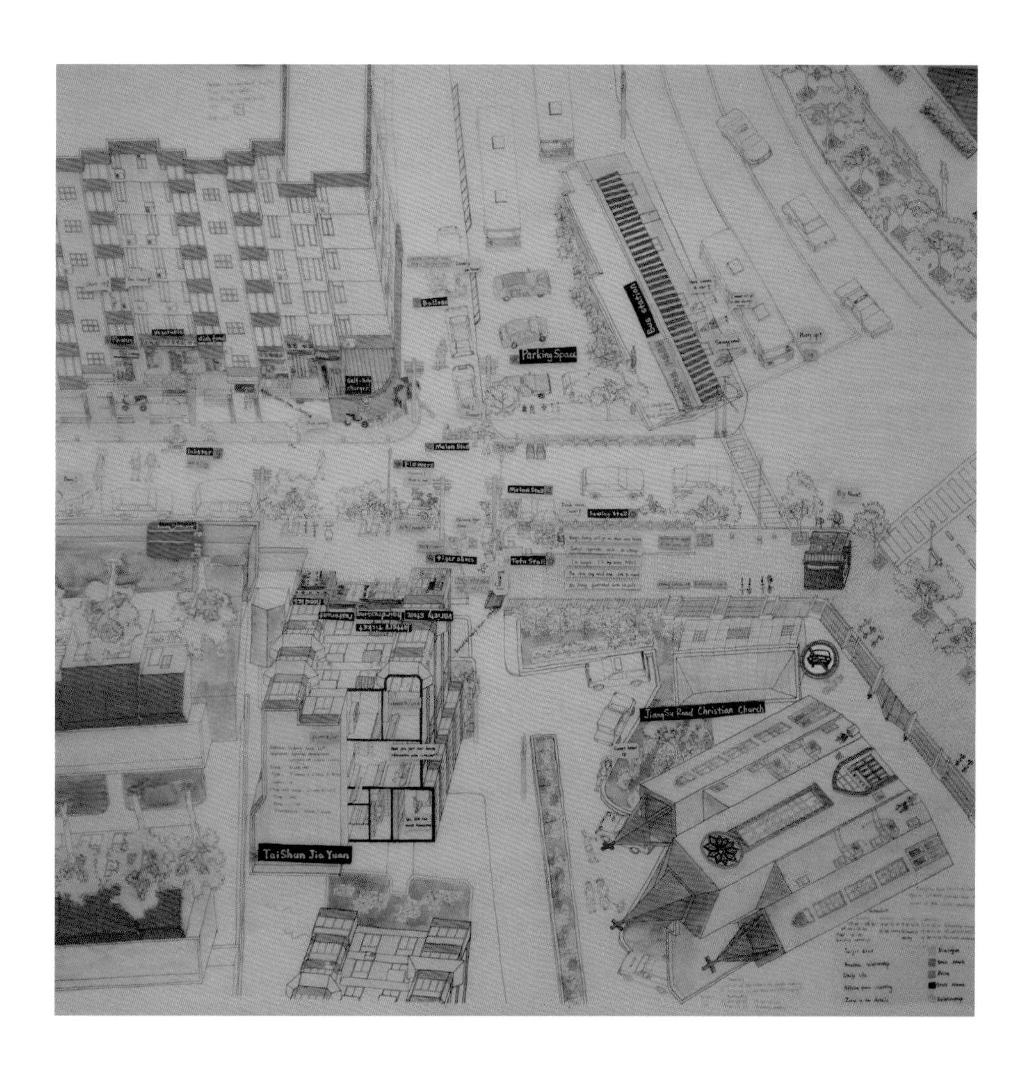

Bus station
Parking Space
Melon Stall
Flowers
Melon Stall
Tofu Stall
JiangSu Road Christian Church
TaiShun Jia Yuan

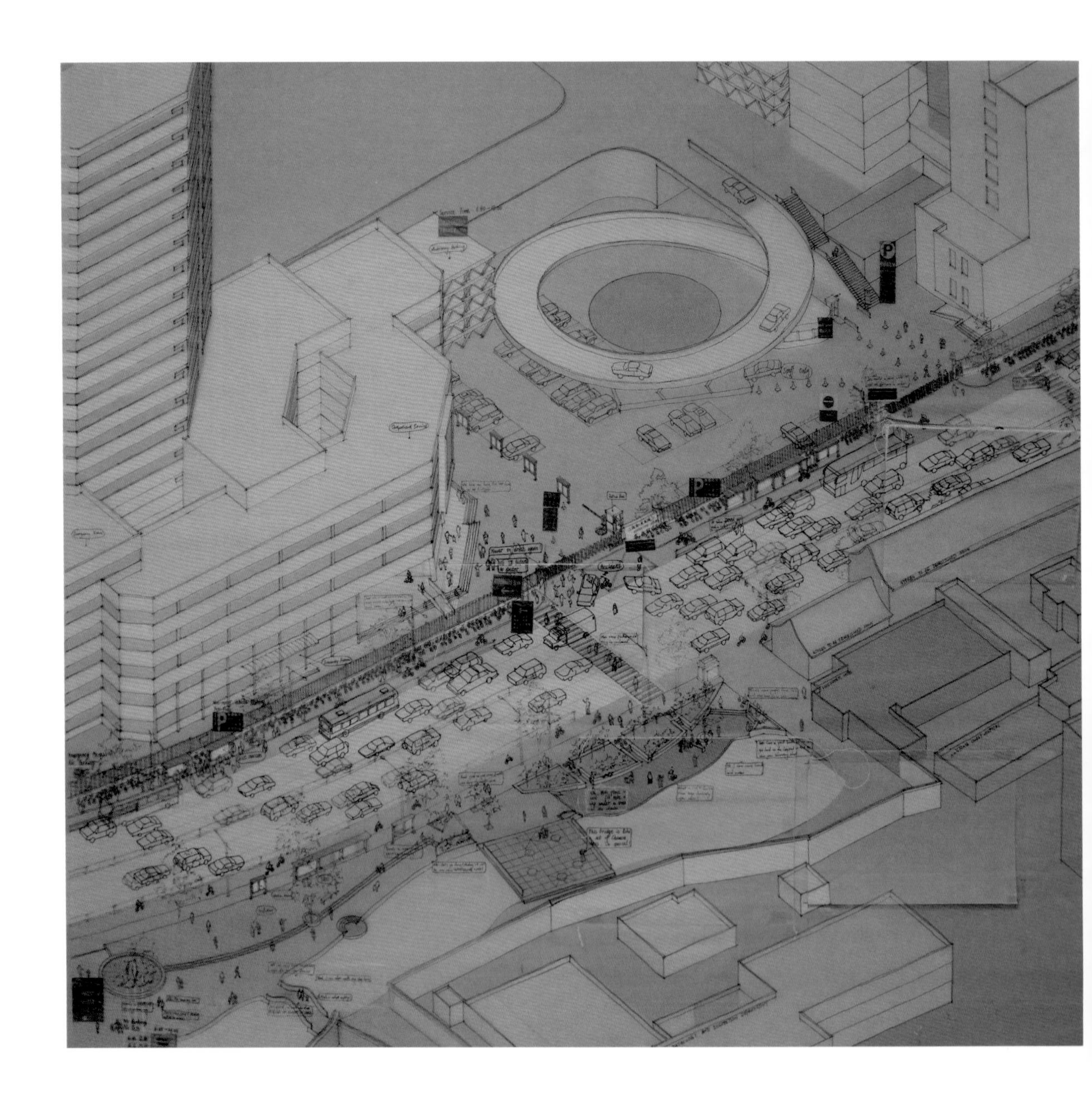

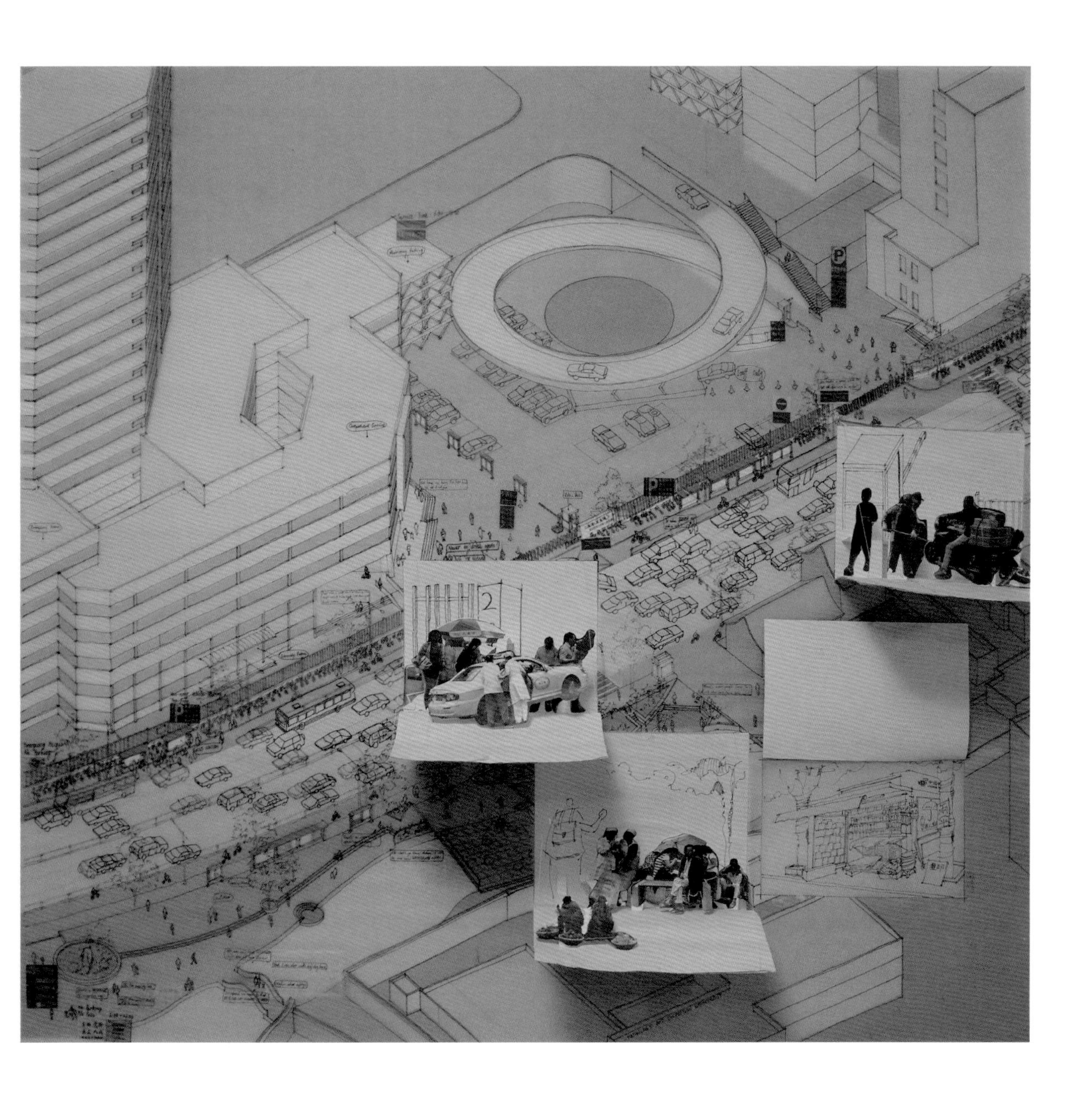

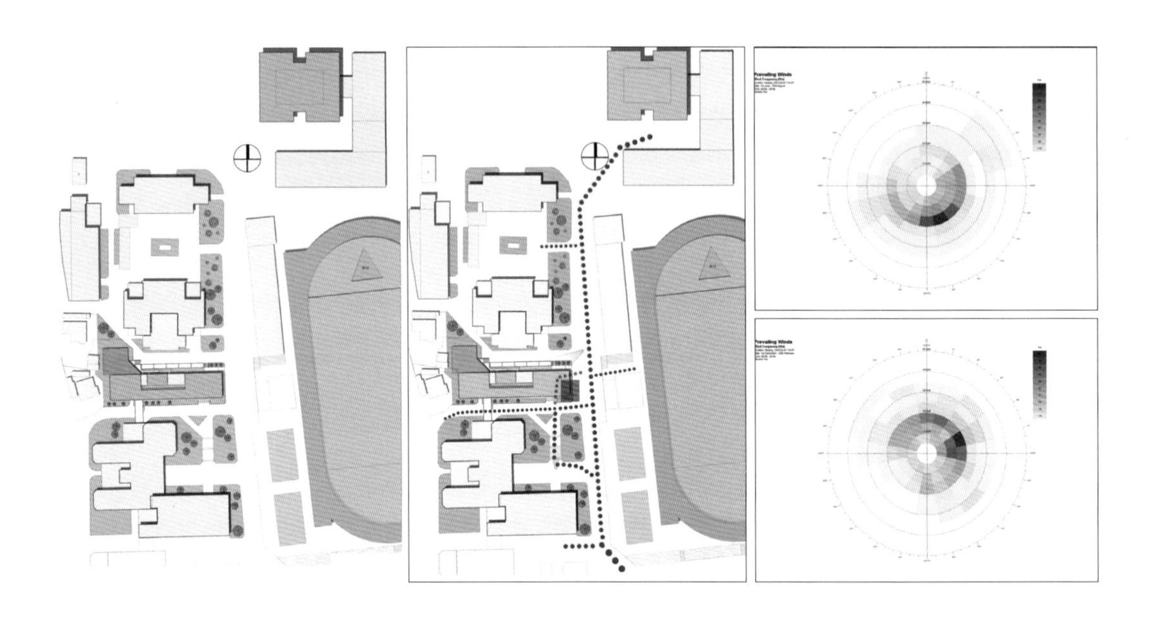

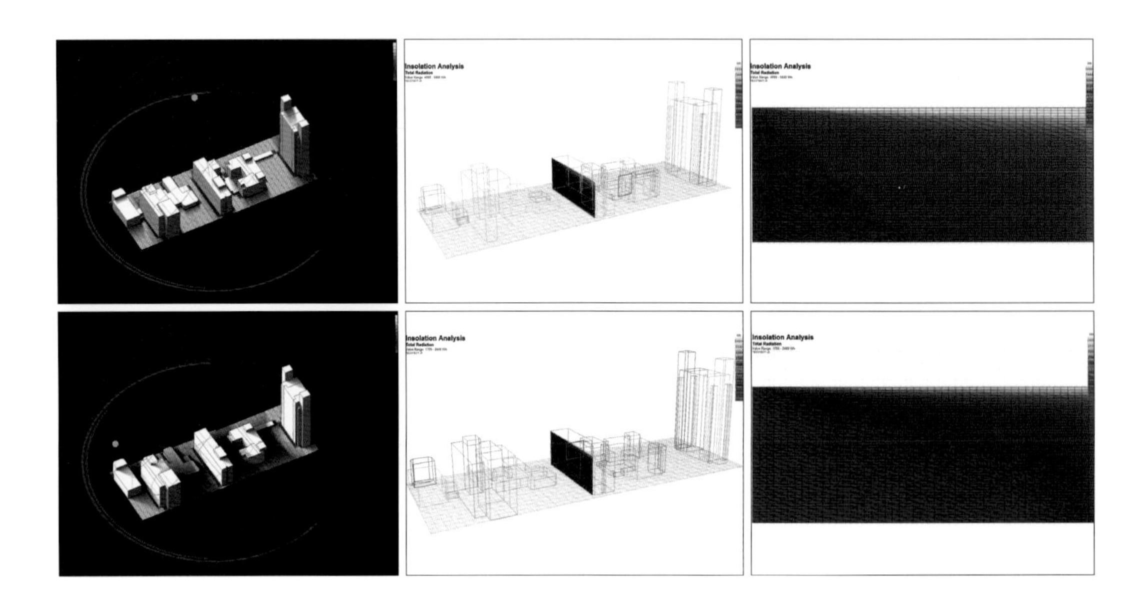

南京大学文科楼气候、场地与日照分析。

Climate, site and insolation analysis of the Liberal Arts Building in NJU.

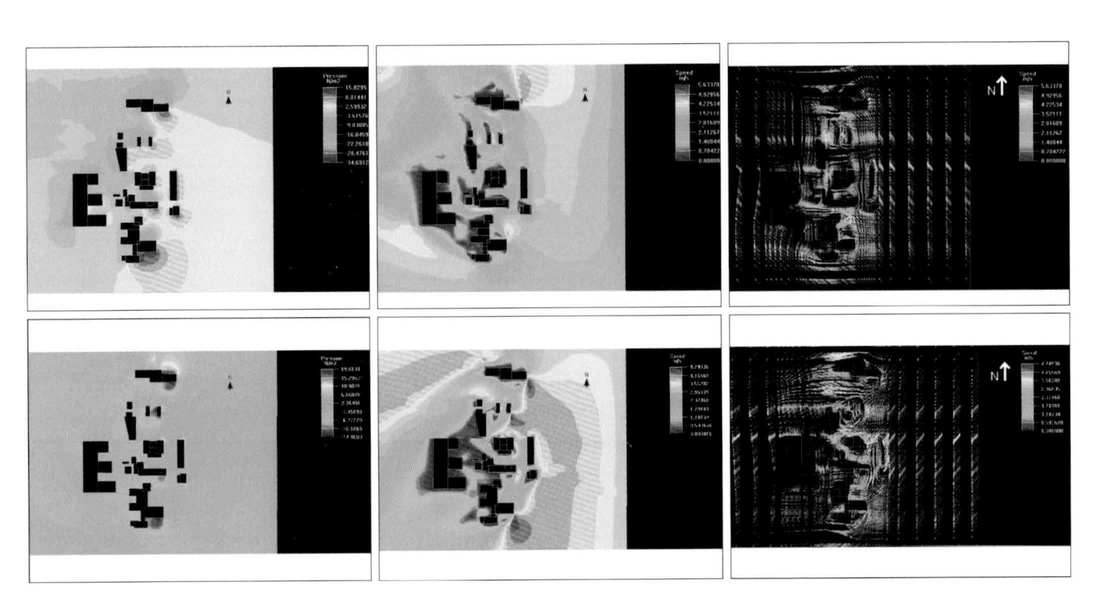

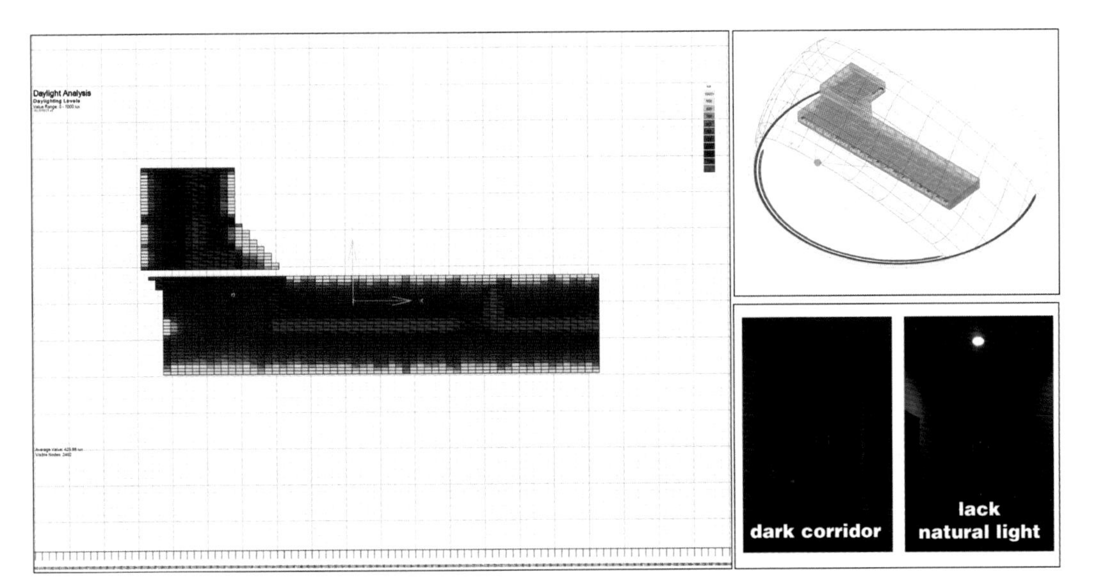

南京大学文科楼风环境与天然采光分析。

Wind environment and daylight analysis of the Liberal Arts Building in NJU.

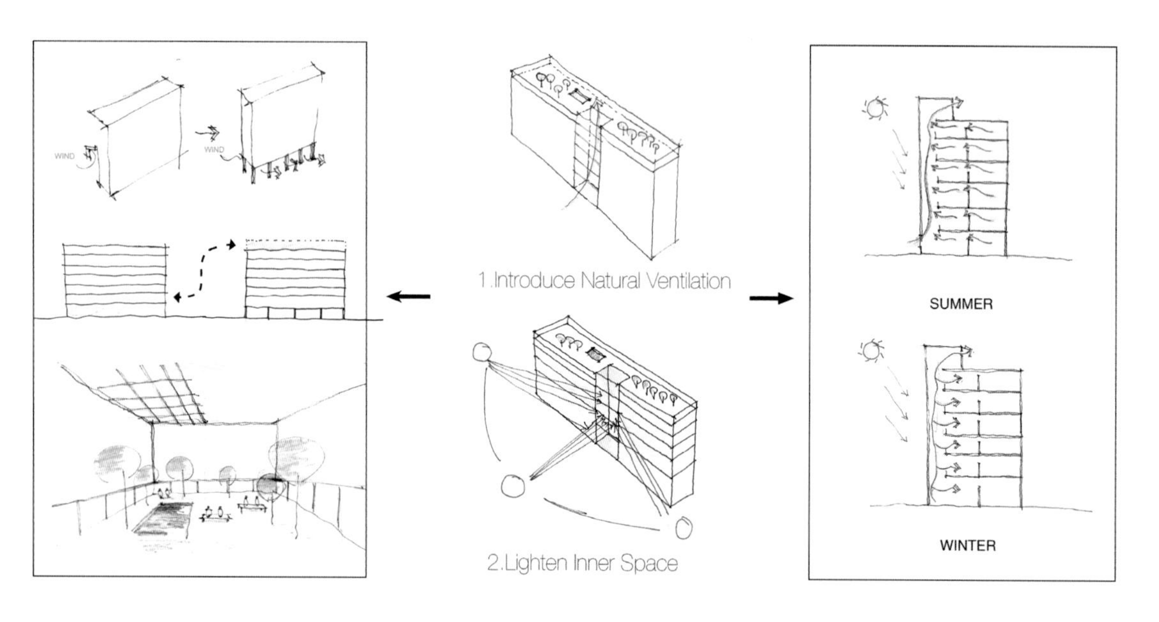

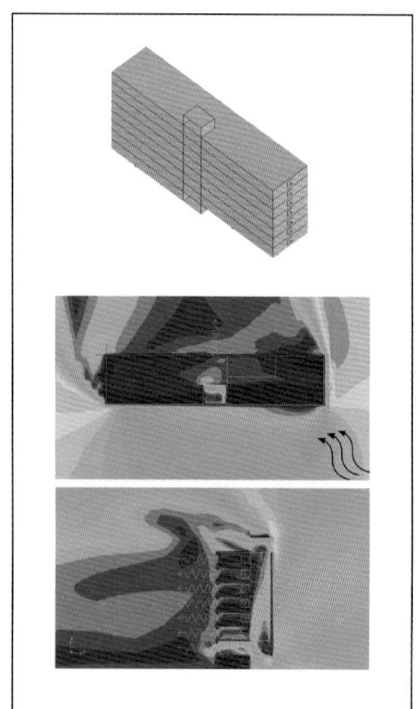

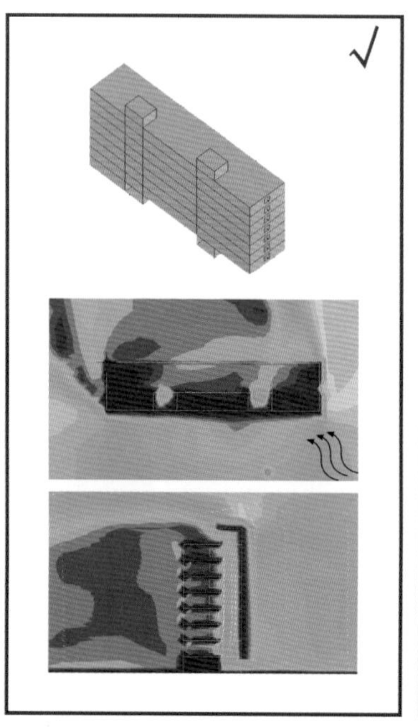

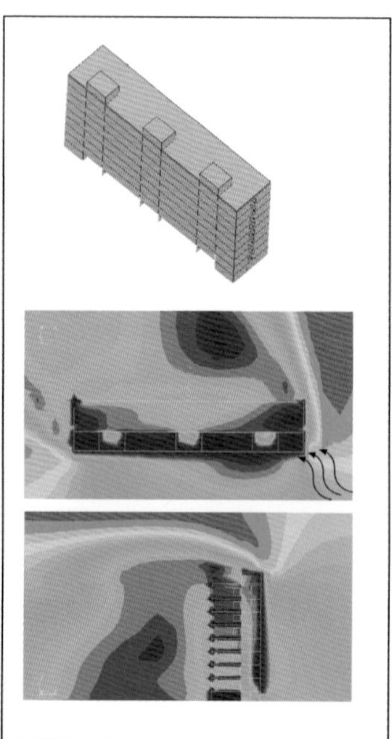

基于自然通风与天然采光分析的方案比较。

Scheme comparison based on natural ventilation and daylight analysis.

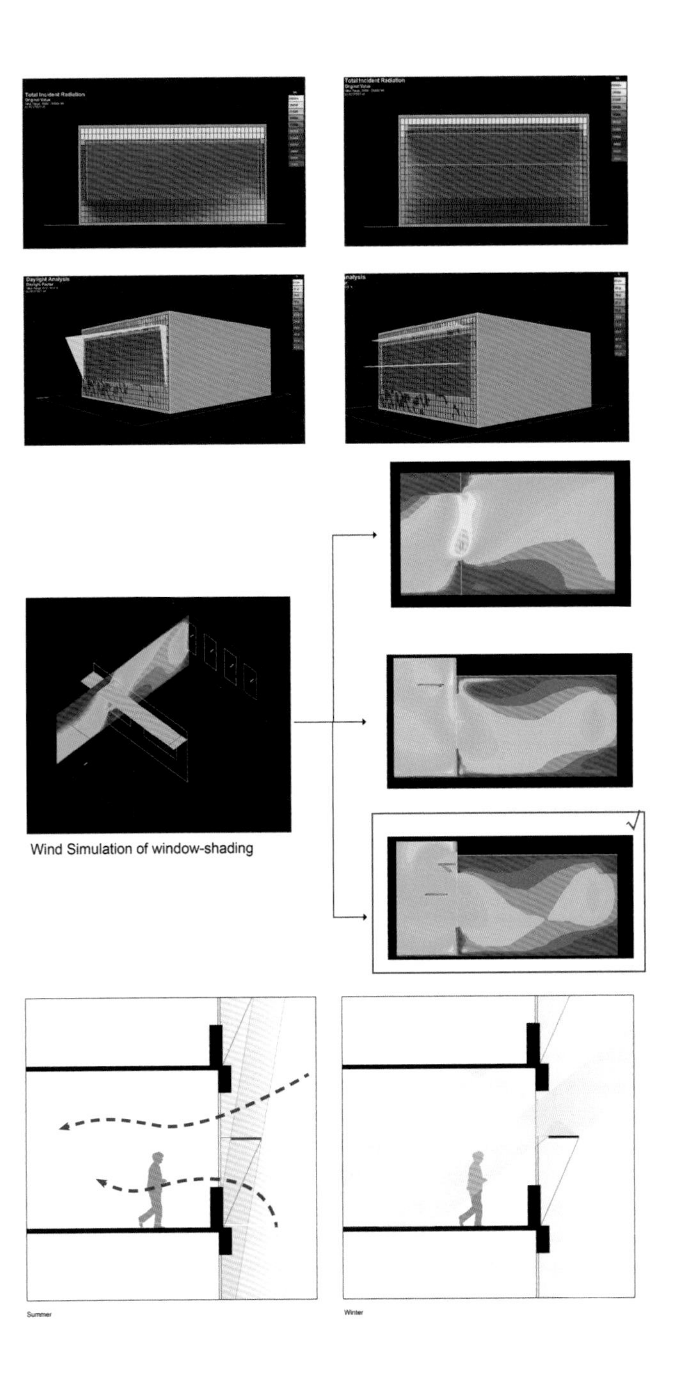

Wind Simulation of window-shading

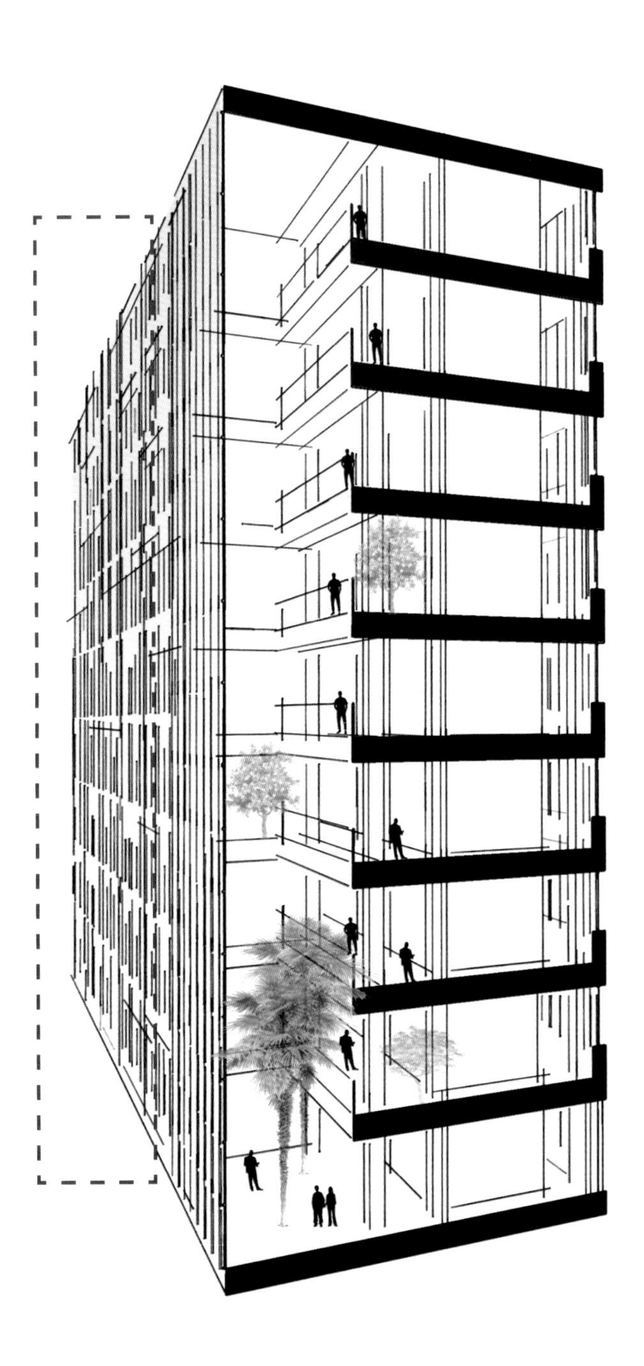

Before

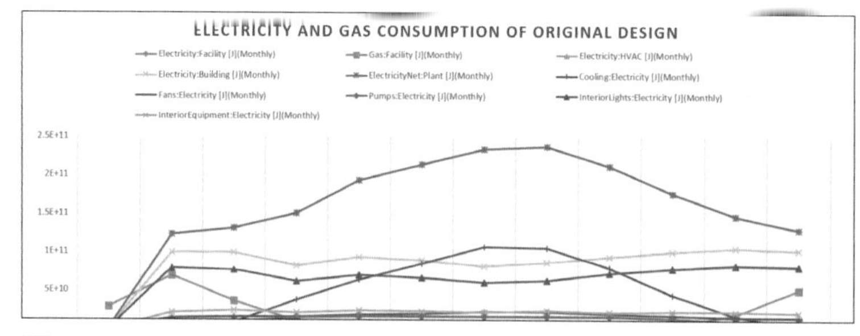

After

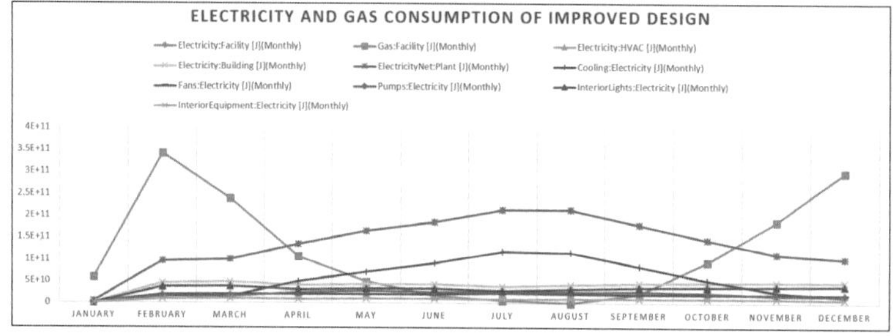

基于自然通风与建筑能耗的优化设计方案。

Optimized design scheme based on natural ventilation and building energy consumption analysis.

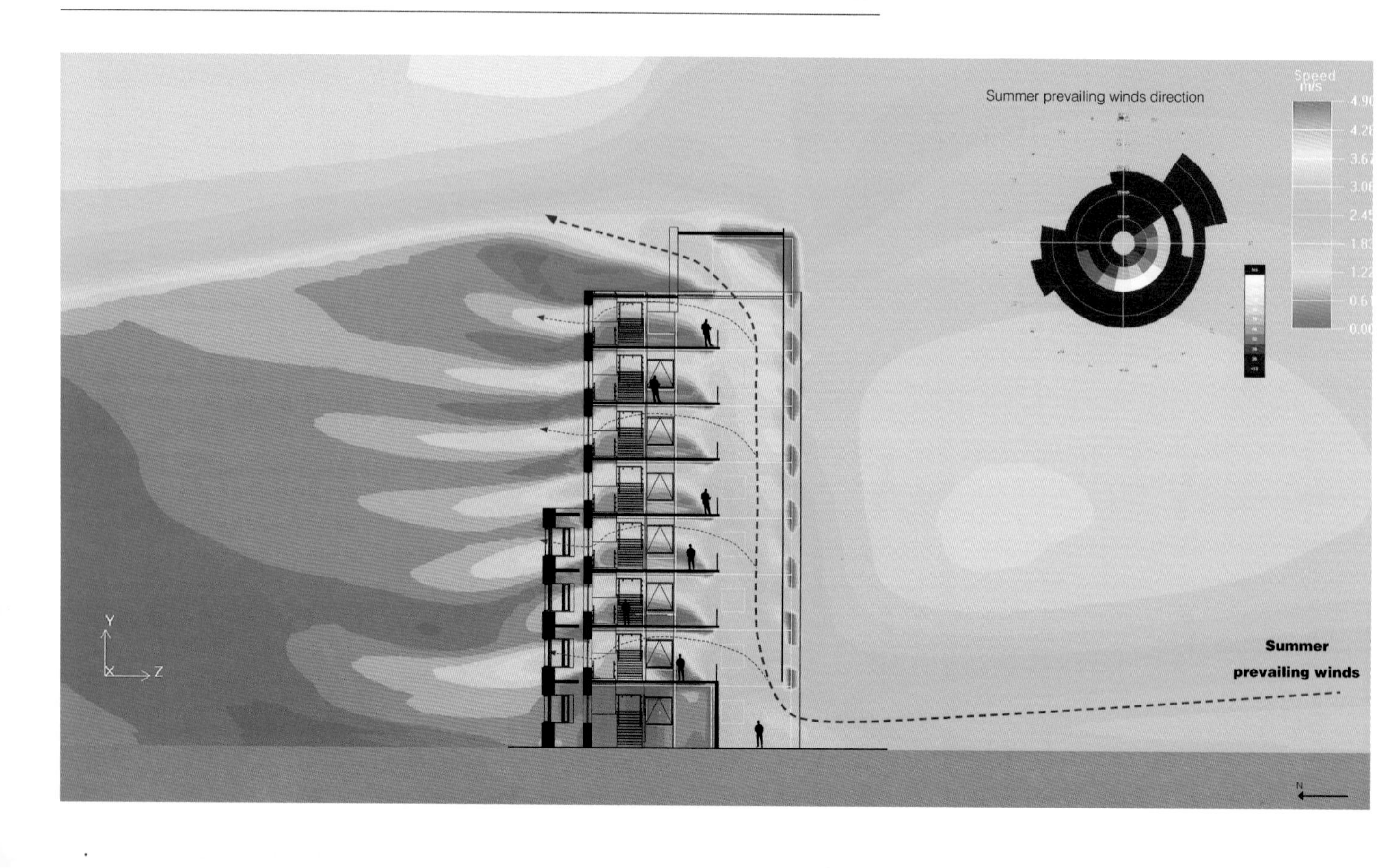

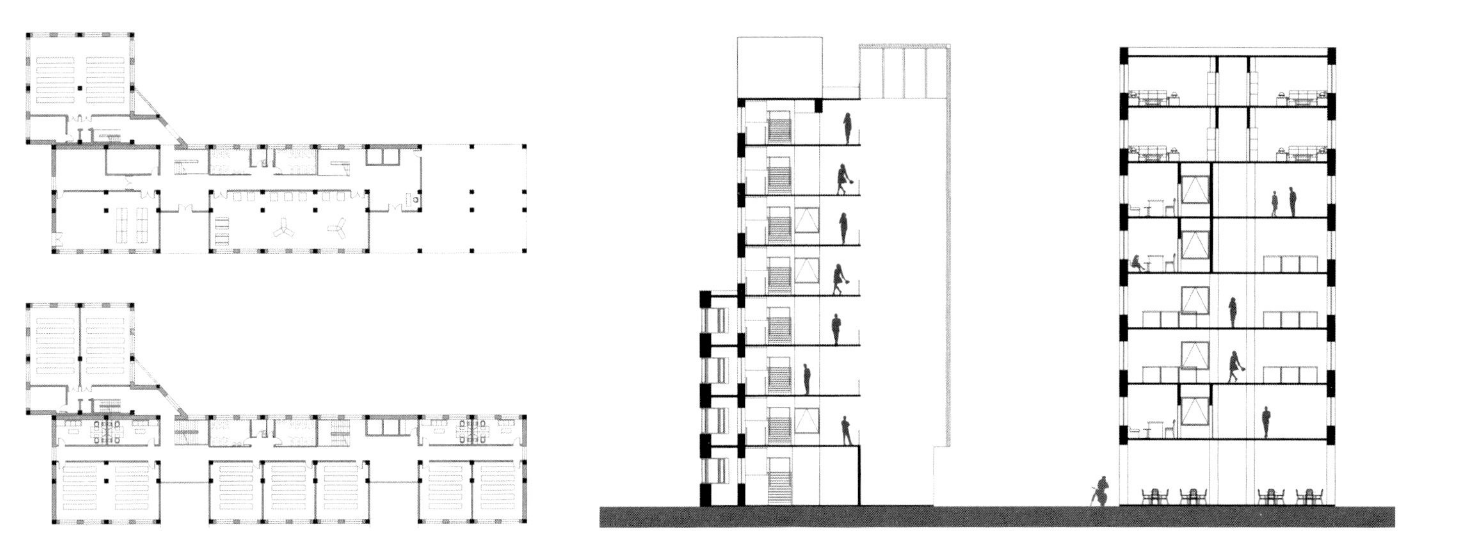

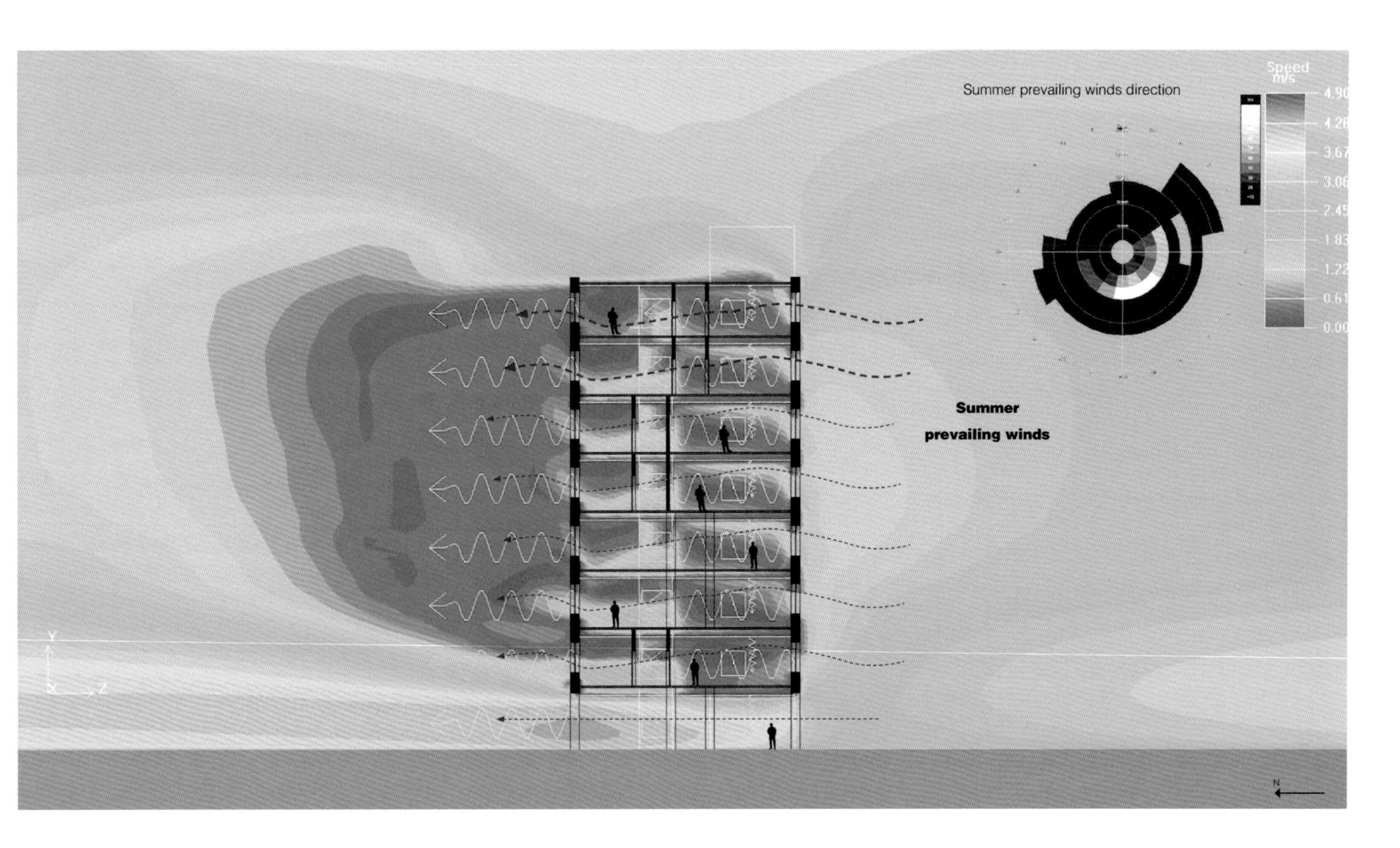
Summer prevailing winds direction
Speed m/s
4.90
4.28
3.67
3.06
2.45
1.83
1.22
0.61
0.00
Summer
prevailing winds
Y
X
Z
N

研究生学生作品索引
Index of Works by Graduate Students

基本建筑建构研究
Constructional Design

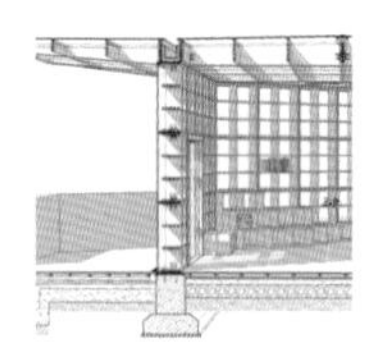

孙雅贤、岳海旭

徐天驹、张海宁

城市空间设计 · 层与界
From Volumn To Space·Multillayer & Interface

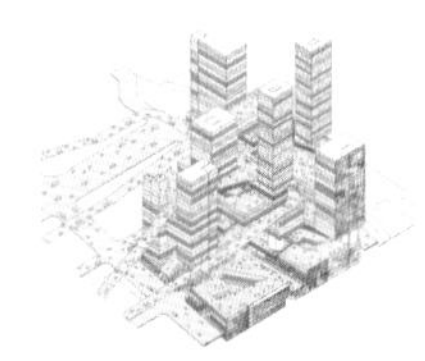

姚梦、刘文沛、徐晏

许文韬、刘晨、尤逸尘

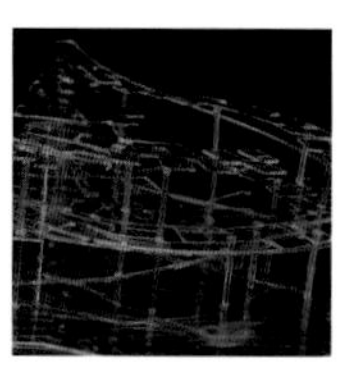

焦宏斌、单泓景、蒋西亚

电影建筑学
Cinematic Atchitecture

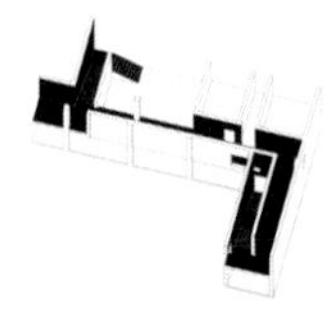

《对一个楼梯的空间注记》
吴书其、林陈、车俊颖、王政、刘思彤

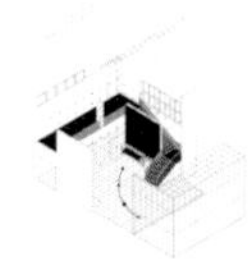

《格网》
程斌、陈晓敏、夏侯蓉、胡任元

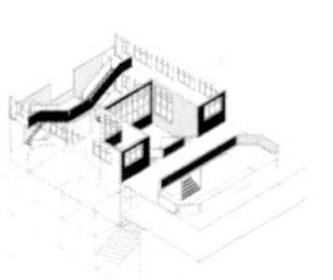

《层叠》
刘芮、陈博宇、陈凌杰、宁凯

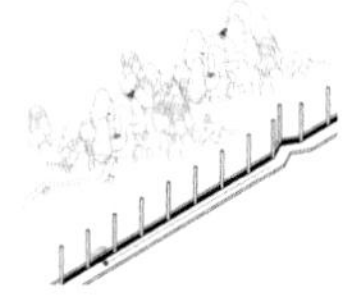

Parallax
吴书其、林陈、车俊颖、王政、刘思彤

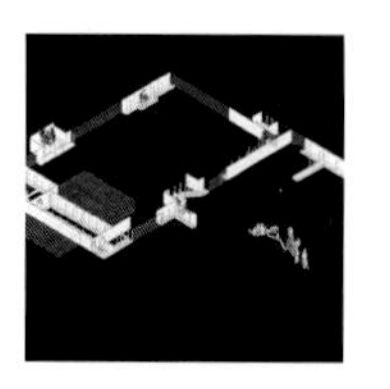

《呓园》
许文韬、张海宁、杨悦、彭蕊寒、韩书园

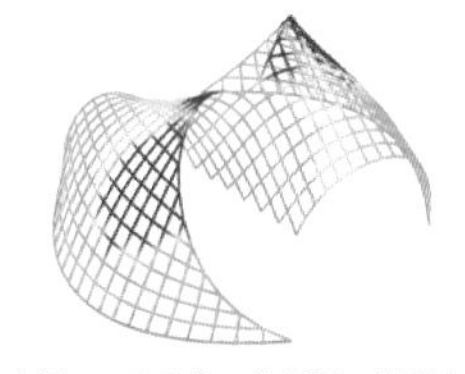

陈修远、吴昇奕、徐思恒、张明杰

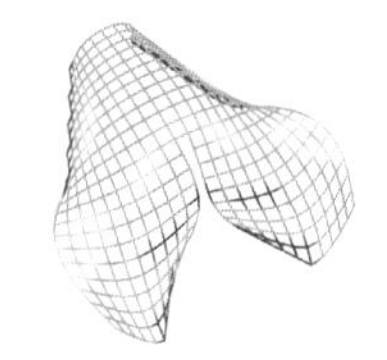

陈凌杰、孙雅贤、谭健、徐晏

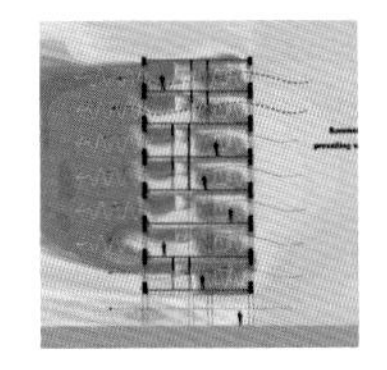

刘晨、张明杰、许文韬、刘宇、谭健

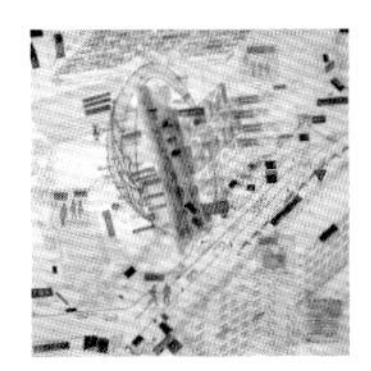

程斌、查新彧、林治

陈博宇、王曙光、刘芮

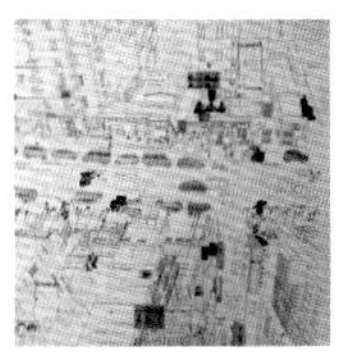

胡任元、张海宁、张强

姚梦、陈小敏、顾一蝶、刘文沛

杨悦、刘思彤、黄广伟

建筑设计课程
ARCHITECTURAL DESIGN COURSES

本科一年级

设计基础（一）

· 季鹏

课程类型：必修

学时学分：64 学时／ 2 学分

Undergraduate Program 1st Year

BASIC DESIGN 1·JI Peng

Type: Required Course

Study Period and Credits: 64 hours/2 credits

教学目标

提升学生感知美、捕捉美和创造美的能力。

研究主题

1. 看到世界的美。
2. 看到形式的美。
3. 理解空间的规则。
4. 理解使用的规则。

教学内容

1. 研究对象的素描表达、黑白灰归纳、拼贴表现和陶土烧造。
2. 针对 Zoom in 概念的训练与纸构成训练。
3. 建筑摄影。
4. 观念与材料的关系。

Training Objective

Improve ability of students to perceive, capture, and create beauty.

Research Subject

1. See beauty of the world.
2. See beauty of forms.
3. Understand rules of space.
4. Understand rules of use.

Teaching Content

1. Sketch presentation, black-white-grey induction, collage expression and pottery making for research objects.
2. Training on the concept of "Zoom in" and paper construction.
3. Architectural photography.
4. Relations between ideas and materials.

本科一年级

设计基础（二）

· 鲁安东　丁沃沃　唐莲

课程类型：必修

学时学分：64 学时／ 2 学分

Undergraduate Program 1st Year

BASIC DESIGN 2 · LU Andong, DING Wowo, TANG Lian

Type: Required Course

Study Period and Credits: 64 hours/2 credits

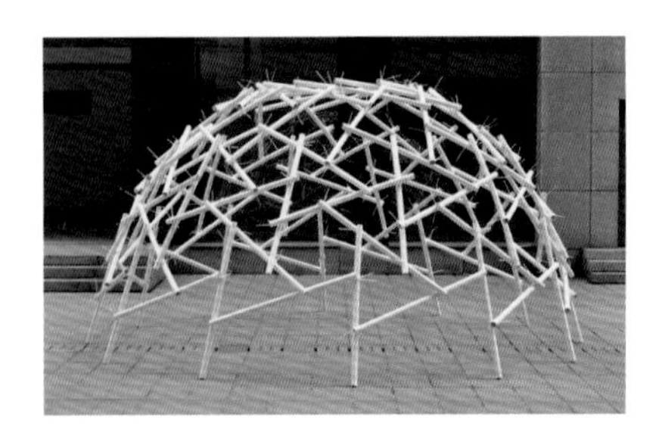

动作—空间分析

通过影像记录人在空间中的动作，挑选关键帧进行动作、尺度、几何与感知分析。目的在于：使学生初步认识身体、尺度与环境的相互影响；学会观察并理解场地；初步认识形式与背后规则的关系；学会发现日常使用中的问题并解决问题；学会使用分析图交流构思。

折纸空间

折叠纸板创造空间，用轴测图、拼贴图再现空间。目的在于：使学生初步掌握二维到三维的转化，初步认识图和空间的再现关系；利用单一材料围合复杂空间，初步认识结构与空间的关系；学会用分析图进行表述。

互承的艺术

运用互承结构原理，搭建人能进入的覆盖空间。目的在于：初步理解结构知识对于构筑空间的意义；在搭建过程中初步建立材料、节点、造价等概念；强化场地意识（包括朝向、环境、流线等）；学会用分析图进行表述。

Action-space Analysis

Record human's actions in space with video and select key frames to conduct analysis on actions, dimensions, geometry and perception. It aims to enable students to preliminarily understand the mutual influence among body, dimensions and environment; learn how to observe and understand the site; preliminarily understand relations between forms and rules behind them; learn how to find out problems in daily use and how to solve these problems; learn how to use analysis charts to exchange ideas.

Folding Space

Fold paperboard to create space, and represent space with axonometric drawing and collage. It aims to enable students to preliminarily grasp the transformation from 2D to 3D, preliminarily understand the reappearance relationship between drawing and space; enclose complex space with single material, and preliminarily understand the relations between structure and space; learn how to express with analysis charts.

Art of Mutually-supporting

Apply principles of mutually-supporting structure to erect an accessible covered space. It aims to preliminarily understand meaning of structure knowledge to construction of space; roughly establish the concepts material, joints, and cost during erection; enhance site awareness (including orientation, environment, streamline, etc.); to learn expression with analysis charts.

本科二年级

建筑设计基础

·刘铨　冷天

课程类型：必修

学时学分：64 学时／ 4 学分

Undergraduate Program 2nd Year

ARCHITECTURAL BASIC DESIGN · LIU Quan, LENG Tian

Type: Required Course

Study Period and Credits: 64 hours/4 credits

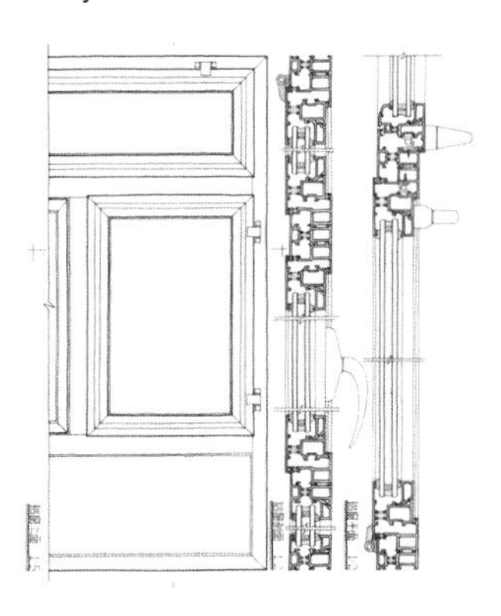

课题内容

认知与表达

教学目标

本课程是建筑学专业本科生的专业通识基础课程。本课程的任务主要是一方面让新生从专业的角度认知与实体建筑相关的基本知识，如主要建筑构件与材料、基本构造原理、空间尺度、建筑环境等知识；另一方面通过学习运用建筑学的专业表达方法来更好地掌握这些建筑基本知识，为今后深入的专业学习奠定基础。

教学内容

1. 认知建筑

（1）立面局部测绘

（2）建筑平、剖面测绘

（3）建筑构件测绘

2. 认知图示

（1）单体建筑图示认知

（2）建筑构件图示认知

3. 认知环境

（1）街道空间认知

（2）建筑肌理类型认知

（3）地形与植被认知

4. 专业建筑表达

（1）建筑图纸表达

（2）建筑模型表达

（3）环境分析图表达

Subject Content

Cognition and presentation

Training Objective

The course is the basic course of general professional knowledge for undergraduates of architecture. Task of the course is, on the one hand, allow students to cognize basic knowledge about physical building from an professional perspective, such as main building members and materials, basic constructional principles, spatial dimensions, and building environment; and on the other hand, to better master such basic architectural knowledge through studying application of professional presentation method of architecture, and to lay down solid foundation for future in-depth study of professional knowledge.

Teaching Content

1. Cognizing building

(1) Surveying and drawing of partial elevation

(2) Surveying and drawing plans, profiles of building

(3) Surveying and drawing building members

2. Cognizing drawings

(1) Cognition to drawings of individual building

(2) Cognition to drawings of building members

3.Cognizing environment

(1) Cognition to street space

(2) Cognition to types of building texture

(3) Cognition to terrain and vegetation

4. Professional architectural presentation

(1) Presentation with architectural drawings

(2) Presentation with architectural models

(3) Presentation with environmental analysis charts

本科二年级

建筑设计（一）：小型公共建筑设计

·刘铨　冷天　王丹丹

课程类型：必修

学时学分：64 学时／ 4 学分

Undergraduate Program 2nd Year

ARCHITECTURAL DESIGN 1: SMALL PUBLIC BUILDING · LIU Quan, LENG Tian,WANG Dandan

Type: Required Course

Study Period and Credits: 64 hours/4 credits

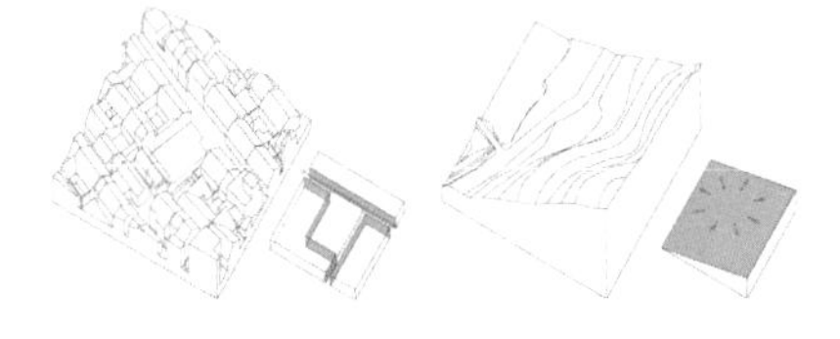

课题内容

老城古玩店设计

教学目标

根据建筑功与外部空间的功能、流线需要，运用水平与垂直构件进行基本的空间限定及造型训练。

研究主题

学习利用水平构件与垂直构件形成并组织室内外空间；学习出入口设置与简单的流线组织；掌握基本的人体活动与空间、构件的尺度关系；注意室内空间与相邻建筑、街道或庭院之间的空间及界面关系。

教学内容

在城市历史街区中设计具有商业与展示功能的小型建筑单体。在地块内现存建筑物已被拆除的假定前提下，独立完成一次完整的小型建筑单体方案的设计任务。建筑功能为具有展示性质的古玩商店，包括展示与销售、顾客接待空间以及值班室、洗手间、垂直交通等辅助功能空间，建筑层数二层，建筑面积 160~200 m^2，不得开挖地下室获取额外的使用空间；楼梯、门窗等建筑构件的设计需满足基本建筑规范的要求。

Subject Content

Design of an Antique Shop in the Historic Urban District

Training Objective

To conduct space limit and modeling exercise with the horizontal and vertical components based on the functions of the building and external space as well as the streamline requirements.

Study Topic

Learn to model and organize the indoor and outdoor spaces with the horizontal and vertical components; study entrances and exits settings and simple streamline organization; understand the scale relation between the basic human activities and the spaces as well as components; and pay attention to the space and interface relations between the indoor space and the adjacent buildings, streets or courtyards.

Teaching Content

Design a small building with commercial and display functions in an urban historic block. Given the precondition that the existing buildings on the plot have been demolished, independently complete a small building design plan. The building is a two-story antique shop with the display function and a building area of 160~200m^2, containing spaces for display, sales and customer reception as well as the auxiliary functional spaces such as the duty room, rest room and vertical communication facilities. Extra using space cannot be acquired by digging a basement; design of the building components such as stairs, doors and windows should meet the requirements of the basic building codes.

本科二年级

建筑设计（二）：小型公共建筑设计

·刘铨　冷天　王丹丹

课程类型：必修

学时学分：64 学时／4 学分

Undergraduate Program 2nd Year

ARCHITECTURAL DESIGN 2: SMALL PUBLIC BUILDING · LIU Quan, LENG Tian,WANG Dandan

Type: Required Course

Study Period and Credits: 64 hours/4 credits

课题内容

风景区茶室设计

教学目标

本课程训练使用空间形式语言进行设计操作。在设定上，教案希望学生进一步认识到“空间形式语言”作为设计工具，去解读、应对场地环境与功能需求并推动设计进展的作用。

研究主题

场地与界面、功能与空间、空间的组织、尺度与感知。

教学内容

1. 本次设计场地是紫金山风景区内的一处坡地，要求学生从场地水平向界面的限定来考虑设计建筑的形体、布局及其最终的空间视觉感受。

2. 本次设计的建筑功能设定为风景区茶室，建筑面积 300m^2，建筑层数 2 层，其中包括入口门厅、容纳 60 人的大厅，容纳 30 人的 2~4 人雅座若干，以及必要的辅助空间。同时，必须进行室外场地的设计。

3. 基于地形的形式化表达，寻求适应建筑空间需求的场地改造形式进行高差的处理，把对地形的理解与公共建筑的基本功能空间组织模式及形态塑造联系起来。

4. 在空间形式处理中注意通过图示表达理解空间构成要素与人的空间体验的关系，主要包括尺度感和围合感，注意景观朝向问题。

Subject Content

Design of a Tea Shop in the Scenic Spot

Training Objective

This course aims to train design operations with the language of spatial forms. In term of course setting, the teaching plan is intended to allow students to further understand the role of “the language of spatial forms”, as a design tool, in interpreting, dealing with field environment and function demand, and promoting design progress.

Research Subject

Site and interface, function and space, organization of space, dimensions and perception.

Teaching Content

1. The design site is a sloped field within the Zijinshan scenic spot, and it requires students to consider the profile, layout and final visual effect of space of the designed building based on limits of horizontal interface of the site.

2. The building to be design will functions as a teahouse at scenic spot, covers an area of 300 m^2, consists of two floors, including an entrance hallway, a public hall with seating capacity of 60 people, several private rooms with seating capacity of 2~4 people which can serve 30 guests in aggregate, as well as other necessary auxiliary spaces. Meanwhile, design for outdoor space must also be conducted.

3. On basis of formalized expression of topography, try to deal with height difference in the form of field modification by accommodating demand of building space, and link understanding of topography to the space organization patterns and form modelling for basic functions of public building.

4. Pay attention to the understanding of relations between factors of space formation and spatial experience of human through graphic expression in the process of handling spatial forms.

本科三年级

建筑设计研究（三）：小型公共建筑设计

·周凌　童滋雨　窦平平

课程类型：必修

学时学分：72 学时／4 学分

Undergraduate Program 3rd Year

ARCHITECTURAL DESIGN 3: SMALL PUBLIC BUILDING · ZHOU Ling, TONG Ziyu,DOU Pingping

Type: Required Course

Study Period and Credits: 72 hours/4 credits

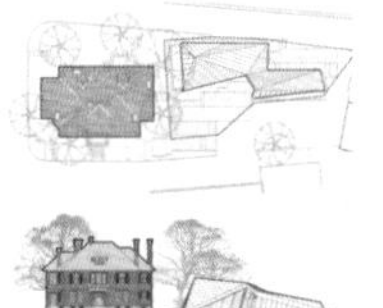

课题内容

赛珍珠纪念馆扩建

教学目标

此课程训练最基本的建造问题，使学生在学习设计的初始阶段就知道房子如何造起来，深入认识形成建筑的基本条件：结构、材料、构造原理及其应用方法，同时课程也面对场地、环境和功能问题。训练核心是结构、材料、场地。在学习组织功能与场地同时，强化认识建筑结构、建筑构件、建筑围护等实体要素。

文脉：充分考虑校园环境、历史建筑、校园围墙以及现有绿化，需与环境取得良好关系。

退让：建筑基底与投影不可超出红线范围。若与主体或相邻建筑连接，需满足防火规范。

边界：建筑与环境之间的界面协调，各户之间界面协调。基底分隔物（围墙或绿化等）不超出用地红线。

户外空间：扩建部分保持一定的户外空间，户外空间可在地下。

地下空间：充分利用地下空间。

教学内容

基地内地面最大可建面积约 100 m^2，地下可建面积 200~300 m^2，总建筑面积约 400~500 m^2，建筑地上 1 层，限高 6 m，地下层数层高不限，展示区域 200~300 m^2，导游处 10 m^2，纪念品部 30 m^2，茶餐厅 60 m^2，厨房区域 >10 m^2，另包括门厅与交通，卫生间。

Subject Content

Extension of Pearl S.Buck's House in Nanjing

Training Objective

This course trains the students to solve the basic construction of architecture. Students should learn how to build an architecture at the very beginning of their studying, understand the basic aspects of architectures: the principles and applications of structure, material and construction. The course also includes the problem of site, enviroment and function. The keypoints of the course include site, structure and material. Students should strengthen the understanding of physical elements including structures, components and façades while learning to organize the function and site.

Context: The enviroment, historical building, the edge of the campus and the green belt around the site should be taken into consideration. The expansion is expected to have a good relationship with the surroudings.

Retreat Distance: The new architecture can't beyond the red line. Fire protection rule should be complied.

Boundary: Both the boundary between different buildings and between building and environment should be harmonized.

Open Space: Open space should be considered, which permitted to be placed underground.

Underground Space: Underground space should be well used.

Teaching Content

The maximum ground can be used in the base area is about 100 m^2, while underground construction area is about 200~300 m^2, and the total floor area of architecture should be about 400~500 m^2.The architecture should be 1 floor above the groud lower than 6 m.The underground levels have no limitation.Exhibition area: 200~300 m^2, information center: 10 m^2, shop: 30 m^2, coffee bar: 60 m^2, kitchen: >10 m^2, lobby & walking space, toilet.

本科三年级

建筑设计（四）：中型公共建筑设计

·周凌　童滋雨　窦平平

课程类型：必修

学时学分：72 学时／ 4 学分

Undergraduate Program 3rd Year

ARCHITECTURAL DESIGN 4: PUBLIC BUILDING · ZHOU Ling, TONG Ziyu,DOU Pingping

Type: Required Course

Study Period and Credits: 72 hours/4 credits

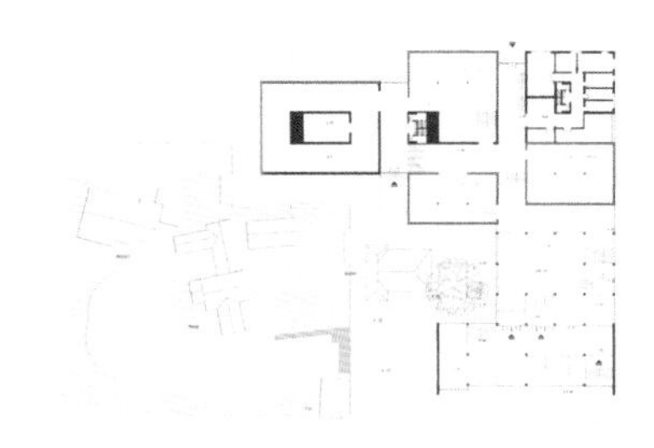

课题内容

傅抱石美术馆

教学目标

课程主题是“空间”(Space) 和“流线”(Circulation)，学习建筑空间组织的技巧和方法，训练空间的效果与表达。空间问题是建筑学的基本问题，课题基于复杂空间组织的训练和学习。从空间秩序入手，安排大空间与小空间，独立空间与重复空间，区分公共与私密空间、服务与被服务空间、开放与封闭空间。同时，空间的串联形成序列，需要有效组织流线，并且充分考虑人在空间中的行为，空间感受。以模型为手段，辅助推敲。设计分阶段体积、空间、结构、围合等，最终形成一个完整的设计。

教学内容

1. 空间组织原则：空间组织要有明确特征，有明确意图，概念要清楚。并且满足功能合理、环境协调、流线便捷的要求。注意三种空间：聚散空间（门厅、出入口、走廊）；序列空间（单元空间）；贯通空间（平面和剖面上均需要贯通，内外贯通、左右前后贯通、上下贯通）。

2. 空间类型：展览陈列空间：3000 m^2；收藏保管空间：700 m^2；技术、研究空间：240 m^2；行政办公空间：150 m^2；休闲服务空间：300 m^2；其他空间：传达室：10 m^2；设备房：200 m^2；交通门厅面积自定；客用、货用电梯各一部；室外停车场。

建筑面积不超过 5000 m^2，高度不超过 18 m。

Subject Content

Fu Baoshi Art Gallery

Training Objective

The course topic is space and circulation, learning techniques and methods of architectural space organization, and training on effect and presentation of space. Space is a basic issue for architecture, and the course is based on training and study on organization of complex spaces. Start from spatial order to arrange large and small spaces, independent space and overlapped space, and to distinguish public and private spaces, serving and served spaces, open and closed spaces. Meanwhile, linking spaces to shape sequence requires effective organizational circulation, as well as full consideration of behaviors, spatial feeling of people in space. Use models as means to assist deliberation. Design stages include volume, space, structure, and enclosure, and shape a complete design in the end.

Teaching Content

1. Space organization principles: Space organization requires distinctive characteristics, explicit intention, and clear concepts. It shall also meet the requirements of reasonable functions, coordinated environment, and convenient circulation. Attention shall be paid to three types of space: converging and diverging space (hallway, entrance and exit, corridor); sequence space (unit space); connecting space (connecting spaces are required on plans and profiles, internal-and-external connection, left-and-right, front-and-rear connections, up-and-down connection).

2. Space type: exhibition & showcase space: 3000 m^2; collection & storage space: 700 m^2; technical, research space: 240 m^2; administrative office space: 150 m^2; leisure service space: 300 m^2; other spaces: janitor's room: 10 m^2; equipment room: 200 m^2; area of traffic hallway to be determined; one guest elevator and one goods elevator; outdoor parking lot.

The floor area shall not exceed 5000 m^2, and the height shall not exceed 18 m.

本科三年级

建筑设计（五六）：大型公共建筑设计

·华晓宁　钟华颖　王铠

课程类型：必修

学时学分：144 学时／ 8 学分

Undergraduate Program 3rd Year

ARCHITECTURAL DESIGN 5-6: COMPLEX BUILDING · HUA Xiaoning, ZHONG Huaying, WANG Kai

Type: Required Course

Study Period and Credits: 144 hours/8 credits

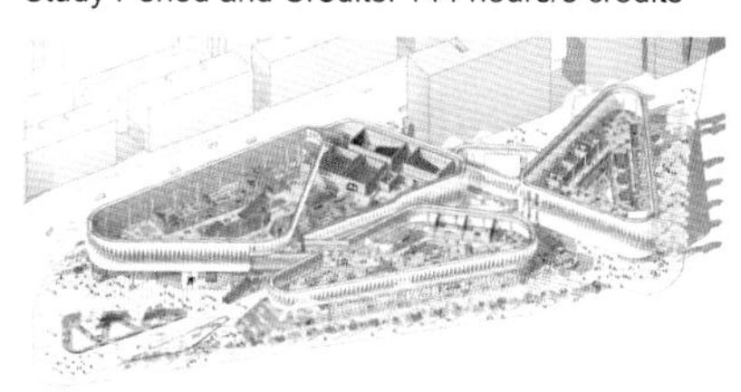

课题内容

城市建筑——社区商业中心＋活动中心设计

研究主题

实与空：关注城市中建筑实体与空间的相互定义、相互显现，将以往习惯上对于建筑本体的过度关注拓展到对于“之间”的空间的关注。

内与外：进一步突破“自身”与“他者”之间的界限，将个体建筑的空间与城市空间视为一个连续统，建筑空间即城市空间的延续，城市空间亦即建筑空间的拓展，两者时刻在对话、互动和融合。

层与流：不同类型的人和物的行为与流动是所有城市与建筑空间的基本框架，当代大都市中不同的流线在不同的高度上层叠交织，构成一个复杂的多维城市。必须首先关注行为和流线的组织，由此才生发出空间的系统和形态。

轴与界：城市纷繁复杂的形态表象之后隐含着秩序和控制性，并将成为新的形态介入。

教学内容

在用地上布置社区商业中心（约 15000 m^2）、社区文体活动中心（约 8000 m^2），并生成相应的城市外部公共空间。

Subject Content

Urban Buildings-Design of Community Business Center and Activity Center

Research Subject

Entity and space: Pay attention to mutual definition, mutual representation of architectural entity and space in cities, and extend traditional excessive attention to the building itself to the space “among them”.

Interior and exterior: Further break through the boundary between “self” and “others”, and consider space of individual buildings and urban space as a continuum.

Stack and flow: Behaviors and flows of different types of people and objects are the basic framework of all urban and building spaces, and a complex multi-dimensional city is formed by stacking up and interweaving of different flow lines at different altitudes in modern metropolis. We must pay attention to organization of behaviors and flow lines first, and then can generate system and morphology of space.

Axis and boundaries: Order and control are concealed behind the morphologic appearance of complexity of cities, which will be involved as new forms.

Teaching Content

Lay out community commercial center (about 15000 m^2) and community recreational and sports activities center (about 8000 m^2) on the land, and generate associated outdoor urban public spaces.

本科四年级

建筑设计（七）高层建筑设计

吉国华　胡友培　尹航

课程类型：必修

学时学分：72 学时／ 4 学分

Undergraduate Program 4th Year

ARCHITECTURAL DESIGN 7: HIGH-RISING BUILDING · JI Guohua, HU Youpei, YIN Hang

Type: Required Course

Study Period and Credits: 72 hours/4 credits

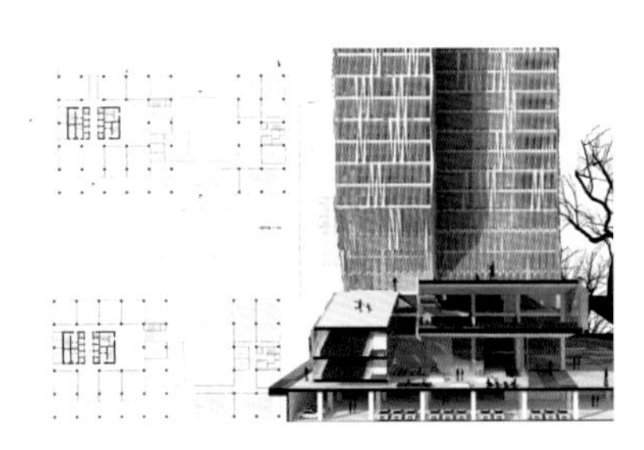

课题内容

高层办公楼设计

教学目标

高层办公建筑设计涉及城市、空间、形体、结构、设备、材料、消防等方面内容，是一项较复杂与综合的任务。本课题采取贴近真实实践的视角，教学重点与目标是帮助学生理解、消化以上涉及各方面知识，提高综合运用并创造性解决问题的技能。

教学内容

建筑容积率 ≤ 5.6，建筑限高 ≤ 100 m，裙房高度 ≤ 24 m，建筑密度 ≤ 40%。需规划合理流线，避免形成交通拥堵。

高层部分为办公楼，设计应兼顾各种办公空间形式。裙房设置会议中心，须设置 400 人报告厅 1 个，200 人报告厅 2 个，100 人报告厅 4 个，其他各种会议形式的中小型会议室若干，以及咖啡茶室、休息厅、服务用房等。会议中心应可独立对外使用。机动车交通独立设置，不得进入校内道路系统。地下部分主要为车库和设备用房。

Subject Content

Design of High-rise Office Building

Training Objective

Design of the high-rise office building is a complicated and comprehensive task, involving city, space, form, structure, equipment, materials and fire control. From a perspective close to the practice, this course focuses on and aims at helping students understand and grasp the knowledge of the above-mentioned aspects and improving their skills of integrated use and creatively solving problems.

Teaching Content

Building plot ratio ≤5.6, building height limit ≤100 m, annex height ≤24 m, building density ≤40%. Reasonable circulation must be planned to avoid traffic jam.

The high-rise part is an office building, so the design must give consideration to various forms of office space.The annex is a conference center, which must include 1 lecture hall of 400 seating capacity, 2 lecture halls of 200 seating capacity, 4 lecture halls of 100 seating capacity, several small and medium-sized meeting rooms for various meetings, and coffee & tea room, lobby, and service quarter. The conference center shall be separated and available for external usage. Motor vehicle traffic routes must be separated from road system within the campus. The underground part is mainly for garage and equipment room.

本科四年级

建筑设计（八）城市设计

吉国华　胡友培　尹航

课程类型：必修

学时学分：72 学时／ 4 学分

Undergraduate Program 4th Year

ARCHITECTURAL DESIGN 8: URBAN DESIGN · JI Guohua,HU Youpei, YIN Hang

Type: Required Course

Study Period and Credits: 72 hours/4 credits

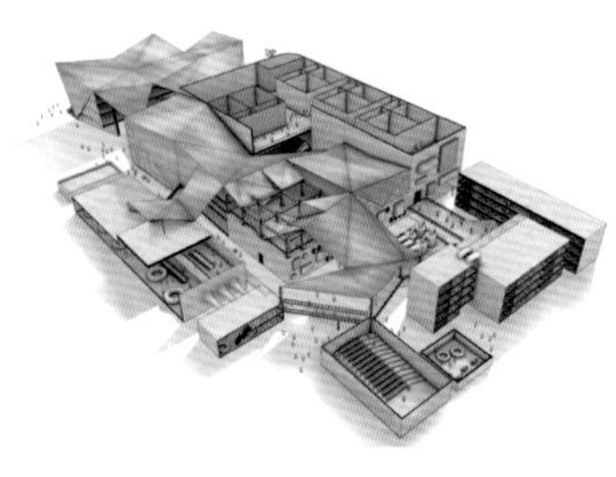

课题内容

南京碑亭巷地块旧城更新城市设计

教学目标

1. 着重训练城市空间场所的创造能力，通过体验认知城市公共开放空间与城市日常生活场所的关联，运用景观环境的策略创造城市空间的特征。

2. 熟练掌握城市设计的方法，熟悉从宏观整体层面处理不同尺度空间的能力，并有效地进行图纸表达。

3. 理解城市更新的概念和价值；通过分析理解城市交通、城市设施在城市体系中的作用。

4. 多人小组合伙，培养团队合作意识和分工协作的工作方式。

教学内容

1. 设计地块位于南京市玄武区，总用地约为 6.20hm^2。地块内国民大会堂旧址、国立美术陈列馆旧址和北侧邮政所大楼可保留，其余地块均需进行更新。地块周边有丰富的博物馆、民国建筑等文化资源，设计应对周边文化环境起到进一步提升作用。地块周边用地情况复杂，设计中需考虑与周边现状的相互影响。

2. 本次设计的总容积率指标为 2.0~2.5，建筑退让、日照等均按相关法规执行。

3. 碑亭巷、石婆婆庵需保留，碑亭巷和太平北路之间现状道路可根据设计进行位置或线形调整。

4. 地下空间除满足单一地块建筑配建的停车需求外，应综合考虑地上、地下城市一体化设计与综合开发。

Subject Content

Urban Design for Old Town Renovation of the Land Parcel of Beiting Lane in Nanjing

Training Objective

1. Emphasize the training on ability of creating urban spatial places, and create features of urban space with the strategy of landscape environment through experiencing and perceiving the links between urban public spaces and urban daily living places.

2. Master methodology of urban design, grasp the ability of handling spaces of different dimension at macro and integral level, and achieve effective representation with drawings.

3. Understand the concept and value of urban renovation; understand the role of urban traffic, urban facilities in the urban system through analysis.

4. Form partnership with several group members to cultivate awareness of teamwork and the working mode of collaboration.

Teaching Content

1.Land parcel of the design is located in Xuanwu District, Nanjing, covering an area of 6.20hm^2 approximately. Site of the former National Assembly Hall, site of the former National Art Gallery and the post office building at north side may be retained, other parts of the land parcel need to be renovated. There are abundant cultural resources such as museums and buildings constructed in the period of the Republic of China around the land parcel, so the design should further improve the cultural environment around it. Land use conditions around the land parcel is very complicated, so mutual influence with surrounding existing conditions must be taken into account in the design.

2.Gross plot ratio of the design is 2.0~2.5, and building setback and sunlight value shall comply with relevant laws and regulations.

3.Beiting Lane and Shipopo Nunnery are to be retained, and location and route of the existing road between Beiting Land and North Taiping Road may be adjusted according to the design.

4.For underground space, besides meeting the associated parking demand of buildings on the single land parcel, overall consideration shall be taken for integrated design and comprehensive development of urban spaces above and under the ground.

本科四年级
毕业设计
· 赵辰
课程类型：必修
学时学分：1 学期 /0.75 学分

Undergraduate Program 4th Year
THESIS PROJECT · ZHAO Chen
Type: Required course
Study Period and Credits: 1 term /0.75 credit

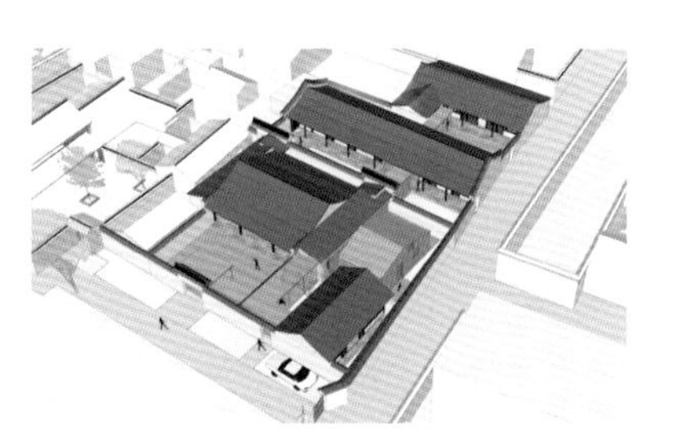

课题内容

北村复兴规划设计项目暨南方传统建筑室内舒适度改善专项

教学目标

掌握建筑设计基本的技能与知识（测绘、建模、调研、分析），并能对特定的地域和历史建筑进行深入的设计研究（内容策划、建筑结构、构造），根据社会发展的需求，提出改造和创造的可能。在选定的村落之现状研究的基础上，进行村落景观空间的整体规划。选择相关重点区域与建筑，进行专项的建筑设计。

研究主题

北村位于福建东北部山区的生态、人文资源都十分优秀的地域，因交通缘故而长期受到发展的限制。随着高速公路与铁路的发展而将迅速得到新的发展机遇，而新的发展必须适应新的低碳、生态型的发展模式，并同时充分保护与发挥地域历史文化与生态的优势。这正是新时代对建筑与乡村规划设计的一种挑战。

Subject Content

Rebirth Project of Beicun and Indoor Comfort Improvement in the Traditional Building of Southern China

Training Objective

Master basic skills and knowledge for architectural design (surveying and drawing, modeling, investigation, analysis), and be cable of conducting in-depth design research (content planning, building structure, construction) on specific region and historical buildings, and put forward possibility of improvement and creation according to the demand of social development. Complete overall planning on landscape space of village based on research of existing conditions of the selected village. And conduct specialized architectural design for selected key areas and buildings.

Research Subject

Beicun Village is located at a mountain area in northeast Fujian, a region with excellent ecological and cultural resources, and its development has been restricted for a long time due to poor traffic. New development opportunities came rapidly along with the development of expressway and railway, while new development must accommodate the new low-carbon, ecological development mode, and give adequate protection and full play to the advantage of regional historical culture and ecology. It is the challenge brought by the new era to architecture and village planning design.

本科四年级
毕业设计
· 窦平平
课程类型：必修
学时学分：1 学期 /0.75 学分

Undergraduate Program 4th Year
THESIS PROJECT · DOU Pingping
Type: Required course
Study Period and Credits: 1 term /0.75 credit

课题内容

外部空间设计：场所记忆与日常性的培育

研究主题

项目基地为南京城西南的一个街区，现规划为低层院落式高级住宅区，本项毕业设计课题为这片街区的步行街巷和节点空间进行外部空间和都市景观建筑设计。该街区整体延续了原先老城的低层高密度城市肌理，街巷结构基本保留了原先的“八爪金龙”街巷格局。

该项目的院落住宅单体户型设计现由李兴钢工作室主持设计，方案已初步完成，尚在调整中。本项目外部空间设计需与住宅单体设计部分进行协调和合作。

教学内容

1. 通过街巷结构的空间建构，在城市尺度上重构场所记忆；
2. 通过住宅与城市、住宅与街巷之间的界面设计，营造公共与私密领域之间的多重性，丰富城市界面和街巷界面的空间层次；
3. 通过景观建筑设计，在微观尺度上关怀日常行为和外部空间使用。

Subject Content

Exterior Space Design: Cultivating the Spirit and Domesticity of a Place

Research Subject

Base of the project is a block in southwest Nanjing City, which is now planned as a low-rise courtyard-style high-grade residential district. This graduation design task is completing architectural design for exterior space and urban landscape for pedestrian lanes and node spaces of the block. The overall block inherited the low-rise high-density city texture of the old town, and its street structure basically kept the original street pattern of “eight-claw golden dragon”.

Individual house layout of the courtyard residence of the project is now undertaken by Li Xinggang Studio, the plan is completeld preliminarily, being under adjustment currently. Exterior space design of the project needs coordination and cooperation with individual residential house design.

Teaching Content

1. Reconstruct spatial memory on city dimensions through spatial construction of street structure;
2. Create multiplicity between public and private areas, and enrich spatial layers between city interface and street interface through design of interfaces between residence and city, residence and street;
3. Take care of daily behavior and use of exterior space on micro dimensions through landscape architectural design.

本科四年级

毕业设计：建筑技术科学专门化设计

· 华晓宁　郜智

课程类型：必修

学时学分：1 学期 /0.75 学分

Undergraduate Program 4th Year

THESIS PROJECT: SPECIALIZED DESIGN OF BUILDING SCIENCE · HUA Xiaoning，GAO Zhi

Type: Required Coure

Study Period and Credits: 1 term /0.75 credit

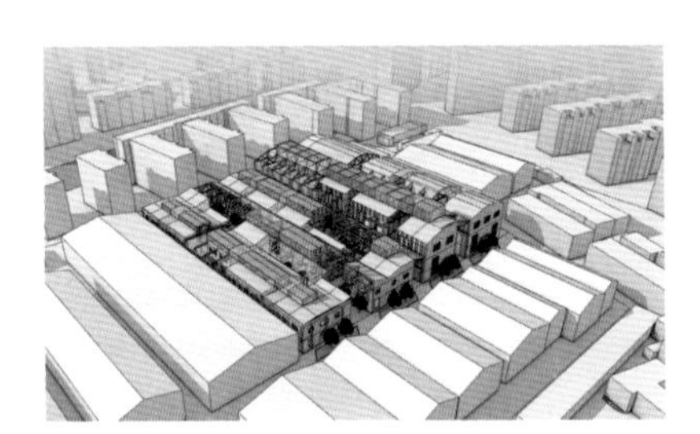

课题内容

江苏省农科院招待所改造设计

教学目标

掌握整合于建筑设计策略中的绿色建筑相关设计理念、知识和方法。

教学内容

旧建筑的改造和再利用是目前我国建筑领域方兴未艾的一个发展方向，其中将节能改造与建筑形态再造和空间再生进行深入整合是一个尤为重要的主题。江苏省农科院招待所一栋 1980 年代的宿舍建筑，在未来将被改造成为江苏省农科院国际学术交流中心的一部分，其使用者及其空间配置要求都将发生彻底的改变。为此，需要在建筑形态和空间的改造中深入整合相关节能技术，以使旧建筑既能满足新的使用要求，同时又能有效降低能耗。

Subject Content

Renovation of the Guesthouse Building in JAAS

Training Objectives

To acquire green building-related design concepts, knowledge and methods integrated into architectural design strategy.

Teaching Content

Improvement and reuse of old buildings is an emerging development direction in China's architecture field, and in-depth combination of energy conservation improvement, reconstruction of building forms and regeneration of space is an especially important topic. The Guesthouse of Jiangsu Academy of Agricultural Sciences is a dormitory building of 1980s, it will be reconstructed as part of the International Academic Exchange Center of JAAS in the future, the users and spatial arrangement will be changed completely. Therefore, related energy conservation technologies must be deeply integrated in the improvement of building forms and spaces, so that the old building can not only meet new use requirement, but also reduce energy consumption effectively.

本科四年级

毕业设计：数字化建造技术专门研究

· 钟华颖

课程类型：必修

学时学分：1 学期 /0.75 学分

Undergraduate Program 4th Year

THESIS PROJECT: SPECIALIZED STUDY ON DIGITAL CONSTRUCTION TECHNOLOGIES · ZHONG Huaying

Type: Required Coure

Study Period and Credits: 1 term /0.75 credit

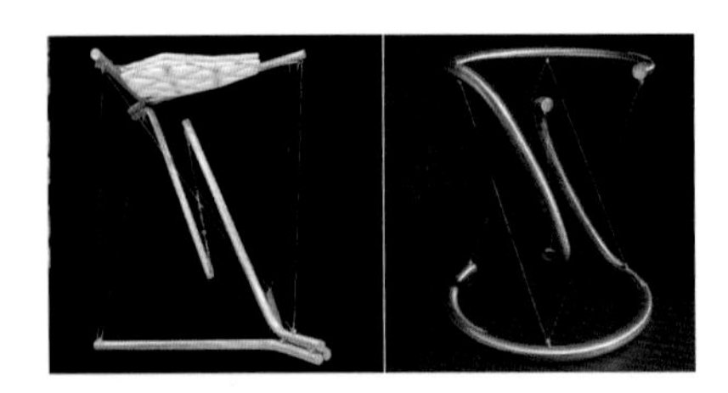

课题内容

基于张拉整体结构的数字化搭建

研究主题

富勒 (R.B.Fuller) 发明的张拉整体结构，是一组不连续的受压构件与一套连续的受拉单元组成的自支承、自应力的空间网格结构。这一结构系统最大限度地利用了材料和截面的特性，用最少的材料营造了最大的跨度和空间。传统的张拉整体结构受压的刚性杆位于系统中央，占据了使用空间，长期以来较少用于建筑设计，多用于桥梁、雕塑。改进的张拉整体结构将结构杆与索置于周边，围合出使用空间，具有了应用于建筑设计的可能性。本次设计旨在探索这一系统的建筑应用。

教学内容

本设计拟在校园内选择某处闲置的场地，搭建一个具有一定使用空间的构筑物，如休息亭、车篷、雨篷等。设计要求基于张拉整体结构，使用数字化手段设计结构形态及外围护系统，并最终完成全尺寸模型的搭建。

Subject Content

Digital Design and Fabrication Based on Tensegrity System

Research Subject

Invented by R. B. Fuller, tensegrity structure is a type of self-supporting, self-stress spatial grid structure consisting of a group of discontinuous compression members and a set of continuous tensile elements. This structure maximizes the utilization of materials and sectional characteristics, and creates the largest span and space with minimum materials. The compressive rigid members of traditional tensegrity structure are located at the center of the system, occupying useful space, which is rarely used in architectural design for a long period, mostly used in bridges and sculptures. In the improved tensegrity structure, structural members and cables are located at perimeter, and a useful space is enclosed, so it has the possibility of being applied in architectural design. The design aims to explore the application of the system in buildings.

Teaching Content

The design aims to construct a structure with certain useful space at a vacant land selected in the campus, such as resting pavilion, car shed, and canopy. The design must be based on tensegrity structure, use digital means to design structural forms and external envelope system, and complete the construction of a full-sized model.

本科四年级

毕业设计：数字化建造技术专门研究

· 童滋雨

课程类型：必修

学时学分：1 学期 /0.75 学分

Senior Year

THESIS PROJECT: SPECIALIZED STUDY ON DIGITAL CONSTRUCTION TECHNOLOGIES · TONG Ziyu

Type: Required Coure

Study Period and Credits: 1 term /0.75 credit

课题内容

基于自承结构体系的数字化搭建

研究主题

互承结构是一种构造独特的空间结构体系，其特点是每根构件都被相邻构件支承，同时自身又支承相邻构件，即构件之间以一种递推的方式相互支承，在几何上和结构上均无主次层次可言，形成一种富于韵律的建筑美感。互承结构通过构件之间的相互支承解决了荷载传递的问题，不仅构造简单，而且可以利用小尺寸构件实现大跨度结构。互承结构体系的这些特点使其在建筑设计中具有特殊的意义和价值。

教学内容

本设计拟在校园内选择某处闲置的场地，搭建一个具有一定跨度的构筑物，如休息亭、车棚、雨篷等。设计要求基于互承结构，使用数字化手段进行模拟和受力验证，并最终完成全尺寸模型的搭建。

Subject Content

Digital Design and Fabrication Based on Reciprocal Frame System

Research Subject

Mutual-supporting structure is a type of spatial structural system with unique construction, it is characterized that each member is supported by adjacent members, and in the meantime, the member supports other adjacent members, that is, members are mutually supported in a recursive way, they are not distinguished as primary or secondary ones in terms of geometry and structure, and shape a kind of architectural beauty enriched with rhythms. Mutual-supporting structure solves the problem of load transmission through mutual support among members, which not only features simple structure, but also can realize large-span structure with small-sized members. Such characteristics of mutual-supporting structure system have particular significance and value in architectural design.

Teaching Content

The design aims to construct a structure with certain useful space at a vacant land selected in the campus, such as resting pavilion, car shed, and canopy. The design must be based on mutual-supporting structure, use digital means to simulate and test verification of the force, and complete the construction of a full-sized model.

本科四年级

毕业设计

· 周凌　胡友培　冯金龙

课程类型：必修

学时学分：1 学期 /0.75 学分

Senior Year

THESIS PROJECT · ZHOU Ling，HU Youpei，FENG Jinlong

Type: Required Coure

Study Period and Credits: 1 term /0.75 credit

课题内容

微筑：预制混凝土构建设与研究

教学目标

结合设计竞赛的要求，通过学习预制混凝土装配建造体系的相关知识，探索预制混凝土装配体系的设计可能与诗意表达。

研究主题

预制混凝土建造体系的多重建筑学意义——建构的美学意义、设计方法论的意义、建构的社会文化意义、人居环境意义。

教学内容

以一个微型建筑为设计对象，通过现场调研和预制工厂考场了解预制混凝土的材料特性，认识预制混凝土建造体系在施工上的精确性、快速性和可控性；展开有关结构体系与形式、预制技术与施工、材料性能与建构表现方面的设计研究。在形成初步设计方案基础上，设计详细的构造大样，进行贴近真实建造层面的深化设计。

Subject Content

Mini-building: Construction and Research of Precast Concrete Structures

Training Objective

Explore design possibility and poetic presentation of precast concrete assembly system through study of knowledge related to precast concrete assembly and construction system, and in combination with the requirements of design contest.

Research Subject

Multiple architectural meanings of precast concrete construction system – aesthetic meaning of construction, meaning of design methodology, social and cultural meanings of construction, and meaning of human settlements.

Teaching Content

By making a mini building as design object, get to know material characteristics of precast concrete through field research and precast plant investigation, understand accuracy, rapidity and controllability of precast concrete construction system in term of construction; conduct design research on structural system and forms, precast technology and construction, material performance and constructional expression. Design detailed constructional details, and conduct in-depth design close to real construction level based on the formed preliminary design scheme.

研究生一年级

建筑设计研究（一）：基本设计

· 傅筱

课程类型：必修

学时学分：40 学时／2 学分

Graduate Program 1st Year

DESIGN STUDIO 1: BASIC DESIGN · FU Xiao

Type: Required Course

Study Period and Credits: 40 hours/2 credits

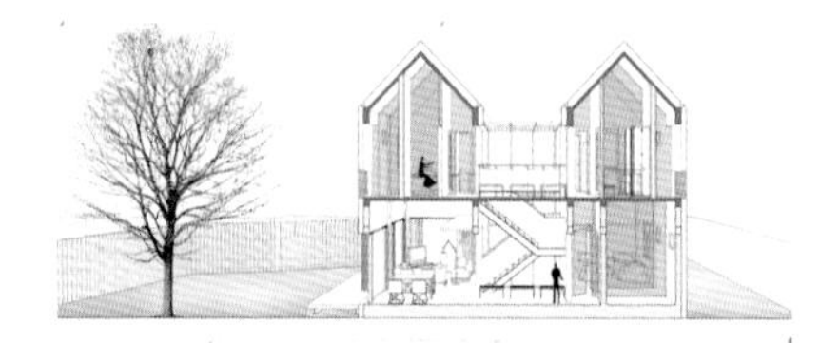

课题内容

宅基地住宅设计

教学目标

课程从“场地、空间、功能、经济性”等建筑的基本问题出发，通过宅基地住宅设计，训练学生对建筑逻辑性的认知，并让学生理解有品质的设计是以基本问题为基础的。

研究主题

基本问题与设计品质

教学内容

在 A、B 两块在宅基地内任选一块进行住宅设计。

Subject Content

Homestead Housing Design

Training Objective

The course starts from fundamental issues of architecture such as “site, space, function, and economical efficiency”, aims to train students to cognize architectural logics, and allow them to understand that quality design is based on such fundamental issues.

Research Subject

Fundamental issues and design quality

Teaching Content

Select one from two homesteads A and B and conduct housing design.

研究生一年级

建筑设计研究（一）：基本设计

· 张雷

课程类型：必修

学时学分：40 学时／2 学分

Graduate Program 1st Year

DESIGN STUDIO 1: BASIC DESIGN · ZHANG Lei

Type: Required Course

Study Period and Credits: 40 hours/2 credits

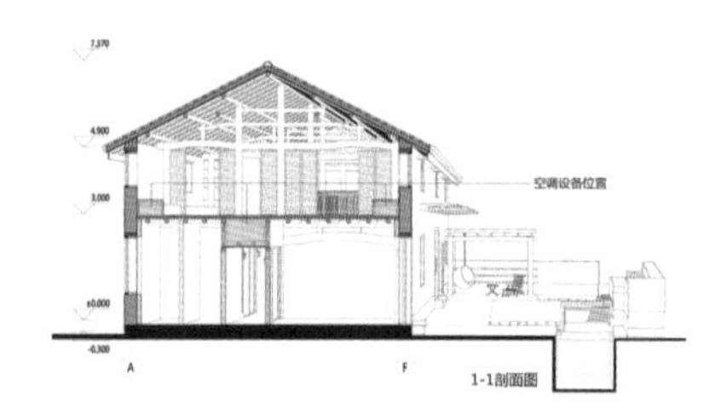

课题内容

传统乡村聚落复兴研究

教学目标

课程从“环境”“空间”“场所”与“建造”等基本的建筑问题出发，对乡村聚落肌理、建筑类型及其生活方式进行分析研究，通过功能置换后的空间再利用，从建筑与基地、空间与活动、材料与实施等关系入手，强化设计问题的分析，强调准确的专业性表达。通过设计训练，达到对地域文化以及建筑设计过程与方法的基本认识与理解。

研究主题

乡土聚落／民居类型／空间再利用／建筑更新／建造逻辑

教学内容

对选定的乡村聚落进行调研，研究功能置换和整修改造的方法和策略，促进乡村传统村落的复兴。

Subject Content

Research on Revitalization of Traditional Rural Settlements

Training Objective

The course starts from fundamental architectural issues such as “environment”, “space”, “site” and “construction”, conducts analysis and research on texture, building type and life style of rural settlements, and aims to strengthen analysis on design problems and emphasize accurate professional presentation on basis of relations between building and base, space and activity, and material and execution, through reutilization of space after function replacement. And obtain basic cognition and understanding on regional culture as well as architectural design process and methodology through design training.

Research Subject

Rural settlement / types of folk house / reutilization of space / building renovation / constructional logic

Teaching Content

Conduct investigation and research on selected rural settlement, study methodology and strategy of function replacement and renovation and improvement, and promote revitalization of traditional rural villages.

研究生一年级

建筑设计研究（一）：概念设计

· 冯路

课程类型：必修

学时学分：40 学时／2 学分

Graduate Program 1st Year

DESIGN STUDIO 1: CONCEPTUAL DESIGN · FENG Lu

Type: Required Course

Study Period and Credits: 40 hours/2 credits

课题内容

半透明性 I

研究主题

以设计为研究，探索“半透明性”在建筑学上的意义和呈现方式。

建筑学长久以来一直追求自身的自治性。它常常通过一种形式和内容的意义结构来完成。相对于“清晰性”而言，这个课程企图尝试探索一种空间的“半透明性”，通过这一切入点重新寻求建筑学和个人之间关系的可能性。

教学内容

以南京大学建筑系所在的蒙民伟楼为设计基地和研究对象。选取蒙民伟楼的某一处或某一类型空间局部作为分析和研究的对象，理解其内在的“清晰性”。在此基础上，进行概念设计，使其空间具有“半透明性”。设计研究将通过绘图和影像等方式来表达。每个设计工作组由三位同学组成。

Subject Content

Translucence I

Research Subject

Conduct research through design to explore architectural meaning and presentation mode of “translucence”.

The architecture has long been pursuing its own autonomy. It is often realized through meaning structure of form and content. Therefore, for the purpose of the “clearness”, this course tries to explore a type of “translucence” of space, and to rediscover the possibility of relations between architecture and individuals from this entry point.

Teaching Content

Apply the Mong Man Wai Building of the Architecture Department, Nanjing University as the design base and research object. Select one space or one type of space of the Mong Man Wai Building as object of analysis and research, and try to understand its inherent “clearness”. Conduct conceptual design on such basis to render the space with “translucence”. Present the design research through ways such as drawing and image. Each design working group will consist of three students.

研究生一年级

建筑设计研究（一）：概念设计

· 鲁安东

课程类型：必修

学时学分：40 学时／2 学分

Graduate Program 1st Year

DESIGN STUDIO 1: CONCEPTUAL DESIGN · LU Andong

Type: Required Course

Study Period and Credits: 40 hours/2 credits

课题内容

扩散：空间营造的流动逻辑

研究主题

以 1920—1930 年代蚕种场建筑为切入点，关注其对于自然元素的创造性的运用，研究光和风从表皮到空间的扩散，探索基于环境因素的空间设计方法，尝试通过设计教学实现研究与设计之间的相互反馈。

教学内容

探讨从立面和剖面入手的设计。空间不再由建筑平面进行组织，而是从 外向内遵循自然的流动逻辑进行营造。

作为设计对象的镇江高资蚕种场， 始建于 1921 年。设计需提出完整的策划方案和单体设计。

Subject Content

Diffusion: Flowing Logic of Space Creation

Research Subject

By taking the silkworm seed station of 1920—1930s as the entry point, focus on the application of the creativity of natural elements, research the diffusion of light and wind from surface to space, explore space design method based on environmental factors, and try to realize mutual feedback between research and design through design teaching.

Teaching Content

Explore design starting from elevation and profile. Spaces are not organized with building plan anymore, but created with the flowing logic following the nature from outside to inside.

As design object, the Zhenjiang Gaozi Silkworm Seed Station was built in 1921. The design must provide complete planning scheme and individual entity design.

研究生一年级

建筑设计研究（二）：建构设计

· 傅筱

课程类型：必修

学时学分：40 学时／2 学分

Graduate Program 1st Year

DESIGN STUDIO 2 : CONSTRUCTIONAL DESIGN · FU Xiao

Type: Required Course

Study Period and Credits: 40 hours/2 credits

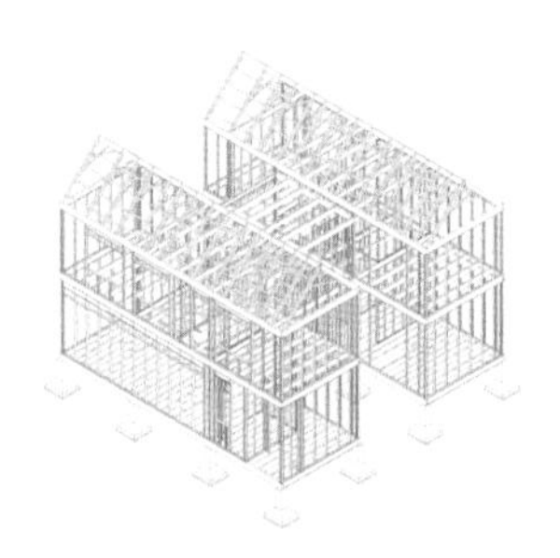

课题内容

"基础设计"的深化与发展

教学目标

1. 训练学生对设计概念与构造设计关联性的认知：

（1）处理好节点的基本工程技术问题

A. 对自然力的抵抗与利用：保温、防水、遮阳……

B. 构造与施工：复杂问题简单化、建造方便性、误差问题……

（2）根据设计概念研究建造材料的选用和节点设计，在满足基本工程技术的前提下，重点研究超越基本工程技术问题的构造设计表达。

2. 训练学生对一个"完整空间形态"建造的认知：

所谓完整空间形态是指包括室外场地、外墙（从屋顶到基础）、设备、装修所构成的空间形态，通过完整空间形态设计，让学生建立构造设计的整体意识。

教学内容

以基本计案例为基础，选取 1~2 个主要设计概念进行深化设计，要求达到节点大样表达深度。

Subject Content

Deepin Development of the Projects of Basic Design

Training Objective

1. Make students understanding the relationship between design concept and detail of construction.

1) Solute the basic technical problems of detail.

a. Resistance and application of natural elements: heat preservation, water resistance, sun shading, etc.

b. Construction and building: simplification of complicated problem, conveniences of construction, deviation, etc.

2) Study the selection of materials and design of details according to the design concepts, especially the presentation of details beyond the basic technical problem.

2. Make students understanding the construction of an integrated space.

The integrated space includes the envelope (from roof to foundation), facilities and decoration. Students will get the integrated awareness of detail design via the design practice of integrated space.

Teaching Content

Deepening design one or two concepts come from the projects of basic design to the detail level.

研究生一年级

建筑设计研究（二）：建构设计

· 郭屹民

课程类型：必修

学时学分：40 学时／2 学分

Graduate Program 1st Year

DESIGN STUDIO 2:CONSTRUCTIONAL DESIGN · GUO Yimin

Type: Required Course

Study Period and Credits: 40 hours/2 credits

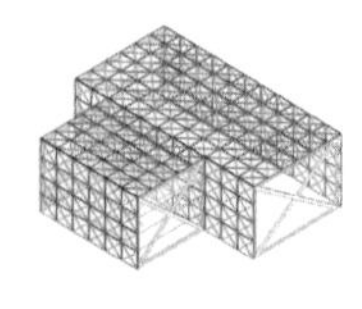

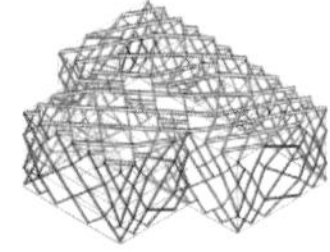

课题内容

南京大学北苑体育活动中心设计

教学目标

1. 掌握结构设计基础知识，并会进行结构分析和结构设计。

2. 了解结构的材料与建造，并会通过材料和建造进行建筑构造设计。

3. 了解结构设计与功能、场地的关系，并会进行与功能相关的建筑设计。

教学内容

北苑校区现有体育馆设施陈旧，功能单一，已无法满足现有师生更加丰富多彩的健身锻炼需求及社交活动要求。现拟在原体育馆建筑用地范围内改扩建体育活动中心。

Subject Content

Design of Beiyuan Sports Center of Nanjing University

Training Objective

1.To grasp the basic knowledge of structural design, and to be able to conduct structural analysis and design.

2.To understand the materials and construction of the structure, and to be able to conduct building structure design with the materials and through construction.

3.To understand the relationships between structural design and the functions as well as sites, and to be able to conduct function-related architectural design.

Teaching Content

There has an existing gymnasium with obsolete facilities and unitary function at Beiyuan campus. It cannot meet the demand of faculty and students on more colorful fitness exercises and social activities. Now it plans to improve and expand the sport center within the land area of the existing gymnasium.

研究生一年级

建筑设计研究（二）：城市设计

· 丁沃沃

课程类型：必修

学时学分：40 学时／ 2 学分

Graduate Program 1st Year

DESIGN STUDIO 2: URBAN DESIGN · DING Wowo

Type: Required Course

Study Period and Credits: 40 hours/2 credits

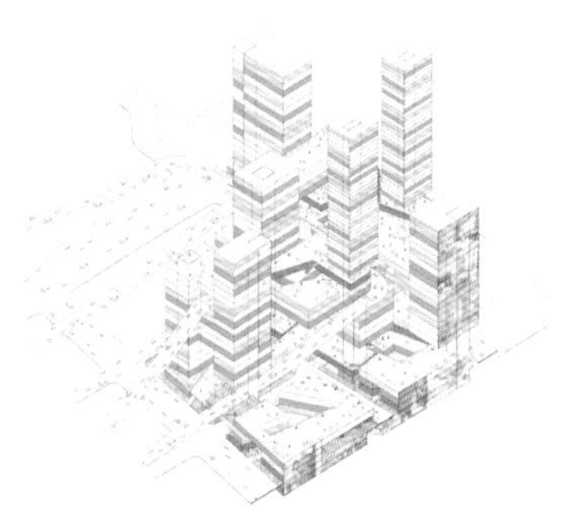

课题内容

城市空间设计

教学目标

本课题拟通过高密度城市建筑与城市空间设计实验，了解城市建筑角色和城市物质空间的本质，初步掌握城市建筑与城市空间塑造之间的关系。此外，通过城市空间设计练习进一步深化空间设计的技能和方法。

研究主题

以“层”“茎”“界”与“空”作为操作载体，通过设计操作，探索高密度城市的构成潜力，认知在高密度城市环境中城市建筑设计元素的转化。

教学内容

1. 以南京下关沿江地块作为设计实验的场所，通过场地地理条件认知、案例研习、设计分析和设计实验，探讨高效城市空间的建构方法和内涵。

2. 通过设计拓展图示技能与表现方法，从空间意向出发建构城市物质空间的层与界面，由“空间意向”“扫描”出城市空间的“层”与“界”。

Subject Content

Urban Space Design

Training Objective

This course aims to understand the nature of roles of urban buildings and urban physical space, and preliminarily grasp relations between urban building and urban space creation through experimental design of high-density urban buildings and urban space. In addition, it aims to further deepen skills and methodology of space design through practices of urban space design.

Research Subject

Explore composition potential of high-density city, and understand transformation of urban building design elements in high-density city environment through design operations by using multilayer, rhizome, interface and opening as operation carriers.

Teaching Content

1. Apply a riverside plot at Xiaguan, Nanjing as the site for the design experiment. Explore tectonic methodology and implication of efficient urban space through cognition to geological field conditions, case study, design analysis, and design experiment.

2. Develop drawing presentation skills and expression method through design, construct multilayer and interface of urban physical space from space image, and “scan” multilayer and interface of urban space from “space image”.

研究生一年级

建筑设计研究（二）：城市设计

· 鲁安东

课程类型：必修

学时学分：40 学时／ 2 学分

Graduate Program 1st Year

DESIGN STUDIO 2 : URBAN DESIGN · LU Andong

Type: Required Course

Study Period and Credits: 40 hours/2 credits

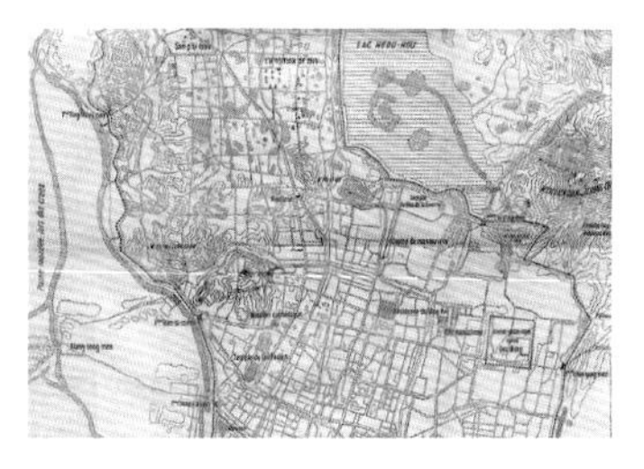

课题内容

湖山之城

教学目标

湖与山是城市空间的边界条件。它们既隔离又聚集着周边的城市空间，并在形态和生态上导致了多样的微观环境。本课程将系统地研究南京城市中的自然山水及其对空间的影响，并选择特征地块进行城市设计。

教学内容

本课程将介绍城市研究的工作方法，帮助设计师来理解、分析和设计城市。本课程将包括三个部分：

1.“系统分析”：我们将通过对环山水地带的详细观察、分析和图示来理解自然山水带来的限制和机会。

2.“生态分析”：我们将使用分析技术来更好地理解城市地块的生态环境，探讨城市发展的多种资源，寻找不同的发展模式，并对生态节点进行设计。

3.“城市设计”：城市设计的目的是重新定义边界条件，在以山水为中心的整体系统中建构新的微观形态和生态。

Subject Content

City of Lakes and Hills

Training Objective

Lakes and hills are boundary conditions of urban space. They isolate and gather urban spaces around, and lead to diversified micro environments morphologically and ecologically. This course will research the natural hills and waters in Nanjing city and their influence on space systematically, and select featured plots to conduct urban design.

Teaching Content

The course will introduce the working method for urban research, and help architects understand, analyze and design the city. The course consists of three parts:

1. “Systematic analysis”: We will understand the restrictions and opportunities brought by natural hills and waters through detailed observation, analysis and drawing presentation of zones surrounded by hills and waters.

2. “Ecological analysis”: We will use analysis techniques to better understand the ecological environment of urban plots, explore various resources for urban development, seek for different development modes, and design ecological nodes.

3. “Urban design”: The urban design aims to redefine boundary conditions, and to construct new micro morphology and ecology in the entire system centering on hills and waters.

建筑理论课程
ARCHITECTURAL THEORY COURSES

本科二年级
建筑导论 • 赵辰等
课程类型：必修
学时/学分：36学时/2学分
Undergraduate Program 2nd Year
INTRODUCTION TO ARCHITECTURE • ZHAO Chen, etc.
Type: Required Course
Study Period and Credits:36 hours / 2 credits

课程内容
1. 建筑学的基本定义
第一讲：建筑与设计/赵辰
第二讲：建筑与城市/丁沃沃
第三讲：建筑与生活/张雷
2. 建筑的基本构成
（1）建筑的物质构成
第四讲：建筑的物质环境/赵辰
第五讲：建筑与节能技术/秦孟昊
第六讲：建筑与生态环境/吴蔚
第七讲：建筑与建造技术/冯金龙
（2）建筑的文化构成
第八讲：建筑与人文、艺术、审美/赵辰
第九讲：建筑与环境景观/华晓宁
第十讲：城市肌体/胡友培
第十一讲：建筑与身体经验/鲁安东
（3）建筑师职业与建筑学术
第十二讲：建筑与表现/赵辰
第十三讲：建筑与几何形态/周凌
第十四讲：建筑与数字技术/吉国华
第十五讲：城市与数字技术/童滋雨
第十六讲：建筑师的职业技能与社会责任/傅筱

Course Content
I Preliminary of architecture
1. Architecture and design / ZHAO Chen
2. Architecture and urbanization / DING Wowo
3. Architecture and life / ZHANG Lei
II Basic attribute of architecture
II-1 Physical attribute
4. Physical environment of architecture / ZHAO Chen
5. Architecture and energy saving / QIN Menghao
6. Architecture and ecological environment / WU Wei
7. Architecture and construction technology / FENG Jinlong
II-2 Cultural attribute
8. Architecture and civilization, arts, aesthetic / ZHAO Chen
9. Architecture and landscaping environment / HUA Xiaoning
10. Urban tissue / HU Youpei
11. Architecture and body / LU Andong
II-3 Architect: profession and academy
12. Architecture and presentation / ZHAO Chen
13. Architecture and geometrical form / ZHOU Ling
14. Architectural and digital technology / JI Guohua
15. Urban and digital technology / TONG Ziyu
16. Architect's professional technique and responsibility / FU Xiao

---

本科三年级
建筑设计基础原理 • 周凌
课程类型：必修
学时/学分：36学时/2学分

Undergraduate Program 3rd Year
BASIC THEORY OF ARCHITECTURAL DESIGN
• ZHOU Ling
Type: Required Course
Study Period and Credits:36 hours / 2 credits

教学目标
本课程是建筑学专业本科生的专业基础理论课程。本课程的任务主要是介绍建筑设计中形式与类型的基本原理。形式原理包含历史上各个时期的设计原则，类型原理讨论不同类型建筑的设计原理。
课程内容
1. 形式与类型概述
2. 古典建筑形式语言
3. 现代建筑形式语言
4. 当代建筑形式语言
5. 类型设计
6. 材料与建造
7. 技术与规范
8. 课程总结
课程要求
1. 讲授大纲的重点内容；
2. 通过分析实例启迪学生的思维，加深学生对有关理论及其应用、工程实例等内容的理解；
3. 通过对实例的讨论，引导学生运用所学的专业理论知识，分析、解决实际问题。

Training Objective
This course is a basic theory course for the undergraduate students of architecture. The main purpose of this course is to introduce the basic principles of the form and type in architectural design. Form theory contains design principles in various periods of history; type theory discusses the design principles of different types of building.
Course Content
1. Overview of forms and types
2. Classical architecture form language
3. Modern architecture form language
4. Contemporary architecture form language
5. Type design
6. Materials and construction
7. Technology and specification
8. Course summary
Course Requirement
1. Teach the key elements of the outline;
2. Enlighten students' thinking and enhance students' understanding of the theories, its applications and project examples through analyzing examples;
3. Guide students using the professional knowledge to analysis and solve practical problems through the discussion of examples.

本科三年级
居住建筑设计与居住区规划原理 • 冷天 刘铨
课程类型：必修
学时/学分：36学时/2学分

Undergraduate Program 3rd Year
THEORY OF HOUSING DESIGN AND RESIDENTIAL PLANNING • LENG Tian, LIU Quan
Type: Required Course
Study Period and Credits:36 hours / 2 credits

课程内容
第一讲：课程概述
第二讲：居住建筑的演变
第三讲：套型空间的设计
第四讲：套型空间的组合与单体设计（一）
第五讲：套型空间的组合与单体设计（二）
第六讲：居住建筑的结构、设备与施工
第七讲：专题讲座：住宅的适应性，支撑体住宅
第八讲：城市规划理论概述
第九讲：现代居住区规划的发展历程
第十讲：居住区的空间组织
第十一讲：居住区的道路交通系统规划与设计
第十二讲：居住区的绿地景观系统规划与设计
第十三讲：居住区公共设施规划、竖向设计与管线综合
第十四讲：专题讲座：住宅产品开发
第十五讲：专题讲座：住宅产品设计实践
第十六讲：课程总结，考试答疑

Course Content
Lect. 1: Introduction of the course
Lect. 2: Development of residential building
Lect. 3: Design of dwelling space
Lect. 4: Dwelling space arrangement and residential building design (1)
Lect. 5: Dwelling space arrangement and residential building design (2)
Lect. 6: Structure, detail, facility and construction of residential buildings
Lect. 7: Adapt ability of residential building, supporting house
Lect. 8: Introduction of the theories of urban planning
Lect. 9: History of modern residential planning
Lect. 10: Organization of residential space
Lect. 11: Traffic system planning and design of residential area
Lect. 12: Landscape planning and design of residential area
Lect. 13: Public facilities and infrastructure system
Lect. 14: Real estate development
Lect. 15: The practice of residential planning and housing design
Lect. 16: Summary, question of the test

研究生一年级
现代建筑设计基础理论 • 张雷
课程类型：必修
学时/学分：18学时/1学分

Graduate Program 1st Year
PRELIMINARIES IN MODERN ARCHITECTURAL DESIGN • ZHANG Lei
Type: Required Course
Study Period and Credits:18 hours/1 credit

课程内容

1. 现代设计思想的演变
2. 基本空间的组织
3. 建筑类型的抽象和还原
4. 材料运用与建造问题
5. 场所的形成及其意义
6. 今天的工作原则与策略

建筑可以被抽象到最基本的空间围合状态来面对它所必须解决的基本的适用问题，用最合理、最直接的空间组织和建造方式去解决问题，以普通材料和通用方法去回应复杂的使用要求，是建筑设计所应该关注的基本原则。

Course Content

1. Transition of the modern thoughts of design
2. Arrangement of basic space
3. Abstraction and reversion of architectural types
4. Material application and constructional issues
5. Formation and significance of sites
6. Nowaday working principles and strategies

Architecture can be abstracted to the most fundamental state of space enclosure, so as to confront all the basic applicable problems which must be resolved. The most reasonable and direct mode of space arrangement and construction shall be applied; ordinary materials and universal methods shall be used as the countermeasures to the complicated application requirement. These are the basic principles on which an architecture design institution shall focus.

研究生一年级
现代建筑设计方法论 • 丁沃沃
课程类型：必修
学时/学分：18学时/1学分

Graduate Program 1st Year
METHODOLOGY OF MODERN ARCHITECTURAL DESIGN • DING Wowo
Type: Required Course
Study Period and Credits:18 hours/1 credit

课程内容

以建筑历史为主线，讨论建筑设计方法演变的动因/理念及其方法论。基于对传统中国建筑和西方古典建筑观念异同的分析，探索方法方面的差异。通过分析建筑形式语言的逻辑关系，讨论建筑形式语言的几何学意义。最后，基于城市形态和城市空间的语境探讨了建筑学自治的意义。

1. 引言
2. 西方建筑学的传统
3. 中国:建筑的意义
4. 历史观与现代性
5. 现代建筑与意识的困境
6. 建筑形式语言的探索
7. 反思与回归理性
8. 结语

Course Content

Along the main line of architectural history,this course discussed the evolution of architectural design motivation ideas and methodology.Due to different concepts between the Chinese architecture and Western architecture Matters. The way for analyzing and exploring has to be studied.By analyzing the logic relationship of architectural form language,the geometrical significance of architectural form language is explored. Finally,within the context of urban form and space,the significance of architectural autonomy has been discussed.

1. Introdution
2. Tradition of western architecture
3. Meaning of architecture in China
4. History and modernity
5. Modern architectural ideology and its dilemma
6. Exploration for architectural form language
7. Re-thinking and return to reason
8. Conclusion

研究生一年级
电影建筑学 • 鲁安东
课程类型：选修
学时/学分：36学时/2学分

Graduate Program 1st Year
CINEMATIC ARCHITECTURE • LU Andong
Type: Elective Course
Study Period and Credits:36 hours/2 credits

课程内容

在本课程中，电影被视做一种独特的空间感知和思想交流的媒介。我们将学习如何使用电影媒介来对建筑和城市空间进行微观的分析与研究。本课程将通过循序渐进的教学帮助学生建立一种新的观察和理解建筑的方式，逐步培养学生的空间感知和空间想象的能力，以及使用电影媒介来交流自己的感觉和观点的能力。

本课程将综合课堂授课、实践操作和讲评讨论三种教学形式。在理论教学上，本课程将通过“普遍的运动”、《存在的直觉》、《组合的空间》和《城市的幻影》等四讲逐步向学生介绍相关的历史和理论，特别是电影和空间的关系，并帮助学生理解相应的课堂练习。在实践操作教学上，本课程将通过四个小练习和三个大练习，帮助学生从镜头练习到空间表达到观点陈述逐步地掌握电影媒介并将其用于对自己设计能力的培养。而讲评讨论环节将向学生介绍电影媒介的技术和方法，并帮助学生对自己的动手经验进行反思。

Course Content

Film, in this course, is seen as a distinctive medium for the perception of space and the communication of thoughts. We shall learn how to use the unique narrative medium of film to conduct a microscopic study on architecture and urbanism. This course will teach the students of a new way of seeing and knowing architecture. Its purpose is not only to teach theories of urbanism and techniques of filmmaking, but also to teach the students, through a complete case study, of a cinematic (visceral and non-abstract) way of thinking, analyzing and presenting ideas.
This course is composed of a series of 4-hour sessions, which gradually lead the students to undertake their own research project and to produce a cinematic essay on a case study of their own choice. The teaching of this course will be conducted in three forms: the lectures will introduce the students to cinematic ways of seeing and understanding architecture; the tutorials will introduce the students to some basic cinematic techniques, including continuity editing, cinematography, storyboard, shooting script, and post-production; the seminars will review and discuss the students' works in several stages.

城市理论课程
URBAN THEORY COURSES

本科四年级
城市设计及其理论 • 丁沃沃 胡友培
课程类型：必修
学时/学分：36学时/2学分

Undergraduate Program 4th Year
THEORY OF URBAN DESIGN • DING Wowo, HU Youpei
Type: Required Course
Study Period and Credits: 36 hours / 2 credits

课程内容

第一讲 课程概述
第二讲 城市设计技术术语：城市规划相关术语；城市形态相关术语；城市交通相关术语；消防相关术语
第三讲 城市设计方法 —— 文本分析：城市设计上位规划；城市设计相关文献；文献分析方法
第四讲 城市设计方法 —— 数据分析：人口数据分析与配置；交通流量数据分析；功能分配数据分析；视线与高度数据分析；城市空间数据模型的建构
第五讲 城市设计方法 —— 城市肌理分类：城市肌理分类概述；肌理形态与建筑容量；肌理形态与开放空间；肌理形态与交通流量；城市绿地指标体系
第六讲 城市设计方法 —— 城市路网组织：城市道路结构与交通结构概述；城市路网与城市功能；城市路网与城市空间；城市路网与市政设施；城市道路断面设计
第七讲 城市设计方法 —— 城市设计表现：城市设计分析图；城市设计概念表达；城市设计成果解析图；城市设计地块深化设计表达；城市设计空间表达
第八讲 城市设计的历史与理论：城市设计的历史意义；城市设计理论的内涵
第九讲 城市路网形态：路网形态的类型和结构；路网形态与肌理；路网形态的变迁
第十讲 城市空间：城市空间的类型；城市空间结构；城市空间形态；城市空间形态的变迁
第十一讲 城市形态学：英国学派；意大利学派；法国学派；空间句法
第十二讲 城市形态的物理环境：城市形态与物理环境；城市形态与环境研究；城市形态与环境测评；城市形态与环境操作
第十三讲 景观都市主义：景观都市主义的理论、操作和范例
第十四讲 城市自组织现象及其研究：城市自组织现象的魅力与问题；城市自组织系统研究方法；典型自组织现象案例研究
第十五讲 建筑学图式理论与方法：图式理论的研究，建筑学图式的概念；图式理论的应用；作为设计工具的图式；当代城市语境中的建筑学图式理论探索
第十六讲 课程总结

Course Content

Lect. 1. Introduction
Lect. 2. Technical terms: terms of urban planning, urban morphology, urban traffic and fire protection
Lect. 3. Urban design methods — documents analysis: urban planning and policies; relative documents; document analysis techniques and skills
Lect. 4. Urban design methods — data analysis: data analysis of demography, traffic flow, public facilities distribution, visual and building height; modelling urban spatial data
Lect. 5. Urban design methods — classification of urban fabrics: introduction of urban fabrics; urban fabrics and floor area ratio; urban fabrics and open space; urban fabrics and traffic flow; criteria system of urban green space
Lect. 6. Urban design methods — organization of urban road network: introduction; urban road network and urban function; urban road network and urban space; urban road network and civic facilities; design of urban road section
Lect. 7. Urban design methods — representation skills of urban Design: mapping and analysis; conceptual diagram; analytical representation of urban design; representation of detail design; spatial representation of urban design
Lect. 8. Brief history and theories of urban design: historical meaning of urban design; connotation of urban design theories
Lect. 9. Form of urban road network: typology, structure and evolution of road network; road network and urban fabrics
Lect. 10. Urban space: typology, structure, morphology and evolution of urban space
Lect. 11. Urban morphology: Cozen School; Italian School; French School; Space Syntax Theory
Lect. 12. Physical environment of urban forms: urban forms and physical environment; environmental study; environmental evaluation and environmental operations
Lect. 13. Landscape urbanism: ideas, theories, operations and examples of landscape urbanism
Lect. 14. Researches on the phenomena of the urban self-organization: charms and problems of urban self-organization phenomena; research methodology on urban self-organization phenomena; case studies of urban self-organization phenomena
Lect. 15. Theory and method of architectural diagram: theoretical study on diagrams; concepts of architectural diagrams; application of diagram theory; diagrams as design tools; theoretical research of architectural diagrams in contemporary urban context
Lect. 16. Summary

研究生一年级
城市形态研究 • 丁沃沃 赵辰 萧红颜
课程类型：必修
学时/学分：36学时/2学分

Graduate Program 1st Year
URBAN MORPHOLOGY • DING Wowo, ZHAO Chen, XIAO Hongyan
Type: Required Course
Study Period and Credits: 36 hours / 2 credits

课程要求

1. 要求学生基于对历史性城市形态的认知分析，加深对中西方城市理论与历史的理解。
2. 要求学生基于历史性城市地段的形态分析，提高对中西方城市空间特质及相关理论的认知能力。

课程内容

第一周：序言 概念、方法及成果
第二周：讲座1 城市形态认知的历史基础 —— 营造观念与技术传承
第三周：讲座2 城市形态认知的历史基础 —— 图文并置与意象构建
第四周：讲座3 城市形态认知的理论基础 —— 价值判断与空间生产
第五周：讲座4 城市形态认知的理论基础 —— 钩沉呈现与特征形塑
第六周：讲座5 历史城市的肌理研究
第七周：讲座6 整体与局部 —— 建筑与城市
第八周：讨论
第九周：讲座7 城市化与城市形态
第十周：讲座8 城市乌托邦
第十一周：讲座9 走出乌托邦
第十二周：讲座10 重新认识城市
第十三周：讲座11 城市设计背景
第十四周：讲座12 城市设计实践
第十五周：讲座13 城市设计理论
第十六周：讲评

Course Requirement

1. Deepen the understanding of Chinese and Western urban theories and histories based on the cognition and analysis of historical urban form.
2. Improve the cognitive abilities of the characteristics and theories of Chinese and Western urban space based on the morphological analysis of the historical urban sites.

Course Content

Week 1. Preface — concepts, methods and results
Week 2. Lect. 1 Historical basis of urban form cognition — Developing concepts and passing of technologies
Week 3. Lect. 2 Historical basis of urban form cognition — Apposition of pictures and text and image construction
Week 4. Lect. 3 Theoretical basis of urban form cognition — Value judgement and space production
Week 5. Lect. 4 Theoretical basis of urban form cognition — History representation and feature shaping
Week 6. Lect. 5 Study on the grain of historical cities
Week 7. Lect. 6 Whole and part: Architecture and urban
Week 8. Discussion
Week 9. Lect. 7 Urbanization and urban form
Week 10. Lect. 8 Urban Utopia
Week 11. Lect. 9 Walk out of Utopia
Week 12. Lect. 10 Have a new look of the city
Week 13. Lect. 11 Background of urban design
Week 14. Lect. 12 Practice of urban design
Week 15. Lect. 13 Theory of urban design
Week 16. Discussions

本科四年级
景观规划设计及其理论 • 尹航
课程类型：选修
学时/学分：36学时/2学分

Undergraduate Program 4th Year
LANDSCAPE PALNNING DESIGN AND THEORY
• YIN Hang
Type: Elective Course
Study Period and Credits: 36 hours / 2 credits

课程介绍
景观规划设计的对象包括所有的室外环境，景观与建筑的关系往往是紧密而互相影响的，这种关系在城市中尤为明显。景观规划设计及理论课程希望从景观设计理念、场地设计技术和建筑周边环境塑造等方面开展课程的教学，为建筑学本科生建立更加全面的景观知识体系，并且完善建筑学本科生在建筑场地设计、总平面规划与城市设计等方面的设计能力。
本课程主要从三个方面展开：一是理念与历史：以历史的视角介绍景观学科的发展过程，让学生对景观学科有一个宏观的了解，初步理解景观设计理念的发展；二是场地与文脉：通过阐述景观规划设计与周边自然环境、地理位置、历史文脉和方案可持续性的关系，建立场地与文脉的设计思维；三是景观与建筑：通过设计方法授课、先例分析作业等方式让学生增强建筑的环境意识，了解建筑的场地设计的影响因素、一般步骤与设计方法，并通过与“建筑设计六”和“建筑设计七”的设计任务书相配合的同步课程设计训练来加强学生景观规划设计的能力。

Course Description
The object of landscape planning design includes all outdoor environments; the relationship between landscape and building is often close and interactive, which is especially obvious in a city. This course expects to carry out teaching from perspective of landscape design concept, site design technology, building's peripheral environment creation, etc. to establish a more comprehensive landscape knowledge system for the undergraduate students of architecture, and perfect their design ability in building site design, master plane planning and urban design and so on.
This course includes three aspects:
1. Concept and history;
2. Site and context;
3. Landscape and building.

本科四年级
东西方园林 • 许浩
课程类型：选修
学时/学分：36学时/2学分

Undergraduate Program 4th Year
GARDEN OF EAST AND WEST • XU Hao
Type: Elective Course
Study Period and Credits: 36 hours / 2 credits

课程介绍
帮助学生系统掌握园林、绿地的基本概念、理论和研究方法，尤其了解园林艺术的发展脉络，侧重各个流派如日式园林、江南私家园林、皇家园林、规则式园林、自由式园林、伊斯兰园林的不同特征和关系，使得学生能够从社会背景、环境等方面解读园林的发展特征，并能够开展一定的评价。

Course Description
Help students systematically master the basic concepts, theories and research methods of gardens and greenbelts, especially understand the evolution of gardening, emphasizing the different features and relationships of various genres, such as Japanese gardens, private gardens by the south of Yangtze River, royal gardens, rule-style gardens, free style gardens and Islamic gardens; enable students to interpret the characteristics of garden development from the point of view of social backgrounds, environment, etc. Furthermore to do the evaluation in depth.

研究生一年级
景观规划进展 • 许浩
课程类型：选修
学时/学分：18学时/1学分

Graduate Program 1st Year
LANDSCAPE PLANNING PROGRESS • XU Hao
Type: Elective Course
Study Period and Credits: 18 hours / 1 credit

课程介绍
生态规划是景观规划的核心内容之一。本课程总结了生态系统、生态保护和生态修复的基本概念。大规模生态保护的基本途径是国家公园体系，而生态修复则是通过人为干涉对破损环境的恢复。本课程介绍了国家公园的价值、分类与成就，并通过具体案例论述了欧洲、澳洲景观设计过程中生态修复的做法。

Course Description
Ecological planning is one of the core contents of landscape planning. This course summarizes the basic concepts of ecological systems, ecological protection and ecological restoration. The basic channel of large-scale ecological protection is the national park system, while ecological restoration is to restore the damaged environment by means of human intervention. This course introduces the values, classification and achievements of national parks, and discusses the practices of ecological restoration in the process of landscape design in Europe and Australia.

研究生一年级
景观都市主义理论与方法 • 华晓宁
课程类型：选修
学时/学分：18学时/1学分

Graduate Program 1st Year
THEORY AND METHOD OF LANDSCAPE URBANISM
• HUA Xiaoning
Type: Elective Course
Study Period and Credits: 18 hours / 1 credit

课程介绍
本课程介绍了景观都市主义思想产生的背景、缘起及其主要理论观点，并结合实例，重点分析了其在不同的场址和任务导向下发展起来的多样化的实践策略和操作性工具。 通过这些内容的讲授，本课程的最终目的是拓宽学生的视野，引导学生改变既往的思维定式，以新的学科交叉整合的思路，分析和解决当代城市问题。

课程内容
第一讲　导论——当代城市与景观媒介
第二讲　生态过程与景观修复
第三讲　基础设施与景观嫁接
第四讲　嵌入与缝合
第五讲　水平性与都市表面
第六讲　城市图绘与图解
第七讲　AA景观都市主义——原型方法
第八讲　总结与作业

Course Description
The course introduces the backgrounds, the generation and the main theoretical opinions of landscape urbanism. With a series of instances, it particularly analyses the various practical strategies and operational techniques guided by various sites and projects. With all these contents, the aim of the course is to widen the students' field of vision, change their habitual thinking and suggest them to analyze and solve contemporary urban problems using the new ideas of the intersection and integration of different disciplines.

Course Content
Lect. 1: Introduction — contemporary cities and landscape medium
Lect. 2: Ecological process and landscape recovering
Lect. 3: Infrastructure and landscape engrafting
Lect. 4: Embedment and oversewing
Lect. 5: Horizontality and urban surface
Lect. 6: Urban mapping and diagram
Lect. 7: AA Landscape Urbanism — archetypical method
Lect. 8: Conclusion and assignment

历史理论课程
HISTORY THEORY COURSES

本科二年级
中国建筑史（古代）・萧红颜
课程类型：必修
学时/学分：36学时/2学分

Undergraduate Program 2nd Year
HISTORY OF CHINESE ARCHITECTURE (ANCIENT)
• XIAO Hongyan
Type: Required Course
Study Period and Credits:36 hours / 2 credits

教学目标

认识中国传统营造的思维特征与技术选择，培养学生理解历史与分析问题的意识。

课程内容

采取有别于传统教学的方式，将专题讲述与类型分析并重，强调多元视角下的问题式教学，力求达成认知与学理的综合效应。

Training Objective

Recognize the thinking characteristics and technology selection of China's traditional construction; develop students' awareness of understanding history and analyzing problems.

Course Content

Use a teaching method different from traditional teaching, concurrently emphasize special lecturing and type analysis as well as the problem-based teaching under diversified visions, attempting to realize the integrated effect of cognition and learnt theory.

本科二年级
外国建筑史（古代）・胡恒
课程类型：必修
学时/学分：36学时/2学分

Undergraduate Program 2nd Year
HISTORY OF WESTERN ARCHITECTURE (ANCIENT)
• HU Heng
Type: Required Course
Study Period and Credits: 36 hours / 2 credits

教学目标

本课程力图对西方建筑史的脉络做一个整体勾勒，使学生在掌握重要的建筑史知识点的同时，对西方建筑史在2000多年里的变迁的结构转折（不同风格的演变）有深入的理解。本课程希望学生对建筑史的发展与人类文明发展之间的密切关联有所认识。

课程内容

1. 概论　2. 希腊建筑　3. 罗马建筑　4. 中世纪建筑
5. 意大利的中世纪建筑　6. 文艺复兴　7. 巴洛克
8. 美国城市　9. 北欧浪漫主义　10. 加泰罗尼亚建筑
11. 先锋派　12. 德意志制造联盟与包豪斯
13. 苏维埃的建筑与城市　14. 1960年代的建筑
15. 1970年代的建筑　16. 答疑

Training Objective

This course seeks to give an overall outline of Western architectural history, so that the students may have an in-depth understanding of the structural transition (different styles of evolution) of Western architectural history in the past 2000 years. This course hopes that students can understand the close association between the development of architectural history and the development of human civilization.

Course Content

1. Generality　2. Greek Architectures　3. Roman Architectures
4. The Middle Ages Architectures
5. The Middle Ages Architectures in Italy　6. Renaissance
7. Baroque　8. American Cities　9. Nordic Romanticism
10. Catalonian Architectures　11. Avant-Garde
12. German Manufacturing Alliance and Bauhaus
13. Soviet Architecture and Cities　14. 1960's Architectures
15. 1970's Architectures　16. Answer Questions

本科三年级
外国建筑史（当代）・胡恒
课程类型：必修
学时/学分：36学时/2学分

Undergraduate Program 3rd Year
HISTORY OF WESTERN ARCHITECTURE (MODERN)
• HU Heng
Type: Required Course
Study Period and Credits:36 hours / 2 credits

教学目标

本课程力图用专题的方式对文艺复兴时期的7位代表性的建筑师与5位现当代的重要建筑师作品做一细致的讲解。本课程将重要建筑师的全部作品尽可能在课程中梳理一遍，使学生能够全面掌握重要建筑师的设计思想、理论主旨、与时代的特殊关联、在建筑史中的意义。

课程内容

1. 伯鲁乃涅斯基　2. 阿尔伯蒂　3. 伯拉孟特
4. 米开朗基罗（1）　5. 米开朗琪罗（2）　6. 罗马诺
7. 桑索维诺　8. 帕拉蒂奥（1）　9. 帕拉蒂奥（2）
10. 赖特　11. 密斯　12. 勒・柯布西耶（1）
13. 勒・柯布西耶（2）　14. 海杜克　15. 妹岛和世
16. 答疑

Training Objective

This course seeks to make a detailed explanation to the works of 7 representative architects in the Renaissance period and 5 important contemporary architects in a special way. This course will try to reorganize all works of these important architects, so that the students can fully grasp their design ideas, theoretical subject and their particular relevance with the era and significance in the architectural history.

Course Content

1. Brunelleschi　2. Alberti　3. Bramante
4. Michelangelo(1)　5. Michelangelo(2)
6. Romano　7. Sansovino　8. Paratio(1)　9. Paratio(2)
10. Wright　11. Mies　12. Le Corbusier(1)　13. Le Corbusier(2)
14. Hejduk　15. Kazuyo Sejima
16. Answer Questions

本科三年级
中国建筑史（近现代）・赵辰
课程类型：必修
学时/学分：36学时/2学分

Undergraduate Program 3rd Year
HISTORY OF CHINESE ARCHITECTURE (MODERN)
• ZHAO Chen
Type: Required Course
Study Period and Credits:36 hours / 2 credits

课程介绍

本课程作为本科建筑学专业的历史与理论课程，是中国建筑史教学中的一部分。在中国与西方的古代建筑历史课程先学的基础上，了解中国社会进入近代，以至于现当代的发展进程。

在对比中西方建筑文化的基础之上，建立对中国近现代建筑的整体认识。深刻理解中国传统建筑文化在近代以来与西方建筑文化的冲突与相融之下，逐步演变发展至今天世界建筑文化的一部分之意义。

Course Description

As the history and theory course for undergraduate students of Architecture, this course is part of the teaching of History of Chinese Architecture. Based on the earlier studying of Chinese and Western history of ancient architecture, understand the evolution progress of Chinese society's entry into modern times and even contemporary age.

Based on the comparison of Chinese and Western building culture, establish the overall understanding of China's modern and contemporary buildings. Have further understanding of the significance of China's traditional building culture's gradual evolution into one part of today's world building culture under conflict and blending with Western building culture in modern times.

研究生一年级
建筑理论研究・王骏阳
课程类型：必修
学时/学分：18学时/1学分

Graduate Program 1st Year
STUDY OF ARCHITECTURAL THEORY • WANG Junyang
Type: Required Course
Study Period and Credits:18 hours / 1 credit

课程介绍

本课程是西方建筑史研究生教学的一部分。主要涉及当代西方建筑界具有代表性的思想和理论，其主题包括历史主义、先锋建筑、批判理论、建构文化以及对当代城市的解读等。本课程大量运用图片资料，广泛涉及哲学、历史、艺术等领域，力求在西方文化发展的背景中呈现建筑思想和理论的相对独立性及关联性，理解建筑作为一种人类活动所具有的社会和文化意义，启发学生的理论思维和批判精神。

课程内容

第一讲 建筑理论概论
第二讲 建筑自治
第三讲 柯林・罗：理想别墅的数学与其他
第四讲 阿道夫・路斯与装饰美学
第五讲 库哈斯与当代城市的解读
第六讲 意识的困境：对现代建筑的反思
第七讲 弗兰普顿的建构文化研究
第八讲 现象学

Course Description

This course is a part of teaching Western architectural history for graduate students. It mainly deals with the representative thoughts and theories in Western architectural circles, including historicism, vanguard building, critical theory, construction culture and interpretation of contemporary cities and more. Using a lot of pictures involving extensive fields including philosophy, history, art, etc., this course attempts to show the relative independence and relevance of architectural thoughts and theories under the development background of Western culture, understand the social and cultural significance owned by architectures as human activities, and inspire students' theoretical thinking and critical spirit.

Course Content

Lect. 1. Introduction to architectural theories
Lect. 2. Autonomous architecture
Lect. 3. Colin Rowe : the mathematics of the ideal villa and others
Lect. 4. Adolf Loos and adornment aesthetics
Lect. 5. Koolhaas and the interpretation of contemporary cities
Lect. 6. Conscious dilemma: the reflection of modern architecture
Lect. 7. Studies in tectonic culture of Frampton
Lect. 8. Phenomenology

研究生一年级
建筑史研究 ・ 胡恒
课程类型：选修
学时/学分：36学时/2学分

Graduate Program 1st Year
ARCHITECTURAL HISTORY RESEARCH • HU Heng
Type: Elective Course
Study Period and Credits:36 hours / 2 credits

教学目标

本课程的目的有二。其一，通过对建筑史研究的方法做一概述，来使学生粗略了解西方建筑史研究方法的总的状况。其二，通过对当代史概念的提出，且用若干具体的案例研究，来向学生展示当代史研究的路数、角度、概念定义、结构布置、主题设定等内容。

课程内容

1. 建筑史方法概述（1）
2. 建筑史方法概述（2）
3. 建筑史方法概述（3）
4. 塔夫里的建筑史研究方法
5. 当代史研究方法 —— 周期
6. 当代史研究方法 —— 杂交
7. 当代史研究方法 —— 阈限
8. 当代史研究方法 —— 对立

Training Objective

This course has two objectives: 1. Give the students a rough understanding of the overall status of the research approaches of the Western architectural history through an overview of them. 2. Show students the approaches, point of view, concept definition, structure layout, theme settings and so on of contemporary history study through proposing the concept of contemporary history and several case studies.

Course Content

1. The overview of the method of architectural history(1)
2. The overview of the method of architectural history(2)
3. The overview of the method of architectural history(3)
4. Tafuri's study method of architectural history
5. The study method of contemporary history — period
6. The study method of contemporary history — hybridization
7. The study method of contemporary history — limen
8. The study method of contemporary history — opposition

研究生一年级
建筑史研究・萧红颜
课程类型：选修
学时/学分：36学时/2学分

Graduate Program 1st Year
ARCHITECTURAL HISTORY RESEARCH • XIAO Hongyan
Type: Elective Course
Study Period and Credits:36 hours / 2 credits

教学目标

本课程尝试从理念与类型两大范畴为切入点，专题讲述中国传统营造基本理念之异变与延续、基本类型之关联与意蕴，强调建筑史应回归艺术史分析框架下阐发相关史证问题及其方法。

课程内容

1. 边角　　2. 堪舆
3. 界域　　4. 传摹
5. 宫台　　6. 池苑
7. 庙墓　　8. 楼亭

Training Objective

This course attempts to start with two areas (concept and type) to state the mutation and continuation, association and implication of basic types of the basic concept of China's traditional construction, emphasizing that the architectural history should return to the art history framework to state relevant history evidence issues and its methods.

Course Content

1. Corner　　2. Geomantic Omen
3. Boundary　　4. Spreading and Copying
5. Table Land of Palace　　6. Pond
7. Temple and Mausoleum　　8. Storied Building Pavilion

建筑技术课程
ARCHITECTURAL TECHNOLOGY COURSES

本科二年级
CAAD理论与实践・童滋雨
课程类型：必修
学时/学分：36学时/2学分

Undergraduate Program 2nd Year
THEORY AND PRACTICE OF CAAD • TONG Ziyu
Type: Required Course
Study Period and Credits: 36 hours / 2 credits

课程介绍

在现阶段的CAD教学中，强调了建筑设计在建筑学教学中的主干地位，将计算机技术定位于绘图工具，本课程就是帮助学生可以尽快并且熟练地掌握如何利用计算机工具进行建筑设计的表达。课程中整合了CAD知识、建筑制图知识以及建筑表现知识，将传统CAD教学中教会学生用计算机绘图的模式向教会学生用计算机绘制有形式感的建筑图的模式转变，强调准确性和表现力作为评价CAD学习的两个最重要指标。

本课程的具体学习内容包括：

1. 初步掌握AutoCAD软件和SketchUP软件的使用，能够熟练完成二维制图和三维建模的操作；

2. 掌握建筑制图的相关知识，包括建筑投影的基本概念，平立剖面、轴测、透视和阴影的制图方法和技巧；

3. 图面效果表达的技巧，包括黑白线条图和彩色图纸的表达方法和排版方法。

Course Description

The core position of architectural design is emphasized in the CAD course. The computer technology is defined as drawing instrument. The course helps students learn how to make architectural presentation using computer fast and expertly. The knowledge of CAD, architectural drawing and architectural presentation are integrated into the course. The traditional mode of teaching students to draw in CAD course will be transformed into teaching students to draw architectural drawing with sense of form. The precision and expression will be emphasized as two most important factors to estimate the teaching effect of CAD course.

Contents of the course include:

1. Use AutoCAD and SketchUP to achieve the 2-D drawing and 3-D modeling expertly.

2. Learn relational knowledge of architectural drawing, including basic concepts of architectural projection, drawing methods and skills of plan, elevation, section, axonometry, perspective and shadow.

3. Skills of presentation, including the methods of expression and lay out using mono and colorful drawings

---

本科三年级
建筑技术 1——结构、构造与施工・傅筱
课程类型：必修
学时/学分：36学时/2学分

Undergraduate Program 3rd Year
ARCHITECTURAL TECHNOLOGY 1 — STRUCTURE, CONSTRUCTION AND EXECUTION • FU Xiao
Type: Required Course
Study Period and Credits:36 hours / 2 credits

课程介绍

本课程是建筑学专业本科生的专业主干课程。本课程的任务主要是以建筑师的工作性质为基础，讨论一个建筑生成过程中最基本的三大技术支撑（结构、构造、施工）的原理性知识要点，以及它们在建筑实践中的相互关系。

Course Description

The course is a major course for the undergraduate students of architecture. The main purpose of this course is based on the nature of the architect's work, to discuss the principle knowledge points of the basic three technical supports in the process of generating construction (structure, construction, execution), and their mutual relations in the architectural practice.

本科三年级
建筑技术 2——建筑物理・吴蔚
课程类型：必修
学时/学分：36学时/2学分

Undergraduate Program 3rd Year
ARCHITECTURAL TECHNOLOGY 2 — BUILDING PHYSICS • WU Wei
Type: Required Course
Study Period and Credits:36 hours / 2 credits

课程介绍

本课程是针对三年级学生所设计，课程介绍了建筑热工学、建筑光学、建筑声学中的基本概念和基本原理，使学生能掌握建筑的热环境、声环境、光环境的基本评估方法，以及相关的国家标准。完成学业后在此方向上能阅读相关书籍，具备在数字技术方法等相关资料的帮助下，完成一定的建筑节能设计的能力。

Course Description

Designed for the Grade-3 students, this course introduces the basic concepts and basic principles in architectural thermal engineering, architectural optics and architectural acoustics, so that the students can master the basic methods for the assessment of building's thermal environment, sound environment and light environment as well as the related national standards. After graduation, the students will be able to read the related books regarding these aspects, and have the ability to complete certain building energy efficiency designs with the help of the related digital techniques and methods.

本科三年级
建筑技术 3——建筑设备・吴蔚
课程类型：必修
学时/学分：36学时/2学分

Undergraduate 3rd Year
ARCHITECTURAL TECHNOLOGY 3 — BUILDING EQUIPMENT • WU Wei
Type: Required Course
Study Period and Credits:36 hours / 2 credits

课程介绍

本课程是针对南京大学建筑与城市规划学院本科学生三年级所设计。课程介绍了建筑给水排水系统、采暖通风与空气调节系统、电气工程的基本理论、基本知识和基本技能，使学生能熟练地阅读水电、暖通工程图，熟悉水电及消防的设计、施工规范，了解燃气供应、安全用电及建筑防火、防雷的初步知识。

Course Description

This course is an undergraduate class offered in the School of Architecture and Urban Planning, Nanjing University. It introduces the basic principle of the building services systems, the technique of integration amongst the building services and the building. Throughout the course, the fundamental importance to energy, ventilation, air-conditioning and comfort in buildings are highlighted.

研究生一年级
传热学与计算流体力学基础・郜志
课程类型：选修
学时/学分：18学时/1学分

Graduate Program 1st Year
FUNDAMENTALS OF HEAT TRANSFER AND COMPUTATIONAL FLUID DYNAMICS • GAO Zhi
Type: Elective Course
Study Period and Credits: 18 hours / 1 credit

课程介绍

本课程的主要任务是使建筑学/建筑技术学专业的学生掌握传热学和计算流体力学的基本概念和基础知识，通过课程教学，使学生熟悉传热学中导热、对流和辐射的经典理论，并了解传热学和计算流体力学的实际应用和最新研究进展，为建筑能源和环境系统的计算和模拟打下坚实的理论基础。教学中尽量简化传热学和计算流体力学经典课程中复杂公式的推导过程，而着重于如何解决建筑能源与建筑环境中涉及流体流动和传热的实际应用问题。

Course Description

This course introduces students majoring in building science and engineering / building technology to the fundamentals of heat transfer and computational fluid dynamics (CFD). Students will study classical theories of conduction, convection and radiation heat transfers, and learn advanced research developments of heat transfer and CFD. The complex mathematics and physics equations are not emphasized. It is desirable that for real-case scenarios students will have the ability to analyze flow and heat transfer phenomena in building energy and environment systems.

研究生一年级
建筑节能与可持续发展 • 秦孟昊
课程类型：选修
学时/学分：18学时/1学分

Graduate Program 1st Year
ENERGY CONSERVATION AND SUSTAINABLE ARCHITECTURE • QIN Menghao
Type: Elective Course
Study Period and Credits:18 hours / 1 credit

课程介绍

随着我国建筑总量的攀升和居住舒适度的提高，建筑能耗急剧上升，建筑节能成为影响能源安全和提高能效的重要因素之一。建筑节能的关键首先是要设计“本身节能的建筑”，建筑师必须从建筑设计的最初阶段，在建筑的形体、结构、开窗方式、外墙选材等方面融入节能设计的定量分析。而这些很难通过传统建筑设计方法达到，必须依靠建筑技术、建筑设备等多学科互动协作才能完成。这已成为世界各大建筑与城市规划学院教学的一个重点。

本课程将采用双语教学，主要面向建筑设计专业学生讲授建筑物理、建筑技术专业关于建筑节能方面的基本理念、设计方法和模拟软件，并指导学生将这些知识互动运用到节能建筑设计的过程中，在建筑设计专业和建筑技术专业之间建立一个互动的平台，从而达到设计“绿色建筑”的目标，并为以后开展交叉学科研究、培养复合型人才奠定基础。

Course Description

With the rising of China's total number of buildings and the need for living comfort, building energy consumption is rising sharply. Building energy efficiency has become one of the key factors influencing the energy security and energy efficiency. The first key for building energy efficiency is to design “a building that conserves energy itself” and architects must carry out planning at the very beginning of building design. However, it is difficult to satisfy them by means of traditional architectural design approaches; it must be realized by interactive collaboration of diversified subjects including construction technology, construction equipment, etc. Strengthening the interaction of architectural design specialties and construction technology specialties in designing has become a key point in this course as well as in the teaching of various large architecture and urban planning colleges around the world.

研究生一年级
建筑环境学 • 郜志
课程类型：选修
学时/学分：18学时/1学分

Graduate Program 1st Year
FUNDAMENTALS OF BUILT ENVIRONMENT • GAO Zhi
Type: Elective Course
Study Period and Credits:18 hours / 1 credit

课程介绍

本课程的主要任务是使建筑学/建筑技术学专业的学生掌握建筑环境的基本概念，学习建筑与城市热湿环境、风环境和空气质量的基础知识。通过课程教学，使学生熟悉城市微气候等理论，并了解人体对热湿环境的反应，掌握建筑环境学的实际应用和最新研究进展，为建筑能源和环境系统的测量与模拟打下坚实的基础。

Course Description

This course introduces students majoring in building science and engineering / building technology to the fundamentals of built environment. Students will study classical theories of built / urban thermal and humid environment, wind environment and air quality. Students will also familiarize urban micro environment and human reactions to thermal and humid environment. It is desirable that students will have the ability to measure and simulate building energy and environment systems based upon the knowledge of the latest development of the study of built environment.

研究生一年级
材料与建造 • 冯金龙
课程类型：必修
学时/学分：18学时/1学分

Graduate Program 1st Year
MATERIAL AND CONSTRUCTION • FENG Jinlong
Type: Required Course
Study Period and Credits:18 hours / 1 credit

课程介绍

本课程将介绍现代建筑技术的发展过程，论述现代建筑技术及其美学观念对建筑设计的重要作用。探讨由材料、结构和构造方式所形成的建筑建造的逻辑方式研究。研究建筑形式产生的物质技术基础，诠释现代建筑的建构理论与研究方法。

Course Description

It introduces the development process of modern architecture technology and discusses the important role played by the modern architecture technology and its aesthetic concepts in the architectural design. It explores the logical methods of construction of the architecture formed by materials, structure and construction. It studies the material and technical basis for the creation of architectural form, and interprets construction theory and research methods for modern architectures.

研究生一年级
计算机辅助技术 • 吉国华
课程类型：选修
学时/学分：36学时/2学分

Graduate Program 1st Year
TECHNOLOGY OF CAAD • JI Guohua
Type: Elective Course
Study Period and Credits:36 hours / 2 credits

课程介绍

随着计算机辅助建筑设计技术的快速发展，当前数字技术在建筑设计中的角色逐渐从辅助绘图转向了真正的辅助设计，并引发了设计的革命和建筑的形式创新。本课程讲授AutoCAD VBA和RhinoScript编程。让学生在掌握“宏”/“脚本”编程的同时，增强以理性的过程思维方式分析和解决设计问题的能力，为数字建筑设计打下必要的基础。

课程分为三个部分：

1. VB语言基础，包括VB基本语法、结构化程序、数组、过程等编程知识和技巧；

2. AutoCAD VBA，包括AutoCAD VBA的结构、二维图形、人机交互、三维对象等，以及基本的图形学知识；

3. RhinoScript概要，包括基本概念、Nurbs概念、VBScript简介、曲线对象、曲面对象等。

Course Description

Following its fast development, the role of digital technology in architecture is changing from computer-aided drawing to real computer-aided design, leading to a revolution of design and the innovation of architectural form. Teaching the programming with AutoCAD VBA and RhinoScript, the lecture attempts to enhance the students' capability of reasoningly analyzing and solving design problems other than the skills of "macro" or "script" programming, to let them lay the base of digital architectural design.

The course consists of three parts:

1. Introduction to VB, including the basic grammar of VB, structural program, array, process, etc.

2. AutoCAD VBA, including the structure of AutoCAD VBA , 2D graphics, interactive methods, 3D objects, and some basic knowledge of computer graphics.

3. Brief introduction of RhinoScript, including basic concepts, the concept of Nurbs, sammary of VBScript, and Rhino objects.

研究生一年级
GIS基础与应用 • 童滋雨
课程类型：选修
学时/学分：18学时/1学分

Graduate Program 1st Year
CONCEPT AND APPLICATION OF GIS • TONG Ziyu
Type: Elective Course
Study Period and Credits:18 hours / 1 credit

课程介绍

本课程的主要目的是让学生理解GIS的相关概念以及GIS对城市研究的意义，并能够利用GIS软件对城市进行分析和研究。

Course Description

This course aims to enable students to understand the related concepts of GIS and the significance of GIS to urban research, and to be able to use GIS software to carry out urban analysis and research.

其他
MISCELLANEA

主讲 卢永毅 教授 博士生导师
同济大学建筑与城市规划学院
主要研究方向：
西方建筑历史与理论研究
历史建筑保护与再生
城市化进程中的
上海近代建筑
近现代中国建筑史课程讲座之四
地点：南京大学鼓楼校区教学楼 120 室
时间：2014 年 12 月 1 日上午 10 点 10 分

WSP
维思平
吴钢
南京大学建筑讲坛
演讲主题
平常建筑·持续创新
WSP 作品思想
CONTINUOUS INNOVATION BEAUTY OF THE ORDINARY
WSP IDEAS AND WORKS

从一张图
到一个城
——玉树重建工作
2014 年 11 月 19 日 晚 7 点
南京大学蒙民伟楼10楼 1003
鞠德东

南京大学建筑与城市规划学院前沿系列讲座第一讲
暨南京大学兼职教授授予仪式
西藏建筑节能研究
刘加平
中国工程院院士
西安建筑科技大学教授、博导
南京大学鼓楼校区科技馆二楼报告厅
二零一四年十一月二十三日星期日下午三点三十分

近现代中国建筑史课程讲座7
历史的重“缝”：
1949年后移居香港的中国近代建筑师
MAINLAND ARCHITECTS IN HONG KONG AFTER 1949,
A BIFURCATED HISTORY OF MODERN CHINESE ARCHITECTURE
王浩娱
香港大学建筑学博士
上海交通大学建筑系讲师
南京大学鼓楼校区教学楼120室
2014年12月15日上午10点10分

《前沿》讲座系列第一讲
读解历史城市
东南大学 陈薇
时间：
二零一五年四月十四日（星期二）
19：00
地点：
南京大学鼓楼校区蒙民伟楼1003室

从 自上而下
到 自下而上

黄印武

沙溪源乡村合作中心

创造新颖而高效的结构和材料

谢亿民

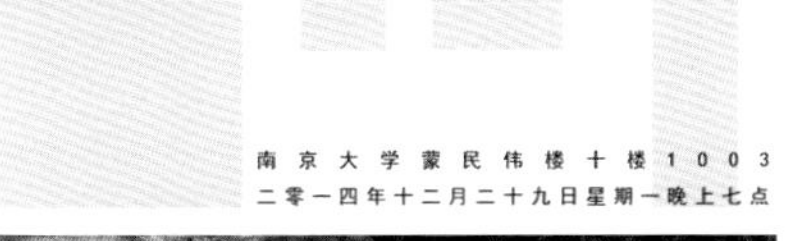

二〇一四至二〇一五学年南大建筑记录系列讲座

平常建筑 · 持续创新
WSP 作品思想

吴钢

维思平建筑设计创始人之一，
主持设计师。

同济大学建筑学硕士， 德国卡尔斯鲁厄大学建筑学硕士

香港中文大学建筑学院教授

亚洲建筑师协会会员

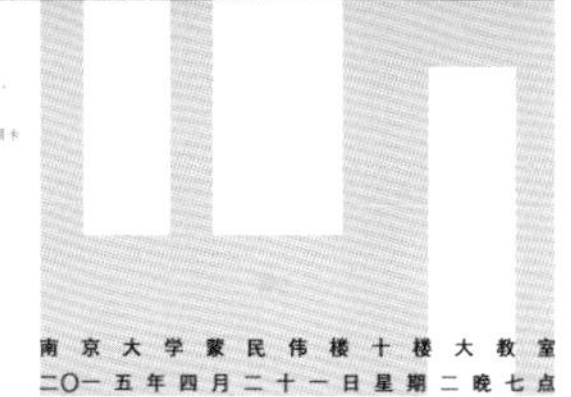

南 京 大 学 蒙 民 伟 楼 十 楼 大 教
二〇一 五 年 六 月 三 日 星 期 三 晚 七

南 京 大 学 蒙 民 伟 楼 十 楼 1 0 0 3
二零一四年十二月二十九日星期一晚上七点

南 京 大 学 蒙 民 伟 楼 十 楼 大 教 室
二〇一 五 年 四 月 二 十 一 日 星 期 二 晚 七 点

對話

刘宇扬

香港大学建筑学院荣誉副教授
上海刘宇扬建筑事务所主持
著名建筑师

张应鹏

哲学博士
苏州九城都市建筑设计有限公司
总建筑师

我对乡村复兴
的理解与原则

赵辰

南京大学教授

我的
乡村梦

任卫中

安吉景坞源生态民居推广中心
中心负责人，民间环保人

南 京 大 学 蒙 民 伟 楼 十 楼 大 教 室
二〇一五年六月十七日星期三下午二点整

南 京 大 学 蒙 民 伟 楼 十 楼 大 教 室
二〇一 五 年 六 月 三 日 星 期 三 晚 八 点

南 京 大 学 蒙 民 伟 楼 十 楼 大 教 室
二〇一五年六月十日星期三晚七点整

硕士学位论文列表
List of Thesis for Master Degree

研究生姓名	研究生论文标题	导师姓名
张成	城市历史文化题材商业街区规划设计研究	张雷
薛晓旸	中国当代建筑中砌筑材料幕墙表皮化研究	张雷
赖友炜	城市化背景下中国当代文化综合体复合化设计研究	张雷
朱鹏飞	我国住宅产业化进程初探——以万科集团为例	冯金龙
俞琳	夏热冬冷地区建筑外遮阳与表皮一体化设计研究	冯金龙
袁亮亮	基于自然通风的多层建筑中庭自动优化设计方法研究	吉国华
樊璐敏	基于天然采光的多层建筑庭院自动优化设计研究	吉国华
周雨馨	对莫拉作品的建造解析	傅筱
李政	工业化介入中国乡村建造的案例解析	傅筱
杨柯	三角形豪式木桁架构造研究——以莫干山蚕种场蚕室屋架为例	周凌
吴黎明	徽州民居营造体系研究——以黟县石亭村黄、齐、吴三宅为例	周凌
赵芹	城市居住地块形态特征研究与表述——以南京为例	丁沃沃
杨浩	街区界面轮廓形态与城市规划指标关系研究	丁沃沃
林肖寅	基于街廓属性解读城市之间平面形态的差异——以南京市为例	丁沃沃
蒋菁菁	城市商业地块与建筑布局模式研究——以南京为例	丁沃沃
孙燕	政策规范与新农村建设中集中建房村庄形态的关联性——以江苏省为例	华晓宁
黄凯熙	基于网络分析和数字高程模型的城市雨洪调蓄开放空间布局优化	华晓宁

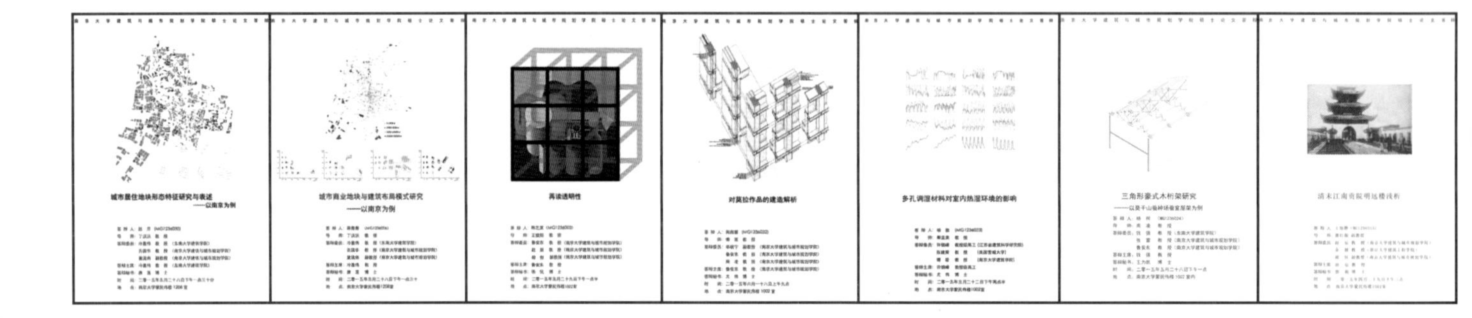

研究生姓名	研究生论文标题	导师姓名
赵潇欣	基于“间架”的中国传统木构架原型及其发展规律研究	赵辰
张方籍	慈城传统住宅的近代化转型研究	赵辰
杨钗芳	当代建筑半透明现象的建构分析	赵辰
武苗苗	慈城公共空间浅论——结合历史城市保护复兴的重新思考	赵辰
王旭静	清末江南贡院明远楼浅析	萧红颜
刘赟俊	1873年江南贡院格局及其构成浅析	萧红颜
胡小敏	建筑形式与建筑结构关系的调查——以当代中国建筑师的20个建筑为例	王骏阳
韩艺宽	再读透明性	王骏阳
杨骏	多孔调湿材料对室内热湿环境的影响	秦孟昊
王洁琼	基于环境功能的近现代苏南地区蚕室建筑类型研究	鲁安东
倪绍敏	南京大学生宿舍天然采光及照明研究	吴蔚
张伟	居住小区绿地布局对微气候影响的模拟研究	郜志、丁沃沃

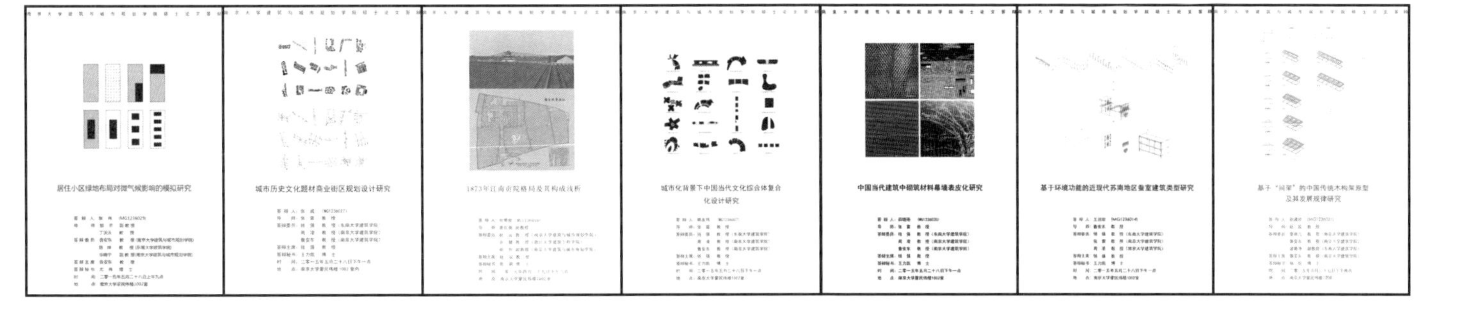

本科生 Undergraduate

2011级学生
Students 2011

崔傲寒 CUI Aohan
冯　琪 FENG Qi
顾聿笙 GU Yusheng
黄凯峰 HUANG Kaifeng
黄雯倩 HUANG Wenqian
贾福龙 JIA Fulong
蒋佳瑶 JIANG Jiayao
蒋建昕 JIANG Jianxin
蒋造时 JIANG Zaoshi
雷朝荣 LEI Zhaorong
黎乐源 LI Leyuan
柳纬宇 LIU Weiyu
缪姣姣 MIAO Jiaojiao
倪若宁 NI Ruoning
彭丹丹 PENG Dandan
宋富敏 SONG Fumin
拓　展 TUO Zhan
王梦琴 WANG Mengqin
王却奁 WANG Quelian
王思绮 WANG Siqi
王潇聆 WANG Xiaoling
王新宇 WANG Xinyu
吴家禾 WU Jiahe
吴结松 WU Jiesong
席　弘 XI Hong
谢忠雄 XIE Zhongxiong
徐亦杨 XU Yiyang
杨益晖 YANG Yihui
张豪杰 ZHANG Haojie
张黎萌 ZHANG Limeng
张人祝 ZHANG Renzhu
周　松 ZHOU Song
周贤春 ZHOU Xianchun
左　思 ZUO Si

2012级学生
Students 2012

陈虹全 CHEN Hongquan
陈思涵 CHEN Sihan
陈　妍 CHEN Yan
从　彬 CONG Bin
段晓昱 DUAN Xiaoyu
高文杰 GAO Wenjie
高祥震 GAO Xiangzhen
葛嘉许 GE Jiaxu
桂　喻 GUI Yu
黄福运 HUANG Fuyun
黄卫健 HUANG Weijian
黄子恩 HUANG Zi'en
季惠敏 JI Huimin
李慧兰 LI Huilan
刘贤斌 LIU Xianbin
刘姿佑 LIU Ziyou
陆怡人 LU Yiren
罗　坤 LUO Kun
钱宇飞 QIAN Yufei
钱雨翀 QIAN Yuchong
全道熏 QUAN Daoxun
沈应浩 SHEN Yinghao
苏　彤 SU Tong
唐林松 TANG Linsong
王　焱 WANG Yan
王一侬 WANG Yinong
吴峥嵘 WU Zhengrong
于明霞 YU Mingxia
臧　倩 ZANG Qian
张馨元 ZHANG Xinyuan
张逸凡 ZHANG Yifan
朱朝龙 ZHU Chaolong
朱凌峥 ZHU Lingzheng
田　甜 TIAN Tian
徐　华 XU Hua
赵媛倩 ZHAO Yuanqian

2013级学生
Students 2013

曹舒琪 CAO Shuqi
陈　露 CHEN Lu
董素宏 DONG Suhong
郭金未 GUO Jinwei
郭　硕 GUO Shuo
贺唯嘉 HE Weijia
胡慧慧 HU Huihui
黄婉莹 HUANG Wanying
黄追日 HUANG Zhuiri
吉雨心 JI Yuxin
贾奕超 JIA Yichao
林之音 LIN Zhiyin
刘稷祺 LIU Jiqi
鲁　晴 LU Qing
罗晓东 LUO Xiaodong
吕　童 Lv Tong
楠田康雄 KUSUDA YASUO
宋宇瑀 SONG Yuxun
谭　皓 TAN Hao
涂成祥 TU Chengxiang
王成阳 WANG Chengyang
王　青 WANG Qing
王秋锐 WANG Qiurui
王　瑶 WANG Yao
王智伟 WANG Zhiwei
武　波 WU Bo
夏凡琦 XIA Fanqi
夏　楠 XIA Nan
徐家炜 XU Jiawei
徐瑜灵 XU Yuling
杨　蕾 YANG Lei
章太雷 ZHANG Tailei
赵　焦 ZHAO Jiao
赵梦娣 ZHAO Mengdi
赵中石 ZHAO Zhongshi
周　怡 ZHOU Yi

2014级学生
Students 2014

蔡英杰 CAI Yingjie
曹　焱 CAO Yan
陈妍霓 CHEN Yanni
杜孟泽杉 DUMENG Zeshan
胡皓捷 HU Haojie
兰　阳 LAN Yang
李雪琦 LI Xueqi
梁晓蕊 LIANG Xiaorui
林　宇 LIN Yu
刘　畅 LIU Chang
刘坤龙 LIU Kunlong
刘宛莹 LIU Wanying
刘为尚 LIU Weishang
卢　鼎 LU Ding
马西伯 MA Xibo
施少鋆 SHI Shaojun
施孝萱 SHI Xiaoxuan
宋　怡 SONG Yi
宋宇宁 SONG Yuning
宋云龙 SONG Yunlong
唐　萌 TANG Meng
完颜尚文 WANYAN Shangwen
夏心雨 XIA Xinyu
谢　峰 XIE Feng
严紫微 Yan Ziwei
杨云睿 YANG Yunrui
杨　钊 YANG Zhao
尹子晗 YIN Zihan
张　俊 ZHANG Jun
张珊珊 ZHANG Shanshan

陈 肯 CHEN Ken
陈 圆 CHEN Yuan
陈 钊 CHEN Zhao
高 菲 GAO Fei
管 理 GUAN Li
韩 梦 HAN Meng
胡 昊 HU Hao
胡鹭鹭 HU Lulu
柯国新 KE Guoxin
黎健波 LI Jianbo
李 港 LI Gang
李恒鑫 LI Hengxin
刘滨洋 LIU Binyang
刘兴渝 LIU Xingyu
刘奕彪 LIU Yibiao
吕 铭 Lv Ming
马 喆 MA Zhe
孟庆忠 MENG Qingzhong
沈周娅 SHEN Zhouya
石延安 SHI Yan'an
王海芹 WANG Haiqin
王力凯 WANG Likai
王亦播 WANG Yibo
吴绉彦 WU Zhouyan
谢智峰 XIE Zhifeng
辛胤庆 XIN Yinqing
徐 睿 XU Rui
杨尚宜 YANG Shangyi
殷 奕 YIN Yi
郁新新 YU Xinxin
袁 芳 YUAN Fang
张 备 ZHANG Bei
张卜予 ZHANG Buyu
曹梦原 CAO Mengyuan
陈 姝 CHEN Shu
陈婷婷 CHEN Tingting
陈 新 CHEN Xin
葛鹏飞 GE Pengfei
胡 曜 HU Yao
金 鑫 JIN Xin
李红瑞 LI Hongrui
李 扬 LI Yang
刘 宇 LIU Yu
吕 程 Lv Cheng
倪力均 NI Lijun
潘 旻 PAN Min
彭文楷 PENG Wenkai
乔 力 QIAO Li
邱金宏 QIU Jinhong
沈均臣 SHEN Junchen
汪 园 WANG Yuan
王 晨 WANG Chen
夏 澍 XIA Shu
徐庆姝 XU Qingshu
姚丛琦 YAO Congqi
于海平 YU Haiping
虞王璐 YU Wanglu
袁金燕 YUAN Jinyan
张 岸 ZHANG An
张 敏 ZHANG Min
张 培 ZHANG Pei
张永雷 ZHANG Yonglei
赵 锐 ZHAO Rui
赵天亚 ZHAO Tianya
周 逸 ZHOU Yi

曹永山 CAO Yongshan
陈 成 CHEN Cheng
陈焕彦 CHEN Huanyan
陈 娟 CHEN Juan
陈 鹏 CHEN Peng
陈中高 CHEN Zhonggao
范丹丹 FAN Dandan
樊璐敏 FAN Lumin
耿 健 GENG Jian
韩艺宽 HAN Yikuan
杭晓萌 HANG Xiaomeng
胡绮玭 HU Qipi
胡小敏 HU Xiaomin
黄凯熙 HUANG Kaixi
黄文华 HUANG Wenhua
黄一庭 HUANG Yiting
贾福有 JIA Fuyou
蒋菁菁 JIANG Jingjing
赖友炜 LAI Youwei
李 政 LI Zheng
林肖寅 LIN Xiaoyin
林中格 LIN Zhongge
刘赟俊 LIU Yunjun
龙俊荣 LONG Junrong
陆 恬 LU Tian
倪绍敏 NI Shaomin
潘 东 PAN Dong
邵一丹 SHAO Yidan
司秉卉 SI Binghui
孙 燕 SUN Yan
陶敏悦 TAO Minyue
王 彬 WANG Bin
王洁琼 WANG Jieqiong
王 凯 WANG Kai
王旭静 WANG Xujing
吴黎明 WU Liming
武苗苗 WU Miaomiao
徐怡雯 XU Yiwen
薛晓旸 XUE Xiaoyang
颜骁程 YAN Xiaocheng
杨 灿 YANG Can
杨钗芳 YANG Chaifang
杨 浩 YANG Hao
杨 骏 YANG Jun
杨 柯 YANG Ke
殷 强 YIN Qiang
余 露 YU Lu
俞 冰 YU Bing
俞 琳 YU Lin
袁亮亮 YUAN Liangliang
岳文博 YUE Wenbo
张 成 ZHANG Cheng
张方籍 ZHANG Fangji
张 伟 ZHANG Wei
张文婷 ZHANG Wenting
赵 芹 ZHAO Qin
赵书艺 ZHAO Shuyi
赵潇欣 ZHAO Xiaoxin
郑国活 ZHENG Guohuo
周 青 ZHOU Qing
周雨馨 ZHOU Yuxin
朱鹏飞 ZHU Pengfei
朱 煜 ZHU Yu

奥珅颖 AO Shenying
陈观兴 CHEN Guanxing
陈相营 CHEN Xiangying
段艳文 DUAN Yanwen
符靓璇 FU Jingxuan
黄龙辉 Huang Longhui
姜伟杰 JIANG Weijie
雷冬雪 LEI Dongxue
李 彤 LI Tong
李招成 LI Zhaocheng
刘 佳 LIU Jia
刘彦辰 LIU Yanchen
吕 航 Lv Hang
毛军列 Mao Junlie
孟文儒 MEHG Wenru
潘柳青 PAN Liuqing
潘幼健 PAN Youjian
沙吉敏 SHA Jimin
谭子龙 TAN Zilong
王淡秋 WANG Danqiu
王冬雪 WANG Dongxue
王珊珊 WANG Shanshan
魏江洋 WEI Jiangyang
吴超楠 WU Chaonan
吴嘉鑫 WU Jiaxin
夏 炎 XIA Yan
肖 霄 XIAO Xiao
徐少敏 XU Shaomin
许伯晗 XU Bohan
许 骏 XU Jun
曹 政 CAO Zheng
陈 逸 CHEN Yi
仇高颖 QIU Gaoying
戴 波 DAI Bo
费日晓 FEI Rixiao
郭 瑛 GUO Ying
郭耘锦 GUO Yunjin
季 萍 JI Ping
贾江南 JIA Jiangnan
姜 智 JIANG Zhi
蒯冰清 KUAI Bingqing
李 昭 LI Zhao
力振球 LI Zhenqiu
刘 莹 LIU Ying
刘玉婧 LIU Yujing
柳筱娴 LIU Xiaoxian
沈康惠 SHEN Kanghui
施 伟 SHI Wei
孙冠成 SUN Guancheng
孙 昕 SUN Xin
谭发兵 TAN Fabing
汤建华 TANG Jianhua
王斌鹏 WANG Binpeng
王 晗 WANG Han
王 倩 WANG Qian
吴 宾 WU Bin
徐 蕾 XU Lei
徐沁心 XU Qinxin
徐婉迪 XU Wandi
张 楠 ZHANG Nan
赵 阳 ZHAO Yang
周荣楼 ZHOU Ronglou

车俊颖 CHE Junying
陈博宇 CHEN Boyu
陈凌杰 CHEN Lingjie
陈 曦 CHEN Xi
陈晓敏 CHEN Xiaomin
陈修远 CHEN Xiuyuan
程 斌 CHENG Bin
高 翔 GAO Xiang
顾一蝶 GU Yidie
韩书园 HAN Shuyuan
胡任元 HU Renyuan
黄广伟 HUANG Guangwei
蒋西亚 JIANG Xiya
焦宏斌 JIAO Hongbin
李天骄 LI Tianjiao
廉英豪 LIAN Yinghao
梁万富 LIANG Wanfu
梁耀波 LIANG Yaobo
林 陈 LIN Chen
林伟圳 LIN Weizhen
林 治 LIN Zhi
刘 晨 LIU Chen
刘 芮 LIU Rui
刘思彤 LIU Sitong
刘文沛 LIU Wenpei
刘 宇 LIU Yu
陆扬帆 LU Yangfan
骆国建 LUO Guojian
宁 凯 NING Kai
彭蕊寒 PENG Ruihan
单泓景 SHAN Hongjing
孙雅贤 SUN Yaxian
谭 健 TAN Jian
田金华 TIAN Jinhua
王冰卿 WANG Bingqing
王 健 WANG Jian
王 琳 WANG Lin
王曙光 WANG Shuguang
王 政 WANG Zheng
武春洋 WU Chunyang
吴昇奕 WU Shengyi
吴书其 WU Shuqi
吴婷婷 WU Tingting
夏候蓉 XIA Hourong
谢锡淡 XIE Xidan
徐 麟 XU Lin
徐思恒 XU Siheng
徐天驹 XU Tianju
许文韬 XU Wentao
徐 晏 XU Yan
杨天仪 YANG Tianyi
杨 悦 YANG Yue
杨玉菡 YANG Yuhan
姚晨阳 YAO Chenyang
姚 梦 YAO Meng
尤逸尘 YOU Yichen
于晓彤 YU Xiaotong
岳海旭 YUE Haixu
查新彧 ZHA Xinyu
张 丛 ZHANG Cong
张海宁 ZHANG Haining
张明杰 ZHANG Mingjie
张 进 ZHANG Jin
张 楠 ZHANG Nan
张 强 ZHANG Qiang
郑 伟 ZHENG Wei

图书在版编目（CIP）数据

南京大学建筑与城市规划学院建筑系教学年鉴.
2014~2015 / 王丹丹编. -- 南京：东南大学出版社，
2015.12
ISBN 978-7-5641-6170-5

Ⅰ. ①南… Ⅱ. ①王… Ⅲ. ①建筑学—教学研究—高
等学校—南京市—2014~2015—年鉴②城市规划—教学研
究—高等学校—南京市—2014~2015—年鉴 Ⅳ. ①TU-42

中国版本图书馆CIP数据核字（2015）第276580号

策　　划：丁沃沃　周　凌　华晓宁
装帧设计：王丹丹　丁沃沃
版面制作：曹舒琪
参与制作：颜骁程　陶敏悦
责任编辑：姜　来　魏晓平

出版发行：东南大学出版社
社　　址：南京市四牌楼 2号
出 版 人：江建中
网　　址：http://www.seupress.com
邮　　箱：press@seupress.com
邮　　编：210096
经　　销：全国各地新华书店
印　　刷：南京新世纪联盟印务有限公司
开　　本：787mm×1092mm　1/20
印　　张：10.5
字　　数：595千
版　　次：2015年12月第1版
印　　次：2015年12月第1次印刷
书　　号：ISBN 978-7-5641-6170-5
定　　价：68.00元